Bacteria–Plant Interactions

Advanced Research and Future Trends

Edited by

Jesús Murillo

Laboratorio de Patología Vegetal
Departamento de Producción Agraria
Universidad Pública de Navarra
Pamplona
Spain

Boris A. Vinatzer

Department of Plant Pathology,
Physiology, and Weed Science
Virginia Tech
Blacksburg, VA
USA

Robert W. Jackson

School of Biological Sciences
University of Reading
Reading
UK

Dawn L. Arnold

Centre for Research in Biosciences
University of the West of England
Bristol
UK

Caister Academic Press

Caister Academic Press
Norfolk, UK

www.caister.com

British Library Cataloguing-in-Publication Data
A catalogue record for this book is available from the British Library

ISBN (hardback): 978-1-908230-58-4
ISBN (ebook): 978-1-910190-00-5

Cover design adapted from Figure 4.1.

Contents

Contributors

Saul Burdman
Department of Plant Pathology and Microbiology; and
Otto Warburg Minerva Center for Agricultural Biotechnology
Robert H. Smith Faculty of Agriculture, Food and Environment
Hebrew University of Jerusalem
Rehovot
Israel

saul.burdman@mail.huji.ac.il

Jeff H. Chang
Department of Botany and Plant Pathology; and
Molecular and Cellular Biology Program; and
Center for Genome Research and Biocomputing
Oregon State University
Corvallis, OR
USA

changj@science.oregonstate.edu

Alan Collmer
Department of Plant Pathology and Plant–Microbe Biology
Cornell University
Ithaca, NY
USA

arc2@cornell.edu

Allison L. Creason
Department of Botany and Plant Pathology; and
Molecular and Cellular Biology Program
Oregon State University
Corvallis, OR
USA

creasona@onid.oregonstate.edu

Edward W. Davis II
Department of Botany and Plant Pathology; and
Molecular and Cellular Biology Program
Oregon State University
Corvallis, OR
USA

davidsed@onid.oregonstate.edu

Darrell Desveaux
Department of Cell & Systems Biology; and
Centre for the Analysis of Genome Evolution & Function
University of Toronto
Toronto, ON
Canada

darrell.desveaux@utoronto.ca

Noam Eckshtain-Levi
Department of Plant Pathology and Microbiology; and
Otto Warburg Minerva Center for Agricultural Biotechnology
Robert H. Smith Faculty of Agriculture, Food and Environment
Hebrew University of Jerusalem
Rehovot
Israel

naom.levi2@mail.huji.ac.il

David S. Guttman
Department of Cell & Systems Biology; and
Centre for the Analysis of Genome Evolution & Function
University of Toronto
Toronto, ON
Canada

david.guttman@utoronto.ca

Monica Höfte
Department of Crop Protection
Laboratory of Phytopathology
Ghent University
Gent
Belgium

monica.hofte@ugent.be

Nicola J. Holden
The James Hutton Institute
Cell and Molecular Sciences
Invergowrie
Dundee
UK

Nicola.Holden@hutton.ac.uk

Ashleigh Holmes
The James Hutton Institute
Cell and Molecular Sciences
Invergowrie
Dundee
UK

Asleigh.Holmes@hutton.ac.uk

Robert W. Jackson
School of Biological Sciences
University of Reading
Whiteknights
Reading
UK

r.w.jackson@reading.ac.uk

Nam Phuong Kieu
Department of Crop Protection
Laboratory of Phytopathology
Ghent University
Gent
Belgium

phuongnam1780@yahoo.com

Britt Koskella
BioSciences
University of Exeter
Cornwall
UK

b.l.koskella@exeter.ac.uk

Amy Huei-Yi Lee
Department of Cell & Systems Biology
University of Toronto
Toronto, ON
Canada

amy.microbiology@gmail.com

Magdalen Lindeberg
Department of Plant Pathology and Plant–Microbe Biology
Cornell University
Ithaca, NY
USA

ml16@cornell.edu

Timothy Lo
Department of Cell & Systems Biology
University of Toronto
Toronto, ON
Canada

timothy.lo@mail.utoronto.ca

R. Ryan McNally
Department of Plant Pathology
University of Minnesota
St. Paul, MN
USA

rmcnally@umm.edu

Heath O'Brien
Department of Cell & Systems Biology
University of Toronto
Toronto, ON
Canada

heath.obrien@gmail.com

Feyisara Eyiwumi Olorunleke
Department of Crop Protection
Laboratory of Phytopathology
Ghent University
Gent
Belgium

feyisaraeyiwumi.olorunleke@ugent.be

Tally Rosenberg
Department of Plant Pathology and Microbiology; and
Otto Warburg Minerva Center for Agricultural Biotechnology
Robert H. Smith Faculty of Agriculture, Food and Environment
Hebrew University of Jerusalem
Rehovot
Israel

tally.rosenberg@mail.huji.ac.il

Yannick Rossez
The James Hutton Institute
Cell and Molecular Sciences
Invergowrie
Dundee
UK

yannick.rossez@hutton.ac.uk

Elizabeth A. Savory
Department of Botany and Plant Pathology
Oregon State University
Corvallis, OR
USA

savorye@science.oregonstate.edu

George W. Sundin
Department of Plant, Soil and Microbial Sciences
Michigan State University
East Lansing, MI
USA

sundin@msu.edu

Tiffany B. Taylor
School of Biological Sciences
University of Reading
Whiteknights
Reading
UK

t.b.taylor@reading.ac.uk

Olivier M. Vandeputte
Laboratoire de Biotechnologie Végétale
Université Libre de Bruxelles
Gosselies
Belgium

ovdeputt@ulb.ac.be

Youfu Zhao
Department of Crop Sciences
University of Illinois
Urbana, IL
USA

zhao888@illinois.edu

Current Books of Interest

Epigenetics: Current Research and Emerging Trends	2015
Advanced Vaccine Research Methods for the Decade of Vaccines	2015
Antifungals: From Genomics to Resistance and the Development of Novel Agents	2015
Aeromonas	2015
Antibiotics: Current Innovations and Future Trends	2015
Leishmania: Current Biology and Control	2015
Acanthamoeba: Biology and Pathogenesis (2nd edition)	2015
Microarrays: Current Technology, Innovations and Applications	2014
Metagenomics of the Microbial Nitrogen Cycle: Theory, Methods and Applications	2014
Pathogenic *Neisseria*: Genomics, Molecular Biology and Disease Intervention	2014
Proteomics: Targeted Technology, Innovations and Applications	2014
Biofuels: From Microbes to Molecules	2014
Human Pathogenic Fungi: Molecular Biology and Pathogenic Mechanisms	2014
Applied RNAi: From Fundamental Research to Therapeutic Applications	2014
Halophiles: Genetics and Genomes	2014
Molecular Diagnostics: Current Research and Applications	2014
Phage Therapy: Current Research and Applications	2014
Bioinformatics and Data Analysis in Microbiology	2014
The Cell Biology of Cyanobacteria	2014
Pathogenic *Escherichia coli*: Molecular and Cellular Microbiology	2014
Campylobacter Ecology and Evolution	2014
Burkholderia: From Genomes to Function	2014
Myxobacteria: Genomics, Cellular and Molecular Biology	2014
Next-generation Sequencing: Current Technologies and Applications	2014
Omics in Soil Science	2014
Applications of Molecular Microbiological Methods	2014
Mollicutes: Molecular Biology and Pathogenesis	2014
Genome Analysis: Current Procedures and Applications	2014
Bacterial Toxins: Genetics, Cellular Biology and Practical Applications	2013
Bacterial Membranes: Structural and Molecular Biology	2014
Cold-Adapted Microorganisms	2013
Fusarium: Genomics, Molecular and Cellular Biology	2013

Full details at www.caister.com

Preface

With the continuing increase in human population and major changes in climate, our relative food prosperity in the 1980s/1990s has reversed to expose a significant threat to global food security. Add to this the continual battle to control existing plant disease problems, emerging diseases and food contamination with human pathogens, and we realise there are still significant challenges ahead for producing adequate, safe food. However, we are witnessing major advances in technology and scientific exploration as we aim to cope with these issues. The aim of this book is to introduce the reader to advances in the field of bacteria–plant interactions, centred on plant pathogens, human pathogen contamination and potential control strategies. Fundamental to this is the need to build up model systems to help inform other studies – these are exemplified in the *Pseudomonas* and *Erwinia* chapters. Emerging experimental systems examining emerging or neglected diseases have been considered in the chapters on *Acidovorax* and Gram-positive bacterial pathogens. The recent outbreak of food poisoning in Europe was a timely reminder for the need to examine human pathogen colonisation of plants. Finally, with the loss of regulated chemicals, there is a clear need to innovate control methods – two chapters consider biocontrol approaches. Bacteria represent a good opportunity to find antimicrobials against fungi and oomycetes, while phage therapy offers a solution to bacterial infections.

Robert W. Jackson, Jesús Murillo, Boris A. Vinatzer and Dawn L. Arnold

Functional Diversification of Phytopathogenic Type III Secreted Effector Proteins

1

Amy Huei-Yi Lee, Heath O'Brien, Timothy Lo, David S. Guttman* and Darrell Desveaux*

Abstract

Bacterial phytopathogens and mutualistic symbionts utilize the type III secretion system (T3SS) to deliver type III secreted effector (T3SE) proteins into host cells in order to manipulate host immunity or cellular processes and integrity, with the ultimate aim of promoting bacterial growth and transmission. A large amount of experimental work has gone into the identification of T3SEs. However, a majority have no known host targets or characterized modes of action. The recent explosion in genome sequences of bacterial phytopathogens has led to the rapid identification of a large number of divergent homologues of known T3SEs, but the majority of functional work has been done on single representatives of a given T3SE family. In cases where multiple homologues have been characterized, they often exhibit striking divergence in targets, cofactor requirements, virulence and immune recognition functions. Comparative genomics and functional analyses of T3SE homologues across bacterial pathogens of both plants and animals will enhance our understanding of host specificity and pathogenesis.

Introduction

Type III secretion systems (T3SSs) are broadly distributed among some of the most agronomically important plant-pathogenic bacteria in the genera *Pseudomonas* (Pseudomonadales), *Xanthomonas* (Xanthomonadales), *Ralstonia* and *Acidovorax* (Burkholderiales) and *Erwinia, Pantoea, Pectobacterium, Brenneria* and *Dickeya* (Enterobacteriales), as well as several genera of mutualistic symbionts (Rhizobiales). *Pseudomonas* is a large genus that includes soil bacteria, opportunistic animal pathogens, and agronomically important plant pathogens. The majority of plant-pathogenic strains were divided among 50 pathogenic varieties (pathovars) within the species *P. syringae*. DNA–DNA hybridization and ribotyping analyses indicate that *P. syringae* comprises at least nine genomospecies (Gardan *et al.*, 1999), while multilocus sequence analysis (MLSA) identifies five major phylogroups (Bull *et al.*, 2011; Sarkar and Guttman, 2004). While the official taxonomy of *P. syringae* is currently in flux, with specific clades being renamed as unique species, the name *P. syringae* is still commonly used to refer to the entire species complex (Baltrus *et al.*, 2011; O'Brien *et al.*, 2011b).

*These authors contributed equally

The genus *Xanthomonas* includes over 150 pathovars that have been divided among 20 different species on the basis of DNA–DNA hybridization and MLSA (Parkinson *et al.*, 2007; Rademaker *et al.*, 2005; Vauterin *et al.*, 1995; Young *et al.*, 2008). Both *P. syringae* and *Xanthomonas* spp. cause a variety of blight, canker and spot diseases on a range of host plants.

Ralstonia solanacearum is a broad host range pathogen that causes bacterial wilt on plants in over 50 families (Denny, 2006). The species is divided among four divergent phylogroups by MLSA (Castillo and Greenberg, 2007; Prior and Fegan, 2005), while the specialist pathogens *R. syzygii* and banana blood disease bacterium (BDB) are nested within one of these phylogroups (Remenant *et al.*, 2011).

The plant-pathogenic strains of *Erwinia* have recently been separated into five different genera that are phylogenetically intermixed with several genera of animal pathogens in the Enterobacteriaceae: *E. amylovora* (fire blight of apple and other Rosaceous plants) and related species, *Pantoea stewartii* (Stewart's disease of corn), and three genera of soft rot pathogens: *Pectobacterium, Brenneria* and *Dickeya* (Hauben *et al.*, 1998; Kwon *et al.*, 1997; Mergaert *et al.*, 1993; Samson *et al.*, 2005).

The genus *Acidovorax* includes several species that cause a wide variety of plant diseases (Schaad *et al.*, 2008). Unfortunately, little work has been done on these species and their complements of type III secreted effectors (T3SEs) have not been characterized.

The first T3SEs to be identified were called avirulence proteins based on their induction of a strong plant immune response when recognized by cognate resistance (R) proteins (Gabriel *et al.*, 1986; Kobayashi *et al.*, 1990; Staskawicz *et al.*, 1987; Swanson *et al.*, 1988). This 'effector-triggered immunity' (ETI) is usually associated with a localized programmed cell death reaction termed the 'hypersensitive response' (HR), which has been used as a robust phenotype to identify T3SEs and their cognate resistance proteins (Innes *et al.*, 2008; Jones and Dangl, 2006). Further progress in the identification and characterization of T3SEs was greatly accelerated by the discovery that the expression of both T3SS structural genes as well as T3SEs are regulated in a coordinated manner by the HrpL transcription factor (Genin *et al.*, 1992; Wengelnik and Bonas, 1996; Xiao and Hutcheson, 1994). This finding has led to the identification of a large number of novel T3SE candidates through expression analyses (Ferreira *et al.*, 2006; Noël *et al.*, 2001; Occhialini *et al.*, 2005; Yang *et al.*, 2010a; Zhao *et al.*, 2005), bioinformatic prediction of promoter sequences (Cunnac *et al.*, 2004; Zwiesler-Vollick *et al.*, 2002) and genetic screens for HrpL-regulated genes (Boch *et al.*, 2002; Chang *et al.*, 2005; Fouts *et al.*, 2002; Mukaihara *et al.*, 2004). The discovery that the secretion and the translocation of T3SEs is dependent on an N-terminal signal that can be separated from the effector domain, and that the N-terminus of one T3SE can drive the secretion and translocation of a heterologous T3SE effector domain (Guttman and Greenberg, 2001; Mudgett *et al.*, 2000) or the calmodulin-dependent adenylate cyclase (Cya) toxin protein from the *Bordetella pertussis* (Casper-Lindley *et al.*, 2002; Schechter *et al.*, 2004) provided new means to identify and validate T3SEs. These reporter constructs have been used for genome-wide functional screens for T3SEs in *P. syringae, X. campestris,* and *R. solanacearum* (Guttman *et al.*, 2002; Mukaihara *et al.*, 2010; Roden *et al.*, 2004b). Additionally, the accumulated data from these functional and *in vitro* screens has allowed researchers to further refine the regulatory signatures associated with T3SS-dependent secretion, and facilitated bioinformatic screens for T3SEs (Furutani *et al.*, 2009; Petnicki-Ocwieja *et al.*, 2002; Wang *et al.*, 2011; Yang *et al.*, 2010b). Together, these approaches have identified over 100 T3SE families in *R. solanacearum* (Genin and Denny, 2012), 60 in *P. syringae* (O'Brien *et al.*, 2011a) and approximately 40 in *Xanthomonas* spp. (White *et al.*, 2009). In contrast, *Erwinia* and related genera have a much smaller complement of T3SEs with only five families discovered to date (Malnoy *et al.*, 2012).

With the dramatic increase in genome sequencing over the last decade, it is becoming increasingly common to identify T3SEs based on their sequence similarity to homologues in well-characterized strains. This has resulted in the rapid increase in the number of putative T3SEs available in databases (Fig. 1.1), though functional confirmation of expression and translocation is lacking in most cases. The number of T3SE homologues per strain varies dramatically among sequenced *P. syringae* strains, from nine in *P. syringae* pv. *japonica* MAFF301072PT (Baltrus *et al.*, 2011) to 36 in *P. syringae* pv. *tomato* (*Pto*) DC3000, though only 29 appear to be expressed to a significant level (Kvitko *et al.*, 2009; Schechter *et al.*, 2006). There are also dramatic differences in T3SE complements even within very closely related *P. syringae* strains of the same pathovar as only 14 T3SEs were found to be in common between *Pto* DC3000 and *Pto* T1 (Almeida *et al.*, 2009). Similarly, two phylogenetically diverged lineages of *P. syringae* pv. *avellanae* have only two putatively functional T3SE families in common and vary in the total number of full-length T3SEs per strain from 10 to 25 (O'Brien *et al.*, 2012).

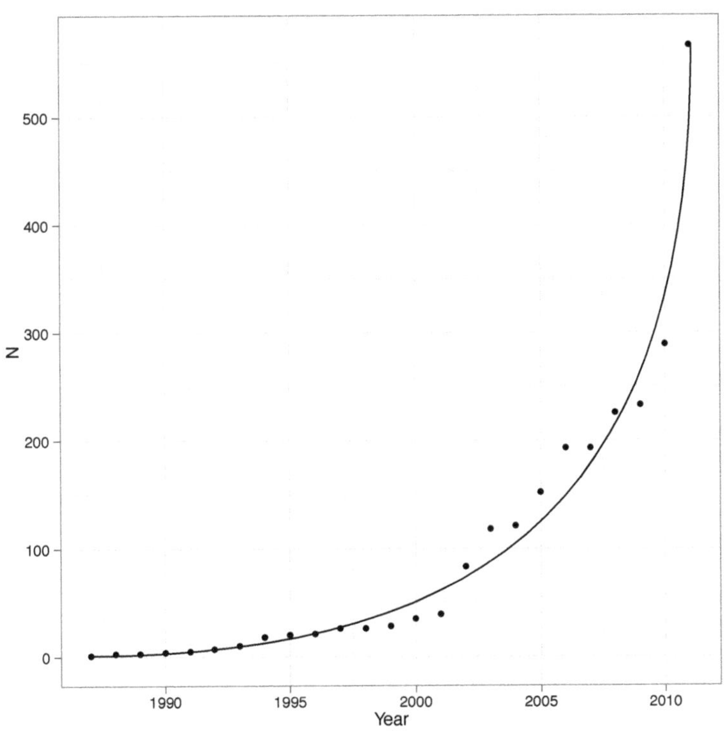

Figure 1.1 Number of *P. syringae* T3SE homologue sequences deposited in GenBank by year (data from pseudomonas-syringae.org).

Most *Xanthomonas* strains possess genes for between 20 and 25 T3SE families and there appears to be much less variability in T3SE complements between strains. Different pathovars of *X. vasculorum* share all but two of ~22 T3SE families (Studholme *et al.*, 2010), while 19 of 26 T3SE families are present in all four sequenced strains of *X. oryzae*, representing two pathovars (Bogdanove *et al.*, 2011). *X. oryzae* also carries 16–26 copies of a large multigene family of transcription activator-like (TAL) DNA-binding T3SEs. This fascinating family of T3SEs has a central domain of 33 to 35 amino acid repeats that determines its DNA-binding specificity (Boch and Bonas, 2010).

R. solanacearum produces the largest number of T3SE homologues, with 60–75 homologues per strain, including five multigene T3SE families with three to seven copies per strain (Genin and Denny, 2012). About 50 T3SE genes are conserved across the *R. solanacearum* species complex, and over 60 are shared among strains of the same phylotype (Genin and Denny, 2012). All sequenced strains of *E. amylovora* have homologues of all five known *Erwinia* T3SE families, while only two are found in the related species *E. pyrifoliae* (Zhao and Qi, 2011).

In addition to understanding the dynamics of T3SE gain and loss, the recent explosion in genome sequencing permits the identification of a core secretome that was likely to be present in the common ancestor of a group of pathogens. This core secretome presumably was essential for the ancestral pathogen to overcome the plant basal immunity and adopt a pathogenic lifestyle. The core secretome is small for both *P. syringae* and *Xanthomonas* relative to the total number of T3SEs found in these pathogens. Seven *P. syringae* T3SEs are genealogically congruent with core genome phylogeny, consistent with vertical inheritance from the common ancestor, although some have been pseudogenized or lost entirely in some lineages (O'Brien *et al.*, 2012). Three of these T3SEs, HopAA1, HopM1 and AvrE1, are localized in the Conserved Effector Locus (CEL) that is adjacent to the T3SS structural gene cluster (Alfano *et al.*, 2000). Two additional T3SE families (HopAB and HopX) have homologues in all or most *P. syringae* strains, but both of these are diverse families comprising multiple functionally diverged subfamilies and their gene trees do not match the species tree (Baltrus *et al.*, 2011; Lindeberg *et al.*, 2005; O'Brien *et al.*, 2011b).

Ten T3SEs were found to be present in each of seven *Xanthomonas* strains, including XopZ1. This is a particularly intriguing finding since XopZ1 is homologous to HopAS1, which is one of the core T3SEs in *P. syringae* (Studholme *et al.*, 2010; White *et al.*, 2009). *P. syringae* strain 642 and *X. campestris* pv. *raphani* 756C are interesting exceptions to the general pattern of T3SE distribution, as they both lack most of the core T3SEs found in all other characterized strains of their respective groups (Bogdanove *et al.*, 2011; Clarke *et al.*, 2010). *X. campestris* pv. *raphani* 756C is the non-vascular counterpart to the well-studied vascular pathogen *X. campestris* pv. *campestris*. *P. syringae* 642 is in a clade of closely related leaf-colonizing strains that appear to be non-pathogenic. This strain lacks the canonical *P. syringae* cluster of loci encoding the T3SS, and carries a highly divergent T3SS as well as T3SEs most similar to the *P. aeruginosa* T3SEs ExoU and ExoY (Clarke *et al.*, 2010).

R. solanacearum has a much larger core secretome complement, with more than 34 of 74 families conserved among 18 analysed strains. An additional 21 families may have been present in the common ancestor of the species complex and lost in some lineages (Guidot *et al.*, 2007; Poueymiro and Genin, 2009).

While the majority of phytopathogen T3SE families lack homologues in other species, a subset of these T3SEs has been horizontally transferred from divergent species. T3SE horizontal gene transfer (HGT) is most pronounced between *P. syringae* and *Xanthomonas* spp., which have 17 homologous T3SE families in common (Table 1.1). This represents almost half of the T3SE families that have been identified in *Xanthomonas* and a third of those identified in *P. syringae* (O'Brien *et al.*, 2011b; White *et al.*, 2009). *R. solanacearum* has slightly fewer T3SE families in common with other phytopathogens, with 14 shared with *Xanthomonas* spp. and ten shared with *P. syringae*, including seven that are found in all three groups (Table 1.1). One of these, the YopJ family, also has a homologue in *E. amylovora*

Table 1.1 Homologous T3SE families in different plant pathogens[1]

Pseudomonas syringae	*Xanthomonas*	*Ralstonia solanacearum*	*Erwinia*	*Pantoea stewartii*
AvrA	AvrBs1			
AvrB	XopAH (AvrXccC)			
AvrE			DspE/A	WtsE
AvrRps4	XopO			
AvrRpt2			AvrRpt2Ea	
HopAA		HopAA1 family		
HopAB (avrPtoB)		AvrPtoB family		
HopAE	XopAA			
HopAF	XopAF			
HopAO	XopH			
HopAS	XopZ			
HopAU	XopN			
HopC		RipT	HopCEa	
HopD (AvrPphD)	XopB	AvrPphD family		
HopF (AvrPphF)		AvrPphF family		
HopG	XopAG	HopG1 family		
HopH	XopG	HopH1 family		
HopK	XopAK			
HopO	XopAI			
HopQ	XopQ	RipB		
HopR1	XopAM	HopR1 family		
HopX (AvrPphE)	XopE	AvrPphE family		
HopZ	XopJ	PopP	Eop1	
	AvrBs3	AvrBs3 family		
	XopAD	SKWP		
	XopAE	PopC		
	XopAL	RSIPO_04353	Eop3	
	XopC	RSp1239		
	XopP	HLK		
	XopX	XopX family		

[1]Data from Malnoy *et al*. (2012), Poueymiro and Genin (2009) and White *et al*. (2009).

(Eop1). All four of the other *Erwinia* T3SEs are also reported to have homologues in *P. syringae* (Malnoy *et al.*, 2012). The *E. amylovora* effector AvrRpt2EA shares 58% amino acid identity to AvrRpt2 from *P. syringae*, but is not found in other closely related *Erwinia* species (Zhao *et al.*, 2006). Eop3, though previously reported as a homologue of *P. syringae* HopX1 (AvrPphE), has no significant sequence similarity to HopX1 and is instead much more similar to XopAL from *X. campestris* and RSIPO_04353 from *R. solanacearum*. DspE/A, which is homologous to AvrE1 from *P. syringae*, also has homologues in all other sequenced plant-associated species of *Erwinia* and *Pseudomonas* (except for the epiphytic species *E. billingiae*), as well as *Pantoea, Pectobacterium, Dickeya* (Araki *et al.*, 2007; Glasner *et al.*, 2011; Holeva *et al.*, 2004; Kube *et al.*, 2008; Mor *et al.*, 2001).

Even though the molecular mechanisms of many T3SEs remain to be characterized, the enzymatic functions and host targets of a subset have been identified. In the following sections, we will focus our discussion on those T3SE families with known host targets and modes of action. Additionally, we will highlight the phylogenetic relationship between different homologues and the functional diversity within each T3SE family.

AvrB

The *P. syringae* AvrB family of T3SEs is homologous to the XopAH (AvrXccC) family from *X. campestris* (Table 1.1). Additionally, AvrB homologues are present in *X. axonopodis* pv. *punicae* str. LMG 859 and the animal pathogen *Fluoribacter dumoffii* Tex-KL (Fig. 1.2). Four subfamilies of AvrB are present in *P. syringae* (Lindeberg *et al.*, 2005), with the *X. campestris* homologue being the most similar to AvrB2. In contrast, the *X. axonopodis* and *F. dumoffii* homologues form lineages distinct from any of the *P. syringae* subfamilies. The founding members, AvrB1$_{PgyR4}$ and AvrB2$_{PgyR4}$, were first identified from the soybean strain of *P. syringae* pv. *glycinea* race 4 (PgyR4) (Staskawicz *et al.*, 1987). Members of the AvrB family are not restricted to soybean pathogens, and have now been identified in *P. syringae* strains that infect a wide range of hosts, including rice and horse chestnut (O'Brien *et al.*, 2011a, 2012).

Although phylogenetically related, AvrB family members do not display the same phenotypes *in planta*. For instance, AvrB1$_{PgyR4}$ triggers HR in *Arabidopsis thaliana* (Innes *et al.*, 1993) while AvrB2$_{PgyR4}$ does not (Wanner *et al.*, 1993). In soybean, AvrB2$_{PgyR4}$ triggers a HR that is distinct from that induced by AvrB1$_{PgyR4}$ (Staskawicz *et al.*, 1987; Tamaki *et al.*, 1988). AvrB2$_{PphNPS3121}$ differs from AvrB2$_{PgyR4}$ by only 2 amino acid substitutions, and elicits a pattern of soybean cultivar reactions that is identical to AvrB2$_{PgyR4}$ (Yucel *et al.*, 1994). Another AvrB2 subfamily member, AvrB2$_{Pph1449B}$, can suppress the HopF1$_{Pph1449B}$-induced HR in the Canadian Wonder cultivar of common bean (Tsiamis *et al.*, 2000). On the other hand, XopAH1, the *X. campestris* pv. *campestris* AvrB family homologue most closely related to the AvrB2 subfamily (Fig. 1.2), has been shown to confer a virulence advantage in cabbage, *Brassica oleracea*, while it reduces bacterial growth on mustard, *Brassica napiformis* L.H. Baily (Wang *et al.*, 2007). AvrB3$_{PsyB728A}$ induces strong immune responses in all tested lettuce cultivars by Wroblewski *et al.*, and variable levels of recognition in different pepper and tomato cultivars (Wroblewski *et al.*, 2009). It is not currently clear if the *in planta* differences between AvrB family members are due to differences in enzymatic activities or *in planta* targets of AvrB. Given that the majority of the molecular and biochemical characterizations

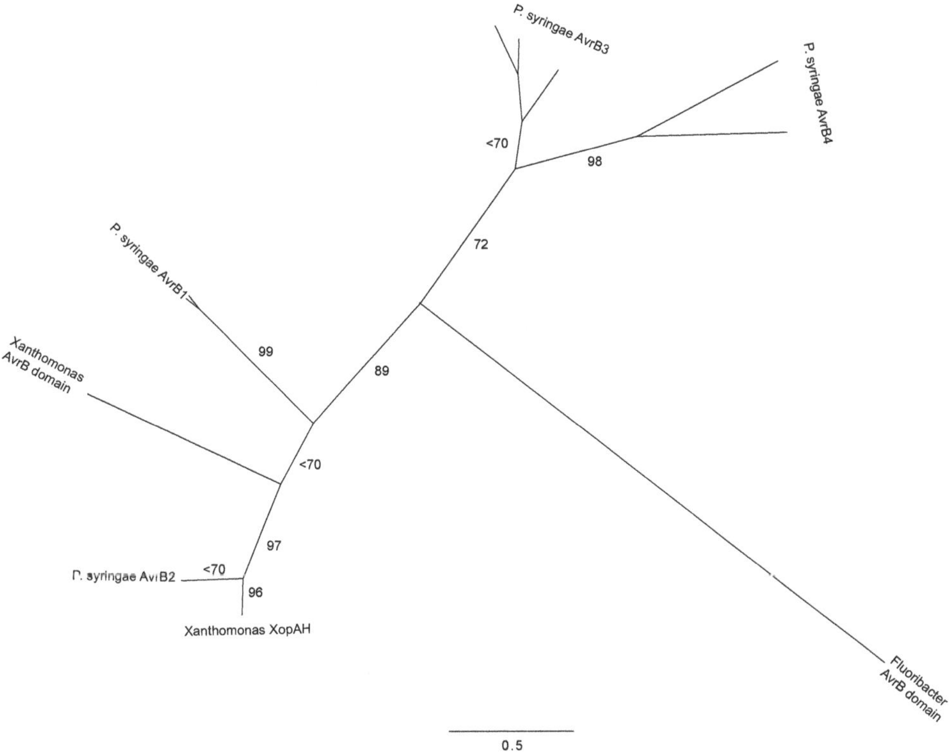

Figure 1.2 Evolutionary relationships among AvrB homologues. Homologues were identified by blasting amino acid sequences from pseudomonas-syringae.com against the non-redundant protein database at GenBank and all DNA sequences coding for hits with E-values $<1\times10^{-10}$ were aligned using MAFFT (Katoh *et al.*, 2002) and TranslatorX (Abascal *et al.*, 2010). Trees were inferred using maximum likelihood with PhyML (Guindon *et al.*, 2010) and visualized with FigTree (tree.bio.ed.ac.uk/software/figtree). Numbers beside branches are approximate likelihood-ratio test (aLRT) values (Anisimova and Gascuel, 2006).

of this family have focused on its founding member, AvrB1$_{PgyR4}$, we will focus the rest of the discussion on this family member.

In *Arabidopsis,* AvrB1$_{PgyR4}$ triggers ETI and an associated HR in plants that contain the dominant allele of the R protein RPM1 (resistance to *P. syringae* pv. *maculicola* 1) (Debener *et al.*, 1991; Innes *et al.*, 1993; Staskawicz *et al.*, 1987). However, RPM1 does not directly bind to AvrB1$_{PgyR4}$. Instead, AvrB1$_{PgyR4}$ indirectly phosphorylates a negative regulator of plant immunity, RIN4 (RPM1-interacting protein 4) and this phosphorylation is detected by RPM1 (Chung *et al.*, 2011; Cui *et al.*, 2010; Liu *et al.*, 2011; Mackey *et al.*, 2002). AvrB1 also activates the *Arabidopsis* TAO1 R protein in a RIN4-independent manner. Both RPM1 and TAO1 additively contribute to AvrB1-induced ETI (Eitas *et al.*, 2008). In *Arabidopsis* plants lacking the dominant allele of RPM1, AvrB1$_{PgyR4}$ can promote bacterial virulence by indirectly phosphorylating RIN4, which negatively regulates plant immunity (Cui *et al.*, 2010; Kim *et al.*, 2005; Shang *et al.*, 2006). AvrB1$_{PgyR4}$-induced phosphorylation of RIN4 promotes pathogen virulence through the manipulation of plant hormone homeostasis, in particular jasmonic acid (Cui *et al.*, 2010).

One of the mechanisms by which AvrB1$_{PgyR4}$ alters RIN4 phosphorylation is by targeting the HSP90 (heat shock protein 90) complex. In plants and animals, the HSP90 complex is an important chaperone for many signalling proteins, stabilizing the maturation and activation of these signalling components (Kadota and Shirasu, 2012). Particularly, the HSP90 complex stabilizes R proteins in plants with the help of the co-chaperone RAR1 [required for MLA12 resistance; (Kadota and Shirasu, 2012)]. AvrB1$_{PgyR4}$ is targeted to the HSP90 complex by directly binding to RAR1 (Cui *et al.*, 2010). Consequently, AvrB1$_{PgyR4}$ activates a mitogen-activated protein kinase (MAPK), MPK4, that associates with the HSP90 complex. MPK4 is a negative regulator of the plant immune signalling in *Arabidopsis* (Bittel and Robatzek, 2007; Petersen *et al.*, 2000). Thus, interaction of AvrB1$_{PgyR4}$ with RAR1 directs AvrB1$_{PgyR4}$ to the HSP90 complex, where it then enhances MPK4 activity (Cui *et al.*, 2010). Since MPK4 can interact with and phosphorylate RIN4, increased MPK4 activity leads to increased phosphorylation of RIN4 (Cui *et al.*, 2010).

Although the molecular mechanism by which AvrB1$_{PgyR4}$ indirectly induces RIN4 phosphorylation has been elucidated, the enzymatic function of AvrB1$_{PgyR4}$ remains unknown (Cui *et al.*, 2010). Analysis of crystal structures of AvrB1$_{PgyR4}$ with and without RIN4 has revealed that AvrB1$_{PgyR4}$ may be related to protein kinases (Desveaux *et al.*, 2007; Lee *et al.*, 2004). However, AvrB1$_{PgyR4}$ has not been shown to phosphorylate MPK4 *in vitro* and thus the enzymatic function of AvrB1$_{PgyR4}$ is still unclear (Cui *et al.*, 2010). Additionally, it is not clear if other members of the AvrB T3SE family function as protein kinases. Phosphorylation is not the only enzymatic function that has been proposed for the AvrB family, as structural analysis of AvrB1$_{PgyR4}$ bound to either a RIN4 peptide or a nucleotide has revealed structural similarity to the fic domain (filamentation induced by cAMP) of the *Vibrio parahaemolyticus* T3SE VopS (Kinch *et al.*, 2009). The VopS fic domain has been demonstrated to display a relatively novel post-translational modification: AMPylation (Kinch *et al.*, 2009). AMPylation involves the covalent addition of AMP, which functions similarly to other posttranslational modifications like phosphorylation or acetylation to alter the activity of substrate proteins (Kinch *et al.*, 2009). However, AMPylation activity has not yet been demonstrated for the AvrB family.

Even though the enzymatic function of the AvrB family has not been determined, mutational analysis of AvrB1$_{PgyR4}$ has revealed a solvent-accessible pocket to be important for the *in planta* function of AvrB1$_{PgyR4}$ (Ong and Innes, 2006). Alignment of eight *P. syringae* AvrB homologues and XopAH from *X. campestris* has identified conserved residues in this predicted enzymatic pocket that are required for the *in planta* functions of AvrB1$_{PgyR4}$ (Ong and Innes, 2006). Mutation in one of these conserved residues, D297, increased the RIN4-binding affinity of AvrB$_{PgyR4}$. Additionally, an upper lobe adjacent to the predicted enzymatic pocket has been suggested to play a role in substrate recognition (Ong and Innes, 2006). However, residues in this upper lobe domain are not conserved among the AvrB homologues and mutations in the upper lobe do not completely abolish AvrB1$_{PgyR4}$ functions (Ong and Innes, 2006). This indicates that despite conservation in the enzymatic pocket between AvrB homologues, there may be functional diversification as a result of differences in the substrate-binding domain. Furthermore, most of the residues in the AvrB1$_{PgyR4}$-RIN4 contact area are not well conserved between different AvrB homologues, indicating that AvrB may have additional targets besides RIN4 (Belkhadir *et al.*, 2004; Desveaux *et al.*, 2007). However, RIN4 has homologues in various plant species including lettuce, common bean and soybean (Jeuken *et al.*, 2009; Selote and Kachroo, 2010). For instance, RIN4-like proteins from soybean interact with $AvrB_{PgyR4}$ and are required for recognition of $AvrB_{PgyR4}$ by the RPG1-B (resistance to *P. syringae* pv. *glycinea* 1) R protein (Selote and Kachroo, 2010; Selote *et al.*, 2012). Thus, it is possible that AvrB family members have diversified to target various RIN4 homologues across different plant species (Chen *et al.*, 2010).

AvrPto

The AvrPto family of T3SEs was first identified from two strains of *P. syringae* pv. *tomato*, *Pto*JL1065 and *Pto*DC3000 (Ronald *et al.*, 1992). The AvrPto family of T3SEs is found exclusively in *P. syringae*, as no homologues have been found in other species. Until 2009, only three closely related AvrPto1 homologues were known for this family. However, the recent explosion in sequence data has revealed the diversity of this family, with eight divergent lineages found in 11 different *P. syringae* pathovars (Nguyen *et al.*, 2010) (Fig. 1.3). The

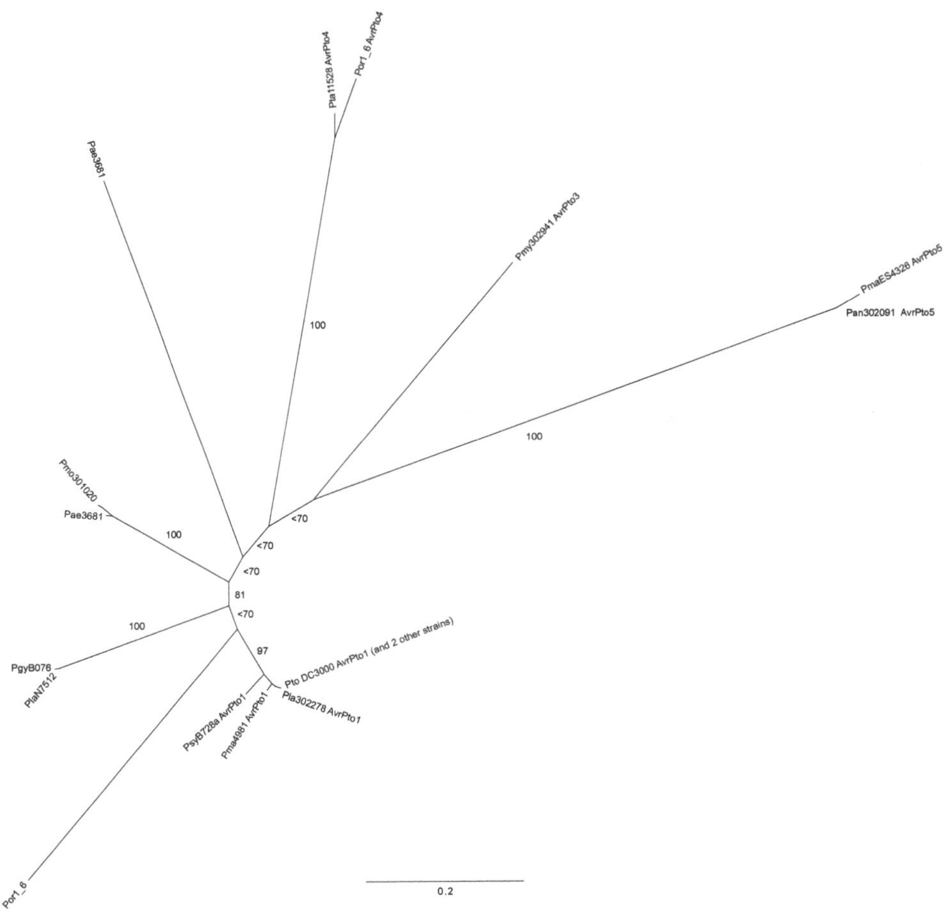

Figure 1.3 Evolutionary relationships among AvrPto homologues. See Fig. 1.2 for details. Strain names are from pseudomonas-syringae.org.

AvrPto family of T3SEs contains two domains with both domains being important for *in planta* functions (Yeam *et al.*, 2010).

The first AvrPto domain contains the CD loop, which is required for the AvrPto family members to directly interact with the tomato kinase, Pto (*Pseudomonas* tomato resistance) (Wulf *et al.*, 2004; Xing *et al.*, 2007). In tomato plants with the cognate R-gene Prf (*Pseudomonas* resistance and fenthion sensitivity), the interaction between Pto and AvrPto$_{\text{PtoJL1065}}$ induces ETI (Chang *et al.*, 2000; Xing *et al.*, 2007). In addition to interacting with the Pto kinase, the AvrPto CD loop also binds to the receptor-like kinase (RLK)/Pelle family of protein kinases (Xiang *et al.*, 2008). Particularly, by interacting with an RLK/Pelle protein kinase in *Arabidopsis* known as BAK1 (BRI1-associated kinase) protein, AvrPto$_{\text{PtoJL1065}}$ can inhibit plant immunity. BAK1 was first shown to be involved in brassinosteroid signalling and has been shown to interact with another RLK, BRI1 (brassinosteroid insensitive 1) (Li *et al.*, 2002; Nam and Li, 2002). In addition, BAK1 plays an important role in plant immunity by interacting with the *Arabidopsis* flagellin receptor, FLS2 (flagellin-sensitive 2) (Heese *et al.*, 2007; Schulze *et al.*, 2010). FLS2 is a receptor kinase that recognizes bacterial flagellin, a microbe-associated molecular pattern (MAMP) (Gomez-Gomez and Boller, 2000). By targeting BAK1 and FLS2, AvrPto1$_{\text{PtoJL1065}}$ inhibits basal immunity, also known as PRR-triggered immunity (PTI), in plants (Shan *et al.*, 2008; Xiang *et al.*, 2008, 2011). However, it is currently unclear whether AvrPto1$_{\text{PtoJL1065}}$ preferentially targets FLS2 or BAK1 (Shan *et al.*, 2008; Xiang *et al.*, 2008, 2011). Furthermore, the molecular mechanism by which AvrPto1$_{\text{PtoJL1065}}$ inhibits MPTI remains unknown. It is possible that AvrPto1$_{\text{PtoJL1065}}$ inhibits MPTI by directly binding to BAK1 to prevent the formation of MPTI signalling complexes (Shan *et al.*, 2008). It is also possible that AvrPto1$_{\text{PtoJL1065}}$ is a protein kinase inhibitor that prevents the kinase activity of FLS2 (Xiang *et al.*, 2008, 2011).

The second domain of the AvrPto family is the C-terminal domain (CTD), which when phosphorylated promotes virulence by an unknown mechanism that is distinct from that of the CD loop (Yeam *et al.*, 2010). In tobacco plants, a putative R protein Rpa (recognition of phosphorylated AvrPto) has been shown to recognize the phosphorylation of the CTD (Yeam *et al.*, 2010). Interestingly, both the CD loop and CTD are required for AvrPto1$_{\text{PtoJL1065}}$ virulence function, as mutations in either domain reduce bacterial virulence (Yeam *et al.*, 2010). In addition, a multiple sequence alignment of AvrPto family members has revealed that conserved residues of both the CD loop and CTD are important for the virulence function of AvrPto1 (Nguyen *et al.*, 2010). For example, mutation of S94, I96 or G99 in the CD loop abolish the virulence advantage conferred by AvrPto1$_{\text{PtoJL1065}}$ (Chang *et al.*, 2001; Shan *et al.*, 2000), while phosphorylation of serine residues within the CTD, S147 and S149, have also been shown to be involved in immune recognition by tobacco species (Yeam *et al.*, 2010). However, there are functional differences between AvrPto family members despite the presence of these conserved virulence residues. For example, AvrPto1$_{\text{PtoJL1065}}$ promotes virulence in the absence of Prf in tomato and strongly inhibits bacterial growth in resistant cultivars of tomato. On the other hand, AvrPto1$_{\text{PmoMAFF301020}}$ and AvrPto1$_{\text{PlaN7512}}$ promote virulence equivalent to AvrPto1$_{\text{PtoJL1065}}$ in the absence of Prf in tomato; yet when expressed in resistant tomato cultivars, these two effectors do not inhibit bacterial growth to the same extent as AvrPto1$_{\text{PtoJL1065}}$ (Nguyen *et al.*, 2010). Even fairly similar AvrPto subfamily members such AvrPto$_{\text{PmoMAFF301020}}$ and AvrPto$_{\text{PlaN7512}}$ (70% and 73% amino acid identity to the AvrPto1$_{\text{PtoJL1065}}$ respectively) have shown reproducible differences in their virulence activities (Nguyen *et al.*, 2010).

The observation that AvrPto1$_{PtoJL1065}$ can interact with multiple RLKs (Shan *et al.*, 2008; Xiang *et al.*, 2008, 2011) suggests that AvrPto1$_{PtoJL1065}$ may have additional virulence targets, as the RLK/Pelle protein kinase family is extremely large in plants, with over 600 members in *Arabidopsis* alone (Lehti-Shiu and Shiu, 2012). Given that RLK/Pelle family members are involved in many signalling processes that affect development, organ differentiation and the response to both abiotic and biotic stresses in plants (Lehti-Shiu and Shiu, 2012; Li *et al.*, 2002; Shan *et al.*, 2008), it raises an interesting possibility that other AvrPto family members may also target the RLK/Pelle protein kinases. Furthermore, it is likely that the divergence of the AvrPto family could be attributed to the targeting of different RLK proteins in the plant hosts. Avoidance of plant immunity may also drive the divergence of the AvrPto family members. For instance, *avrPto1* alleles collected from virulent and hyper-virulent *P. syringae* tomato strains have been shown to contain a single amino acid change that prevented the interaction of AvrPto1 with the Pto kinase in plants, and thus consequently abolished ETI (Kunkeaw *et al.*, 2010). Importantly, these AvrPto1 alleles still maintain their virulence functions as they can promote bacterial growth (Kunkeaw *et al.*, 2010).

Divergent members of the AvrPto family, such as AvrPto$_{Por36_1}$ and AvrPto3$_{PmyMAFF302941}$, which share 47% and 36% amino acid identity with AvrPto1$_{PtoJL1065}$ respectively, could potentially target novel host targets (Nguyen *et al.*, 2010). AvrPto3$_{PmyMAFF302941}$ lacks conserved residues in the CD loop and only has a functional CTD domain. In contrast, AvrPto$_{Por36_1}$ does not have conserved residues in either domain, and is not recognized by the R proteins, Prf in tomato or Rpa in tobacco (Nguyen *et al.*, 2010). However, AvrPto$_{Por36_1}$ elicits a resistance response that was not dependent on either Pto or Rpa, suggesting that AvrPto$_{Por36_1}$ is recognized by a novel resistance protein (Nguyen *et al.*, 2010). Additionally, AvrPto$_{Por36_1}$ does not have a virulence function in tomato (Nguyen *et al.*, 2010). Thus, AvrPto$_{Por36_1}$ raises intriguing questions about the enzymatic functions of divergent AvrPto family members.

HopAB (AvrPtoB)

Previous names of the members of this family have included AvrPtoB, VirPphA and HopPmaL, and these effectors have now been grouped into the diverse HopAB family of T3SEs (Guttman *et al.*, 2002; Jackson *et al.*, 1999; Kim *et al.*, 2002; Lindeberg *et al.*, 2005). The HopAB family is restricted to *P. syringae*, with a distant homologue in *R. solanacearum* strain MolK2 (Fig. 1.4). This family is very diverse and is present in the genomes of all *P. syringae* strains with a canonical T3SS (O'Brien *et al.*, 2011b). HopAB has been divided into three subfamilies, though HopAB2 homologues do not cluster together phylogenetically (Fig. 1.4). Comparisons among six HopAB family members revealed only 52% to 66% amino acid identity, emphasizing the diversity within this T3SE family (Lin *et al.*, 2006). Different family members have been shown to elicit different ETI responses in lettuce, pepper and tomato (Wroblewski *et al.*, 2009). For instance, most HopAB1 subfamily members induce cell death in almost all lettuce genotypes, except HopAB1$_{Ppi}$, which does not elicit any cell death (Wroblewski *et al.*, 2009). The HopAB2 subfamily also elicits cell death in almost all lettuce genotypes, while the sole HopAB3 family member resembled HopAB1$_{Ppi}$ with no cell death observed for any lettuce genotypes tested (Wroblewski *et al.*, 2009). Although the first member of this family, HopAB1$_{Pph1449B}$, was discovered in the bean pathogen *P. syringae Pph*1449B, the most intensely studied member of this family is HopAB2$_{PtoDC3000}$, also

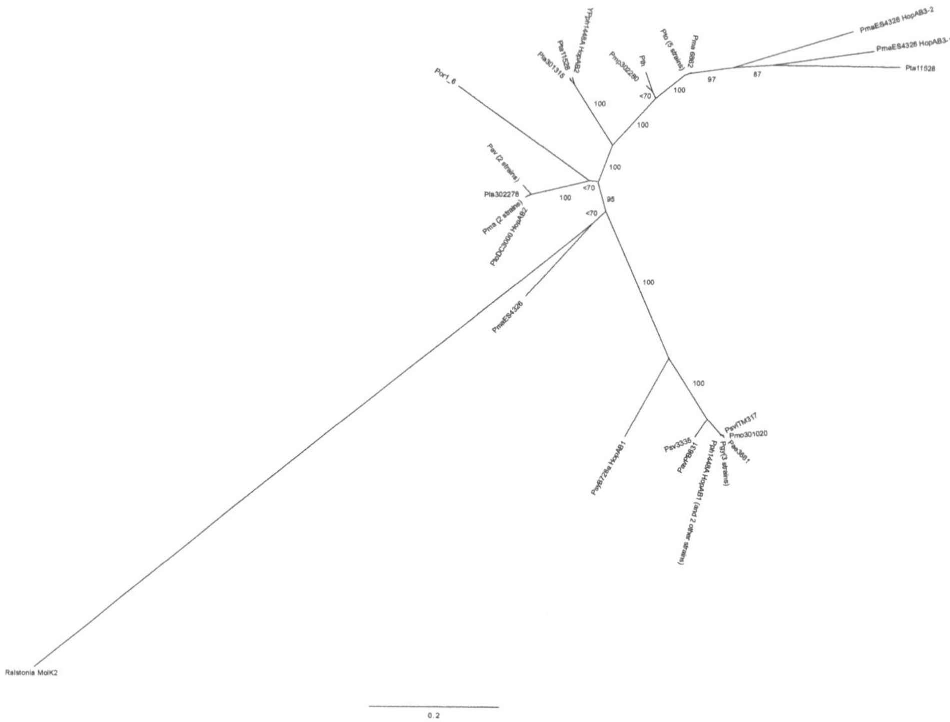

Figure 1.4 Evolutionary relationships among HopAB homologues. See Fig. 1.2 for details. Strain names are from pseudomonas-syringae.org.

known as AvrPtoB. (Jackson *et al.*, 1999; Kim *et al.*, 2002). HopAB2$_{\text{PtoDC3000}}$ was first identified as an avirulence factor in tomato that triggers ETI when it interacts with the tomato kinase Pto (Abramovitch and Martin, 2005). Since then, HopAB2$_{\text{PtoDC300}}$ has been shown to display a variety of functions including the suppression of plant basal resistance (He *et al.*, 2006), promotion of ethylene biosynthesis (Xiao *et al.*, 2007) and the ubiquitination of the tomato Fen kinase (Rosebrock *et al.*, 2007).

There are two functional domains identified in HopAB2$_{\text{PtoDC3000}}$. The N-terminal region contains a multifunctional domain whereby HopAB2$_{\text{PtoDC3000}}$ amino acids 1 to 307 are sufficient for promoting ethylene synthesis (Xiao *et al.*, 2007) and pathogen virulence (Xiao *et al.*, 2007; Zeng *et al.*, 2012). In tomato cultivars containing the protein kinase Pto and the corresponding R protein Prf (Abramovitch and Martin, 2005), this region is sufficient for the elicitation of ETI. The N-terminal domain of HopAB2$_{\text{PtoDC3000}}$ has been shown to interact with multiple plant proteins, including *Arabidopsis* BAK1 (Shan *et al.*, 2008), CERK1 (chitin elicitor receptor kinase 1) (Gimenez-Ibanez *et al.*, 2009) and Bti9 (AvrPtoB tomato-interacting 9), the tomato homologue of CERK1 (Zeng *et al.*, 2012). All of the aforementioned proteins belong to the RLK/Pelle family of protein kinases and interaction with these RLKs allows HopAB2$_{\text{PtoDC3000}}$ to inhibit plant basal resistance signalling and promote *P. syringae* virulence (Gimenez-Ibanez *et al.*, 2009; Shan *et al.*, 2008; Zeng *et al.*,

2012). However, the molecular mechanism by which HopAB2$_{PtoDC3000}$ manipulates RLKs is not fully understood.

The C-terminal domain (CTD) of HopAB2$_{PtoDC3000}$ is an E3 ubiquitin ligase that targets the tomato kinase Fen for degradation to prevent ETI (Janjusevic *et al.*, 2006) and allows the N-terminal domains of HopAB2$_{PtoDC3000}$ to promote virulence in tomato (Abramovitch *et al.*, 2006; Rosebrock *et al.*, 2007). Evidence both for and against the involvement of the CTD in the suppression of plant basal immunity have been documented and currently it is still unclear whether the CTD is truly required for plant basal immunity suppression (Gimenez-Ibanez *et al.*, 2009; Gohre *et al.*, 2008; Zeng *et al.*, 2012). In the case of CERK1, the CTD of HopAB2$_{PtoDC3000}$ is required to ubiquitinate CERK1, which consequently targets CERK1 for degradation (Gimenez-Ibanez *et al.*, 2009). On the other hand, the HopAB2$_{PtoDC3000}$ CTD is not required for targeting Bti9, it is instead postulated that HopAB2$_{PtoDC3000}$ inhibits the kinase activity of Bti9 (Zeng *et al.*, 2012). It is possible HopAB2$_{PtoDC3000}$ has evolved multiple mechanisms to manipulate a variety of RLKs.

The interaction of Bti9 with HopAB2$_{PtoDC3000}$ has been shown to be structurally similar, though not identical, to its interaction with Pto (Zeng *et al.*, 2012). This has led to the hypothesis that the tomato kinase Pto may have evolved as a molecular mimic of the protein kinase domain of LysM (lysin motif)-containing RLKs (Zeng *et al.*, 2012). LysM is a protein motif that has been shown to bind bacterial peptidoglycans (Buist *et al.*, 2008) as well as to fungal chitin (Iizasa *et al.*, 2010; Petutschnig *et al.*, 2010). A number of diverse HopAB family members, including HopAB1$_{PsyB728A}$, HopAB3$_{PtoT1}$ and HopAB3–1$_{PmaES4326}$ have been shown to interact with Pto. Thus, it is plausible that these family members may also interact with Bti9 and potentially other LysM-RLKs (Lin *et al.*, 2006). Furthermore, the variable amino acid substitutions that exist may allow other HopAB family members, such as HopAB1$_{PsyB728A}$ and HopAB3$_{PtoT1}$, to target alternate RLKs to suppress PTI (Lin *et al.*, 2006). In support of this, HopAB2$_{PtoDC3000}$ can interact with Bti9-related RLKs in tomato, SlLyk11, SlLyk12 and SlLyk13 (Zeng *et al.*, 2012).

The diverse HopAB family includes a number of truncated family members, such as HopAB3-1$_{PmaES4326}$ and HopAB3-2$_{PmaES4326}$, that do not contain all the domains mentioned above (Guttman *et al.*, 2002). HopAB3-1$_{PmaES4326}$ is 385 amino acids long, while HopAB3-2$_{PmaES4326}$ is 155 amino acids long, with both T3SEs lacking the canonical CTD found in HopAB2$_{PtoDC3000}$ (Guttman *et al.*, 2002; Lin *et al.*, 2006). However, HopAB3-1$_{PmaES4326}$ can still promote virulence in susceptible tomato cultivars (Lin *et al.*, 2006). HopAB3-1$_{PmaES4326}$ contains two domains: the N-terminal domain, which corresponds to amino acids 1 to 307 of HopAB2$_{PtoDC3000}$ and a novel C-terminal domain (Singer *et al.*, 2012). The N- and C-terminal domains share structural similarity despite no sequence similarity, suggesting that they may target different plant host proteins (Singer *et al.*, 2012). Although the CTD of HopAB3-1$_{PmaES4326}$ partially contains the BAK1-binding domain of HopAB2$_{PtoDC3000}$, HopAB3-1$_{PmaES4326}$ does not bind to BAK1, which provides further support that HopAB3-1$_{PmaES4326}$ has alternative plant host targets (Singer *et al.*, 2012).

Although HopAB2$_{PtoDC3000}$ has been shown to enhance levels of both ethylene (Cohn and Martin, 2005; Xiao *et al.*, 2007) and abscisic acid (ABA) (de Torres-Zabala *et al.*, 2007) to promote disease, the molecular mechanism by which HopAB2$_{PtoDC3000}$ affects hormone homeostasis remains unclear. Ethylene and ABA are both plant hormones that have been shown to be important for many aspects of plant immunity as well as many developmental processes (Bleecker and Kende, 2000; Cao *et al.*, 2011). Ethylene plays a role in the progression of disease symptoms including disease-related cell death (Cohn and Martin, 2005), while ABA can regulate stomata, the entry point for many pathogens, and can suppress salicylic acid-mediated responses (Cao *et al.*, 2011). Given that $HopAB2_{PtoDC3000}$ can interact with numerous plant proteins, it is likely that these novel targets are the key to understanding the manipulation of plant hormones by $HopAB2_{PtoDC3000}$. The role of other HopAB family members on plant hormone levels remains to be tested and it will be interesting to see whether HopAB family members can manipulate other hormone pathways.

OspF/HopAI

The OspF family of T3SEs is found in a broad swath of animal pathogens in addition to *P. syringae* (Cui and Shao, 2011). The animal pathogen homologues all cluster together and, with the exception of *Morganella morganii* KT, they share > 50% amino acid identity, while the *P. syringae* homologues and those from two genera of marine bacteria are much more diverged (< 35% amino acid identity; Fig. 1.5). There is also high sequence divergence among *P. syringae* homologues, with sequence identity as low as 63% between *P. syringae* homologues.

The founding member of this family, OspF from *Shigella flexneri*, has been shown to modulate host innate immunity by inhibiting the MAPK pathway (Li *et al.*, 2007). Specifically, *Shigella* uses OspF to prevent the phosphorylation of multiple MAPKs involved in innate immunity. A heterologous yeast screen also identified the MAPK pathway as a target of OspF (Kramer *et al.*, 2007). The mechanism by which OspF dephosphorylates its MAPK targets is different from canonical phosphatases (Li *et al.*, 2007). Instead, OspF is a

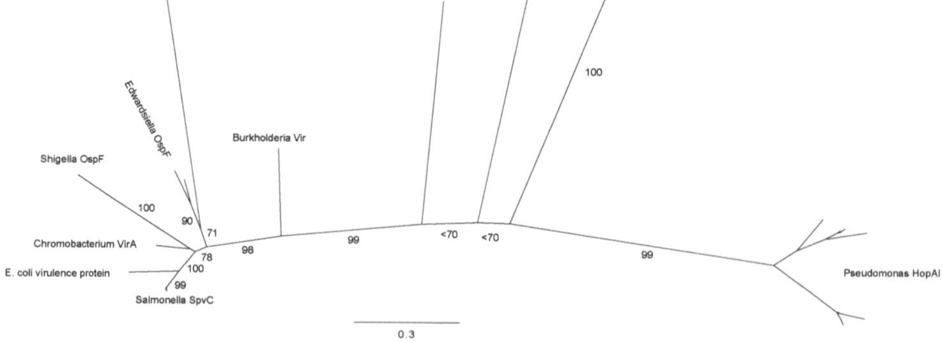

Figure 1.5 Evolutionary relationships among OspF/HopAI homologues. See Fig. 1.2 for details.

nuclear-localized phosphothreonine lyase that irreversibly inactivates MAPKs by removing phosphate groups from phosphothreonine residues and leaving behind a non-reactive methyl group (Li *et al.*, 2007). Mutations in the conserved K102, H104 or K134 residues of OspF inactivate its phosphothreonine lyase activity. However, these OspF mutants can still bind to the MAPK targets, indicating that K102, H104 and K134 residues are important for the enzymatic activity of OspF. The OspF homologue in non-typhoid *Salmonella* strains, SpvC, also functions as a phosphothreonine lyase to irreversibly inactivate the same MAPK targets (Li *et al.*, 2007; Mazurkiewicz *et al.*, 2008). Structural studies on *Salmonella* SpvC indicate that recognition of the substrates is dependent on the presence of two phosphorylated residues on the MAPKs (Chen *et al.*, 2008). Consequently, OspF- or SpvC-modified MAPKs can no longer be phosphorylated, and the host MAPK pathways are effectively disabled (Li *et al.*, 2007).

Interestingly, the distantly related *P. syringae* OspF homologue, HopAI1$_{\mathrm{PtoDC3000}}$ (37% amino acid identity), also targets the MAPK pathway (Zhang *et al.*, 2007). Transgenic expression of HopAI1$_{\mathrm{PtoDC3000}}$ in plants completely suppresses the accumulation of ROS, the activation of the MAPK pathway and cell wall-based defences in response to flagellin (Zhang *et al.*, 2007). Specifically, a co-immunoprecipitation experiment shows that HopAI1$_{\mathrm{PtoDC3000}}$ directly interacts with MPK3 and MPK6 *in planta* (Zhang *et al.*, 2007). HopAI1$_{\mathrm{PtoDC3000}}$-modified MPK3 and MPK6 are covalently inactivated and cannot be re-phosphorylated (Zhang *et al.*, 2007). Like the *Shigella* OspF, HopAI1$_{\mathrm{PtoDC3000}}$ is also a phosphothreonine lyase that irreversibly dephosphorylates the MAPK targets and shuts down the innate immune responses (Li *et al.*, 2007; Zhang *et al.*, 2007). Furthermore, the conserved H104 residue required for the phosphothreonine lyase activity of OspF is critical for HopAI1$_{\mathrm{PtoDC3000}}$ *in planta* function, as mutation of the corresponding HopAI1$_{\mathrm{PtoDC3000}}$ H102 residue abrogates the suppression of basal defence (Zhang *et al.*, 2007). Thus, the OspF family of T3SEs appears to be an exception to the pattern of extensive T3SE functional diversification because OspF, SpvC, and HopAI1$_{\mathrm{PtoDC3000}}$ all function as phosphothreonine lyases that modify MAPK targets despite their sequence divergence.

YopJ/HopZ

The YopJ family of effectors is widely distributed in bacterial pathogens of both plants and animals (Lewis *et al.*, 2011; Ma *et al.*, 2006). YopJ homologues are present in both plant-associated pathogenic and symbiotic species, including *Pseudomonas*, *Xanthomonas*, *Erwinia*, *Ralstonia* and *Rhizobium* (Lewis *et al.*, 2011; Ma *et al.*, 2006). All of the YopJ homologues from plant-associated bacteria group together in an unrooted phylogeny and have much more sequence divergence (< 30% amino acid identity) than the animal pathogen homologues, which are all > 50% amino acid identity (Fig. 1.6). They also differ from the animal pathogens in that sequences are often more similar to homologues from other genera than from other strains of the same pathogen. In particular, HopZ3 from *P. syringae* is closely related to Eop1 from *E. amylovora* suggesting relatively recent horizontal transfer between these species (Ma *et al.*, 2006), while HopZ2 and HopZ4 from *P. syringae*, PopP2 from *R. solanacearum* and homologues from *Acidovorax*, *Rhizobium* and *Mesorhizobium* are all nested within a cluster that includes four homologues in *Xanthomonas* spp., suggesting that this lineage originated in *Xanthomonas* and has recently been acquired by each of these other groups (Fig. 1.6).

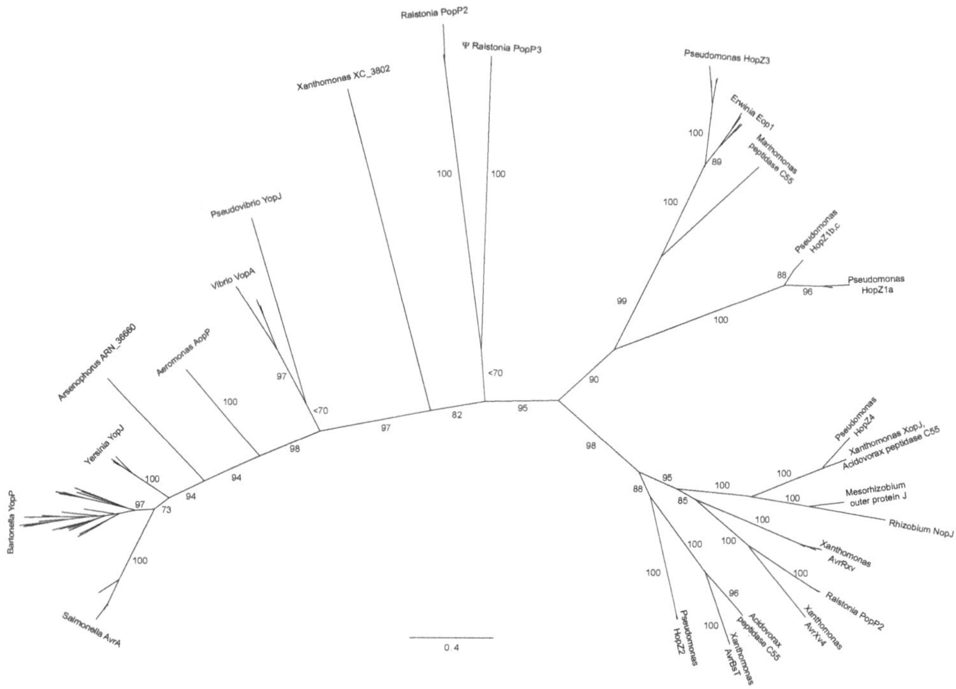

Figure 1.6 Evolutionary relationships among YopJ homologues. See Fig. 1.2 for details.

As seen with OspF, a major goal for animal pathogens is to hijack host signalling pathways, such as the MAPK and the NFκB pathways to perturb immune responses. The founding member of the YopJ family, YopJ from *Yersinia pestis,* also targets the MAPK and NFκB signalling pathways (Mittal *et al.*, 2006; Monack *et al.*, 1997; Mukherjee *et al.*, 2006; Orth *et al.*, 1999, 2000). In particular, YopJ inhibits immune signal transduction by inhibiting the phosphorylation of MAPKKs and IKK (Mukherjee *et al.*, 2006; Orth *et al.*, 1999, 2000). Early work predicted that YopJ prevents the activation of these kinases via its cysteine protease activity (Orth *et al.*, 2000). This prediction was based on the similarity between the catalytic domain of YopJ and the adenovirus protease (AVP) (Orth *et al.*, 2000). Indeed, YopJ appears to cleave the ubiquitin-like protein, SUMO-1, from unidentified host targets (Orth *et al.*, 2000).

Subsequently, two independent groups demonstrated that YopJ is an acetyltransferase (Mittal *et al.*, 2006; Mukherjee *et al.*, 2006). YopJ acetylates one of its MAPKK targets, MKK6, on the serine and threonine residues in the activation loop of the kinase (Mittal *et al.*, 2006; Mukherjee *et al.*, 2006). YopJ requires its catalytic cysteine residue (that is part of the conserved catalytic triad of histidine, glutamate, and cysteine) in order to modify its MAPKK targets (Mittal *et al.*, 2006; Mukherjee *et al.*, 2006; Yoon *et al.*, 2003). In the presence of acetyl-CoA and YopJ, MKK6 and IκBβ can no longer be phosphorylated by the upstream MAPKKK, thus providing an elegant mechanistic explanation for how YopJ inhibits the MAPK and NFκB signalling pathways (Mittal *et al.*, 2006; Mukherjee *et al.*, 2006).

Given the similarity to cysteine proteases, Mukherjee *et al.* subsequently proposed that YopJ modifies its substrates using a 'ping-pong' mechanism (Mukherjee *et al.*, 2007). Typically, cysteine proteases utilize the catalytic triad to form an acyl-enzyme covalent intermediate and then use water to attack the intermediate in a hydrolase reaction (Mukherjee *et al.*, 2007). Similarly, YopJ utilizes its catalytic triad to first form an acetyl-enzyme covalent intermediate, followed by the transfer of the acetyl group to the serine or threonine residues of the substrates, the MAPKKs (Mukherjee *et al.*, 2007). In addition, YopJ appears to require a eukaryotic co-factor inositol hexakisphosphate (IP6, also known as phytic acid) for full activation of its acetyltransferase activity (Mittal *et al.*, 2010). Upon activation, YopJ acetylates itself and its host MAPKK targets (Mittal *et al.*, 2010). However, the identity of the autoacetylation site and the function of autoacetylation (if any) remain to be determined.

Recent work has shown that in addition to acetylating its MAPKK targets, YopJ also acetylates kinases, such as RICK and TAK1, which are upstream of the MAPK and NFκB signalling pathways (Meinzer *et al.*, 2012; Paquette *et al.*, 2012). Using similar mechanisms as for the MAPKKs, YopJ acetylates RICK and TAK1 at critical serine and threonine residues within the kinase activation loop, consequently inhibiting these kinases (Meinzer *et al.*, 2012; Paquette *et al.*, 2012). Thus, by targeting multiple kinases within the same pathway, YopJ effectively disables the host immune responses (Jones and Neish, 2012; Krachler *et al.*, 2011).

Interestingly, other YopJ homologues in animal pathogens have been shown to target different components of the MAPK and NFκB signalling pathways (Collier-Hyams *et al.*, 2002; Fehr *et al.*, 2006; Krachler *et al.*, 2011; Wu *et al.*, 2010). For instance, VopA, a homologue of YopJ in *Vibrio parahaemolyticus*, targets the MAPK signalling pathways in both mammalian and yeast cells (Trosky *et al.*, 2004). However, unlike YopJ, VopA does not inhibit the NFκB pathway (Trosky *et al.*, 2004). Furthermore, even though VopA is an acetyltransferase that targets MAPKK6 (MKK6), the mechanism of VopA inhibition is different from YopJ. Specifically, VopA not only acetylates in the activation loop of MKK6, but also acetylates a conserved lysine residue that is important for ATP binding of MKK6 (Trosky *et al.*, 2007). Consequently, VopA-acetylated MKK6 cannot bind ATP and the host MAPK signalling cascade is disabled (Trosky *et al.*, 2007). It is not known if VopA requires IP6 activation.

The AvrA effector from *Salmonella typhimurium* requires IP6 in order to activate its acetyltransferase activity (Mittal *et al.*, 2010), which inhibits both the MAPK and NFκB pathways by acetylating MKK4 and MKK7 (Du and Galan, 2009; Jones *et al.*, 2008). On the other hand, AopP from *Aeromonas salmonicida* targets the NFκB pathway specifically and does not inhibit the MAPK pathway (Fehr *et al.*, 2006; Jones *et al.*, 2012). In particular, AopP inhibits the phosphorylation of IKKβ (Jones *et al.*, 2012).

One of the best-characterized YopJ homologues in phytopathogens is the HopZ family from the *P. syringae* species complex. There are currently four HopZ subfamilies (HopZ1, HopZ2, HopZ3, and HopZ4) identified, with no isolate of *P. syringae* carrying more than one functional allele (Baltrus *et al.*, 2011; Ma *et al.*, 2006; O'Brien *et al.*, 2012). Interestingly, *Xanthomonas* also has four known YopJ homologues (AvrBsT, AvrRxv, AvrXv4 and XopJ), but unlike *Pseudomonas*, each *Xanthomonas* pathovar can carry multiple YopJ homologues (Ciesiolka *et al.*, 1999; Szczesny *et al.*, 2010). Similarly, the bacterial wilt pathogen *Ralstonia solanacearum* can also carry multiple YopJ homologues (PopP1, PopP2, or PopP3) within a particular pathovar (Lavie *et al.*, 2002, 2004). These effectors all have the conserved catalytic triad (histidine, glutamate and cysteine) necessary for cysteine protease or acetyltransferase activities (Ma *et al.*, 2006; Mukherjee *et al.*, 2006). Furthermore, this conserved catalytic triad is required for *in planta* functions of many YopJ homologues in bacterial phytopathogens (Lewis *et al.*, 2011). However, unlike the YopJ homologues in animal pathogens, the host targets, cellular localization and functions of YopJ family members in plant pathogens are highly diverse (Lewis *et al.*, 2011). In the following sections, we will discuss the functional diversification of the YopJ T3SE family in bacterial phytopathogens.

Early work characterizing the YopJ homologues in plant pathogens predominantly focused on examining the plant immune responses towards different homologues. For instance, AvrRxv from *X. campestris* pv. *vesicatoria* (*Xcv*) induces a strong immune HR response in a wide variety of plant hosts, including common bean, soybean, cow pea, alfalfa, maize, tomato and cotton (Whalen *et al.*, 1988, 1993). The *Xanthomonas* AvrXv4 is recognized in *Solanum pennellii* and *Nicotiana benthamiana* (Astua-Monge *et al.*, 2000; Roden *et al.*, 2004a). The *Ralstonia* PopP1 elicits HR in *Petunia* (Lavie *et al.*, 2002) while PopP2 causes HR in *Arabidopsis* (Deslandes *et al.*, 2002). HopZ1a$_{PsyA2}$ from *P. syringae* elicits a strong defence response in a wide range of plant species, including *Arabidopsis*, *N. benthamiana* and soybean (Ma *et al.*, 2006). On the other hand, the closely related HopZ1b$_{PgyBR1}$ (72.1% amino acid identity to HopZ1a) is weakly recognized in *Arabidopsis*, strongly recognized in *N. benthamiana* and not recognized in soybean (Ma *et al.*, 2006; Zhou *et al.*, 2009). Given that many phytopathogen YopJ homologues induce HRs, a number of groups have tackled the functional characterization of these T3SEs by identifying their cognate R proteins.

The first R protein shown to recognize a phytopathogen YopJ homologue was RRS1 from the *Arabidopsis* ecotype Nd-1, which recognizes the *Ralstonia* PopP2 (Deslandes *et al.*, 2002). The recognition of PopP2 by RRS1 (resistant to *Ralstonia solanacearum* 1) requires direct interaction between PopP2 and RRS1 in the nucleus (Deslandes *et al.*, 2003; Tasset *et al.*, 2010). RRS1 recognizes PopP2 specifically as RRS1 does not interact with the other *Ralstonia* YopJ homologue, PopP1 (Deslandes *et al.*, 2003). RRS1-mediated immunity requires catalytically active PopP2 as it is dependent on the conserved catalytic cysteine (C321) residue (Tasset *et al.*, 2010). Furthermore, the RRS1-mediated resistance requires the presence of an *Arabidopsis* cysteine protease, RD19 (response to dehydration 19), which co-localizes with PopP2 in the nucleus (Bernoux *et al.*, 2008). Using liquid chromatography/tandem mass spectrometry (LC-MS/MS), Tasset *et al.* (2010) also demonstrate that PopP2 is an acetyltransferase that autoacetylates on a conserved lysine (K383) residue. Interestingly, the autoacetylation mutant PopP2(K383R) is enzymatically inactive and behaves like the catalytic null PopP2(C321A) mutant (Tasset *et al.*, 2010). Consequently, PopP2(K383R) is not recognized by the cognate R protein, RRS1 (Tasset *et al.*, 2010). Therefore, Tasset *et al.* (2010) hypothesize that the conserved K383 residue may function as an acetyl-CoA binding site. Given that PopP2 does not require a eukaryotic cofactor for activation of its autoacetylation activity, the mechanisms by which PopP2 functions as an acetyltransferase may be different from YopJ. Additionally, PopP2 does not appear to acetylate RRS1 or RD19, thus the *in planta* host target of PopP2 acetyltransferase activity is not known (Tasset *et al.*, 2010). PopP2 is also recognized by another *Arabidopsis* R protein, RPS4 (resistant to *Pseudomonas syringae* 4) (Narusaka *et al.*, 2009). However, the mechanism by which RPS4 recognizes PopP2 remains to be determined.

In *Arabidopsis*, HopZ1a$_{PsyA2}$ and the closely related HopZ1b$_{PgyBR1}$ also induce allele-specific responses. Particularly, HopZ1a$_{PsyA2}$ triggers a strong HR response in *Arabidopsis* while HopZ1b$_{PgyBR1}$ triggers a weak HR in only ~25% of plants (Lewis *et al.*, 2008). The R protein that recognizes HopZ1a$_{PsyA2}$ in *Arabidopsis* is an R protein that belongs to the CC-NBS-LRR class, ZAR1 (HopZ-activated resistance) (Lewis *et al.*, 2010). Importantly, ZAR1 does not recognize HopZ1b$_{PgyBR1}$, indicating that HopZ1b$_{PgyBR1}$ has acquired mutations that result in loss of *Arabidopsis* ZAR1 immune detection (Lewis *et al.*, 2008, 2010). The sequence diversification responsible for the HopZ1b$_{PgyBR1}$ evasion of *Arabidopsis* immunity is currently unknown. However, the recognition of HopZ1a$_{PsyA2}$ by *Arabidopsis* is dependent on the catalytic cysteine residue (C216) and myristoylation site (G2) required for membrane-localization (Lewis *et al.*, 2008, 2010). In contrast to the direct interaction between RRS1 and PopP2, there is no current evidence that suggests ZAR1 interacts directly with HopZ1a$_{PsyA2}$. Instead, it is proposed that ZAR1 recognizes HopZ1a$_{PsyA2}$ indirectly by detecting HopZ1a$_{PsyA2}$ enzymatic activities and potentially the HopZ1a$_{PsyA2}$-mediated modification of membrane-localized host proteins (Lewis *et al.*, 2008, 2010). However, beyond the requirement for the catalytic cysteine and the myristoylation glycine residues, the mechanisms by which HopZ1a$_{PsyA2}$ triggers ZAR1-mediated defences are unknown. In the absence of ZAR1, HopZ1a$_{PsyA2}$ promotes the growth of a non-host *P. syringae* in a catalytic-dependent manner, indicating that HopZ1a$_{PsyA2}$ has a virulence function in *Arabidopsis* (Lewis *et al.*, 2010).

In soybean, different *hopZ* alleles also induce host-specific responses; HopZ1a$_{PsyA2}$ induces a strong HR while HopZ1b$_{PgyBR1}$ is not recognized and promotes *P. syringae* growth (Morgan *et al.*, 2010; Zhou *et al.*, 2009). While the soybean R protein that recognizes HopZ1a$_{PsyA2}$ has yet to be identified, the effector sequence diversification responsible for differences in host recognition has been identified. Specifically, Morgan *et al.* reasoned that the allelic specificity of *hopZ1* might be due to positively selected regions within HopZ1 effectors that may confer fitness advantages to the pathogen (Morgan *et al.*, 2010). Remarkably, replacing the positively selected C141 residue of HopZ1a$_{PsyA2}$ to the corresponding lysine residue in HopZ1b$_{PgyBR1}$ abolishes HopZ1a$_{PsyA2}$-induced HR in soybean (Morgan *et al.*, 2010). HopZ1a$_{PsyA2}$(C141K) still triggers an HR in *N. benthamiana*, indicating that the loss of HR in soybean is not due to the loss of enzymatic activity in HopZ1a$_{PsyA2}$(C141K) (Morgan *et al.*, 2010). Using structural modelling, Morgan *et al.* (2010) speculate that the C141 residue may be involved in HopZ1a$_{PsyA2}$ substrate binding and may provide an explanation for the allelic specificity in soybean.

Another approach used to characterize phytopathogen YopJ homologues has been to identify YopJ homologues that suppress host immunity. For instance, the *Xanthomonas* AvrBsT has been shown to suppress the HR that is induced by an unrelated T3SE, AvrBs1 (Szczesny *et al.*, 2010). Similarly, *P. syringae* HopZ3$_{PsyB728a}$ has been shown to inhibit the *N. benthamiana* HR that is induced by various effectors, including AvrPto1$_{PsyB728a}$, HopAA1$_{PsyB728a}$, HopM1$_{PsyB728a}$ and HopAE1$_{PsyB728a}$ (Vinatzer *et al.*, 2006). Additionally, HopZ3$_{PsyB728a}$ can suppress the HopZ1b$_{PgyBR1}$-triggered HR, but not the HopZ1a$_{PsyA2}$-triggered HR in *N. benthamiana*, providing additional evidence of functional diversification between different *hopZ* alleles (Zhou *et al.*, 2009). Given that HopZ3$_{PsyB728a}$ can suppress immunity triggered by a wide range of unrelated T3SEs, it is possible that HopZ3$_{PsyB728a}$ may target an upstream signalling component in plant immunity.

To gain molecular insight into the function of phytopathogen YopJ homologues, recent work has focused on identifying the host targets of these T3SEs. One strategy has been yeast two-hybrid (Y2H) screens to identify plant proteins that directly interact with these phytopathogen T3SEs (Lewis *et al.*, 2012; Szczesny *et al.*, 2010; Whalen *et al.*, 2008; Zhou *et al.*, 2011). For example, the *Xanthomonas* AvrRxv has been shown to interact with a 14-3-3 protein from tomato in both Y2H and *in vitro* assays (Whalen *et al.*, 2008). Mutations in AvrRxv that disrupt the interaction with 14-3-3 abolished AvrRxv-induced HR, indicating that the 14-3-3 binding motif is important for the *in planta* function of AvrRxv (Whalen *et al.*, 2008). Another *Xanthomonas* YopJ homologue, AvrBsT, has been shown to interact with a SNF1-related kinase 1 (SnRK1) in Y2H and *in planta* (Szczesny *et al.*, 2010). Even though the catalytic triads of AvrRxv and AvrBsT are essential for *in planta* functions, the enzymatic activities by which AvrRxv and AvrBsT modify their plant interactors are not known (Orth *et al.*, 2000; Whalen *et al.*, 2008).

In soybean, both HopZ1a$_{PsyA2}$ and HopZ1b$_{PgyBR1}$ interact with a protein involved in isoflavonoid biosynthesis, known as GmHID1 (or enzyme 2-hydroxyisoflavanone dehydratases) (Zhou *et al.*, 2011). Transient expression of HopZ1a$_{PsyA2}$ or HopZ1b$_{PgyBR1}$ in *N. benthamiana* induces a reduction in GmHID1 protein levels (Zhou *et al.*, 2011). GmHID1 is important for basal defence, as silencing of *gmhid1* transcription promotes *P. syringae* virulence on soybean (Zhou *et al.*, 2011). In contrast, HopZ2$_{Ppi895A}$, which interacts with the *Arabidopsis* MLO2 protein in a modified Y2H screen, requires MLO2 for its virulence function (Lewis *et al.*, 2012). However, even though GmHID1 and MLO2 are important for HopZ functions *in planta*, the molecular mechanisms by which HopZ1 or HopZ2 modify these proteins has yet to be characterized.

Recently, HopZ1a$_{PsyA2}$ has been shown to be an acetyltransferase (Lee *et al.*, 2012). Using a heterologous system to identify interacting proteins of HopZ1a$_{PsyA2}$ in human cells, tubulin was identified as an interactor of HopZ1a$_{PsyA2}$ (Lee *et al.*, 2012). Like its YopJ homologue, HopZ1a$_{PsyA2}$ requires a eukaryotic cofactor, IP6, to activate its acetyltransferase activities whereby activated HopZ1a$_{PsyA2}$ acetylates itself and tubulin *in vitro* (Lee *et al.*, 2012). Most importantly, HopZ1a$_{PsyA2}$ binds to plant tubulin and causes destruction of plant microtubule networks in *Arabidopsis* (Lee *et al.*, 2012). However, it is unclear whether HopZ1a$_{PsyA2}$ manipulates the plant microtubule networks directly or indirectly. Nevertheless, HopZ1a$_{PsyA2}$ requires its acetyltransferase activity to cause microtubule destruction, inhibit secretion and consequently block cell wall-based defences in *Arabidopsis* (Lee *et al.*, 2012). Interestingly, the *Xanthomonas* XopJ has also been shown to block secretion and cell wall-based defences (Bartetzko *et al.*, 2009). However, it is not known whether XopJ also targets the host cytoskeleton or inhibits the plant secretory pathways via a similar mechanism as HopZ1a$_{PsyA2}$. Studying the different YopJ homologues has provided insights into the various ways pathogens utilize a conserved catalytic triad to manipulate host processes (Jones and Neish, 2012; Krachler *et al.*, 2011; Lewis *et al.*, 2011).

Conclusions

High-throughput screens and next-generation sequencing have identified a large number of T3SEs in diverse groups of bacterial species. However, functional validation and characterization of these T3SEs lag far behind their discovery. While it is tempting to predict the function and molecular mechanisms of newly discovered effectors based on the characterized homologues, many T3SE homologues demonstrate enormous functional diversification within any particular family. One of the few exceptions to this observation is the OspF family, where distantly related homologues maintain the same enzymatic activity as a phosphothreonine lyase to modify their host MAPK targets. As discussed in this book chapter, multidomain T3SEs of the AvrPto or the HopAB family display varied *in planta* function by virtue of diverse host targets that they can interact with. Even T3SEs with conserved catalytic triad and potential enzymatic activities, such as members of the AvrB or the YopJ family, possess diverse function and host targets. These examples illustrate the importance of molecular characterization of multiple homologues within any particular T3SE family. Ultimately, homologous T3SEs that have been independently acquired by distantly unrelated bacteria can serve as promising targets for understanding how pathogens modulate their interactions with their hosts.

References

Abascal, F., Zardoya, R., and Telford, M.J. (2010). TranslatorX: multiple alignment of nucleotide sequences guided by amino acid translations. Nucleic Acids Res. *38*, W7–13.

Abramovitch, R.B., and Martin, G.B. (2005). AvrPtoB: a bacterial type III effector that both elicits and suppresses programmed cell death associated with plant immunity. FEMS Microbiol. Lett. *245*, 1–8.

Abramovitch, R.B., Janjusevic, R., Stebbins, C.E., and Martin, G.B. (2006). Type III effector AvrPtoB requires intrinsic E3 ubiquitin ligase activity to suppress plant cell death and immunity. Proc. Natl. Acad. Sci. U.S.A. *103*, 2851–2856.

Alfano, J.R., Charkowski, A.O., Deng, W.L., Badel, J.L., Petnicki-Ocwieja, T., van Dijk, K., and Collmer, A. (2000). The *Pseudomonas syringae* Hrp pathogenicity island has a tripartite mosaic structure composed of a cluster of type III secretion genes bounded by exchangeable effector and conserved effector loci that contribute to parasitic fitness and pathogenicity in plants. Proc. Natl. Acad. Sci. U.S.A. *97*, 4856–4861.

Almeida, N.F., Yan, S., Lindeberg, M., Studholme, D.J., Schneider, D.J., Condon, B., Liu, H., Viana, C.J., Warren, A., Evans, C., *et al.* (2009). A draft genome sequence of *Pseudomonas syringae* pv. *tomato* T1 reveals a type III effector repertoire significantly divergent from that of *Pseudomonas syringae* pv. *tomato* DC3000. Mol. Plant Microbe Interact. *22*, 52–62.

Anisimova, M., and Gascuel, O. (2006). Approximate likelihood-ratio test for branches: a fast, accurate, and powerful alternative. Syst. Biol. *55*, 539–552.

Araki, H., Innan, H., Kreitman, M., and Bergelson, J. (2007). Molecular evolution of pathogenicity-island genes in *Pseudomonas viridiflava*. Genetics *177*, 1031–1041.

Astua-Monge, G., Minsavage, G.V., Stall, R.E., Vallejos, C.E., Davis, M.J., and Jones, J.B. (2000). Xv4-vrxv4: a new gene-for-gene interaction identified between *Xanthomonas campestris* pv. *vesicatoria* race T3 and wild tomato relative *Lycopersicon pennellii*. Mol. Plant Microbe Interact. *13*, 1346–1355.

Baltrus, D.A., Nishimura, M.T., Romanchuk, A., Chang, J.H., Mukhtar, M.S., Cherkis, K., Roach, J., Grant, S.R., Jones, C.D., and Dangl, J.L. (2011). Dynamic evolution of pathogenicity revealed by sequencing and comparative genomics of 19 *Pseudomonas syringae* isolates. PLoS Pathog. 7, e1002132.

Bartetzko, V., Sonnewald, S., Vogel, F., Hartner, K., Stadler, R., Hammes, U.Z., and Bornke, F. (2009). The *Xanthomonas campestris* pv. *vesicatoria* type III effector protein XopJ inhibits protein secretion: evidence for interference with cell wall-associated defense responses. Mol. Plant Microbe Interact. *22*, 655–664.

Belkhadir, Y., Nimchuk, Z., Hubert, D.A., Mackey, D., and Dangl, J.L. (2004). *Arabidopsis* RIN4 negatively regulates disease resistance mediated by RPS2 and RPM1 downstream or independent of the NDR1 signal modulator and is not required for the virulence functions of bacterial type III effectors AvrRpt2 or AvrRpm1. Plant Cell *16*, 2822–2835.

Bernoux, M., Timmers, T., Jauneau, A., Briere, C., de Wit, P.J., Marco, Y., and Deslandes, L. (2008). RD19, an *Arabidopsis* cysteine protease required for RRS1-R-mediated resistance, is relocalized to the nucleus by the *Ralstonia solanacearum* PopP2 effector. Plant Cell *20*, 2252–2264.

Bittel, P., and Robatzek, S. (2007). Microbe-associated molecular patterns (MAMPs) probe plant immunity. Curr. Opin. Plant Biol. *10*, 335–341.

Bleecker, A.B., and Kende, H. (2000). Ethylene: a gaseous signal molecule in plants. Annu. Rev. Cell Dev. Biol. *16*, 1–18.

Boch, J., and Bonas, U. (2010). *Xanthomonas* AvrBs3 family-type III effectors: discovery and function. Annu. Rev. Phytopathol. *48*, 419–436.

Boch, J., Joardar, V., Gao, L., Robertson, T.L., Lim, M., and Kunkel, B.N. (2002). Identification of *Pseudomonas syringae* pv. *tomato* genes induced during infection of *Arabidopsis thaliana*. Mol. Microbiol. *44*, 73–88.

Bogdanove, A.J., Koebnik, R., Lu, H., Furutani, A., Angiuoli, S.V., Patil, P.B., Van Sluys, M.-A., Ryan, R.P., Meyer, D.F., Han, S.-W., *et al.* (2011). Two new complete genome sequences offer insight into host and tissue specificity of plant pathogenic *Xanthomonas* spp. J. Bacteriol. *193*, 5450–5464.

Buist, G., Steen, A., Kok, J., and Kuipers, O.P. (2008). LysM, a widely distributed protein motif for binding to peptidoglycans. Mol. Microbiol. *68*, 838–847.

Bull, C.T., Clarke, C.R., Cai, R., Vinatzer, B.A., Jardini, T.M., and Koike, S.T. (2011). Multilocus sequence typing of *Pseudomonas syringae* Sensu Lato confirms previously described genomospecies and permits rapid identification of *P. syringae* pv. *coriandricola* and *P. syringae* pv. *apii* causing bacterial leaf spot on parsley. Phytopathology *101*, 847–858.

Cao, F.Y., Yoshioka, K., and Desveaux, D. (2011). The roles of ABA in plant–pathogen interactions. J. Plant Res. *124*, 489–499.

Casper-Lindley, C., Dahlbeck, D., Clark, E.T., and Staskawicz, B.J. (2002). Direct biochemical evidence for type III secretion-dependent translocation of the AvrBs2 effector protein into plant cells. Proc. Natl. Acad. Sci. U.S.A. *99*, 8336–8341.

Castillo, J.A., and Greenberg, J.T. (2007). Evolutionary dynamics of *Ralstonia solanacearum*. Appl. Environ. Microbiol. *73*, 1225–1238.

Chang, J.H., Rathjen, J.P., Bernal, A.J., Staskawicz, B.J., and Michelmore, R.W. (2000). *avrPto* enhances growth and necrosis caused by *Pseudomonas syringae* pv. *tomato* in tomato lines lacking either Pto or Prf. Mol. Plant Microbe Interact. *13*, 568–571.

Chang, J.H., Tobias, C.M., Staskawicz, B.J., and Michelmore, R.W. (2001). Functional studies of the bacterial avirulence protein AvrPto by mutational analysis. Mol. Plant Microbe Interact. *14*, 451–459.

Chang, J.H., Urbach, J.M., Law, T.F., Arnold, L.W., Hu, A., Gombar, S., Grant, S.R., Ausubel, F.M., and Dangl, J.L. (2005). A high-throughput, near saturating screen for type III effector genes from *Pseudomonas syringae*. Proc. Natl. Acad. Sci. U.S.A. *102*, 2549–2554.

Chen, L., Wang, H., Zhang, J., Gu, L., Huang, N., Zhou, J.M., and Chai, J. (2008). Structural basis for the catalytic mechanism of phosphothreonine lyase. Nat. Struct. Mol. Biol. *15*, 101–102.

Chen, N.W., Sevignac, M., Thareau, V., Magdelenat, G., David, P., Ashfield, T., Innes, R.W., and Geffroy, V. (2010). Specific resistances against *Pseudomonas syringae* effectors AvrB and AvrRpm1 have evolved differently in common bean (*Phaseolus vulgaris*), soybean (*Glycine max*), and *Arabidopsis thaliana*. New Phytol. *187*, 941–956.

Chung, E.-H., da Cunha, L., Wu, A.-J., Gao, Z., Cherkis, K., Afzal, A.J., Mackey, D., and Dangl, J.L. (2011). Specific threonine phosphorylation of a host target by two unrelated type III effectors activates a host innate immune receptor in plants. Cell Host Microbe *9*, 125–136.

Ciesiolka, L.D., Hwin, T., Gearlds, J.D., Minsavage, G.V., Saenz, R., Bravo, M., Handley, V., Conover, S.M., Zhang, H., Caporgno, J., *et al.* (1999). Regulation of expression of avirulence gene *avrRxv* and identification of a family of host interaction factors by sequence analysis of *avrBsT*. Mol. Plant Microbe Interact. *12*, 35–44.

Clarke, C.R., Cai, R., Studholme, D.J., Guttman, D.S., and Vinatzer, B.A. (2010). *Pseudomonas syringae* strains naturally lacking the classical *P. syringae hrp/hrc* locus are common leaf colonizers equipped with an atypical type III secretion system. Mol. Plant Microbe Interact. *23*, 198–210.

Cohn, J.R., and Martin, G.B. (2005). *Pseudomonas syringae* pv. *tomato* type III effectors AvrPto and AvrPtoB promote ethylene-dependent cell death in tomato. Plant J. *44*, 139–154.

Collier-Hyams, L.S., Zeng, H., Sun, J., Tomlinson, A.D., Bao, Z.Q., Chen, H., Madara, J.L., Orth, K., and Neish, A.S. (2002). Cutting edge: *Salmonella* AvrA effector inhibits the key proinflammatory, anti-apoptotic NF-kappa B pathway. J. Immunol. *169*, 2846–2850.

Cui, H., Wang, Y., Xue, L., Chu, J., Yan, C., Fu, J., Chen, M., Innes, R.W., and Zhou, J.M. (2010). *Pseudomonas syringae* effector protein AvrB perturbs *Arabidopsis* hormone signaling by activating MAP kinase 4. Cell Host Microbe *7*, 164–175.

Cui, J., and Shao, F. (2011). Biochemistry and cell signaling taught by bacterial effectors. Trends Biochem. Sci. *36*, 532–540.

Cunnac, S., Occhialini, A., Barberis, P., Boucher, C., and Genin, S. (2004). Inventory and functional analysis of the large Hrp regulon in *Ralstonia solanacearum*: identification of novel effector proteins translocated to plant host cells through the type III secretion system. Mol. Microbiol. *53*, 115–128.

Debener, T., Lehnackers, H., Arnold, M., and Dangl, J.L. (1991). Identification and molecular mapping of a single *Arabidopsis thaliana* locus determining resistance to a phytopathogenic *Pseudomonas syringae* isolate. Plant J. *1*, 289–302.

Denny, T. (2006). Plant pathogenic *Ralstonia* species. In Plant-associated Bacteria, S.S. Gnanamanickam, ed. (Springer, Dordrecht, The Netherlands), 573–644.

Deslandes, L., Olivier, J., Theulieres, F., Hirsch, J., Feng, D.X., Bittner-Eddy, P., Beynon, J., and Marco, Y. (2002). Resistance to *Ralstonia solanacearum* in *Arabidopsis thaliana* is conferred by the recessive RRS1-R gene, a member of a novel family of resistance genes. Proc. Natl. Acad. Sci. U.S.A. *99*, 2404–2409.

Deslandes, L., Olivier, J., Peeters, N., Feng, D.X., Khounlotham, M., Boucher, C., Somssich, I., Genin, S., and Marco, Y. (2003). Physical interaction between RRS1-R, a protein conferring resistance to bacterial wilt, and PopP2, a type III effector targeted to the plant nucleus. Proc. Natl. Acad. Sci. U.S.A. *100*, 8024–8029.

Desveaux, D., Singer, A.U., Wu, A.J., McNulty, B.C., Musselwhite, L., Nimchuk, Z., Sondek, J., and Dangl, J.L. (2007). Type III effector activation via nucleotide binding, phosphorylation, and host target interaction. PLoS Pathog. *3*, e48.

Du, F., and Galan, J.E. (2009). Selective inhibition of type III secretion activated signaling by the *Salmonella* effector AvrA. PLoS Pathog. *5*, e1000595.

Eitas, T.K., Nimchuk, Z.L., and Dangl, J.L. (2008). *Arabidopsis* TAO1 is a TIR-NB-LRR protein that contributes to disease resistance induced by the *Pseudomonas syringae* effector AvrB. Proc. Natl. Acad. Sci. U.S.A. *105*, 6475–6480.

Fehr, D., Casanova, C., Liverman, A., Blazkova, H., Orth, K., Dobbelaere, D., Frey, J., and Burr, S.E. (2006). AopP, a type III effector protein of *Aeromonas salmonicida*, inhibits the NF-kappaB signalling pathway. Microbiology *152*, 2809–2818.

Ferreira, A.O., Myers, C.R., Gordon, J.S., Martin, G.B., Vencato, M., Collmer, A., Wehling, M.D., Alfano, J.R., Moreno-Hagelsleb, G., Lamboy, W.F., *et al.* (2006). Whole-genome expression profiling defines the HrpL regulon of *Pseudomonas syringae* pv. *tomato* DC3000, allows *de novo* reconstruciton of the Hrp *cis* element, and identifies novel coregulated genes. Mol. Plant Microbe Interact. *19*, 1167–1179.

Fouts, D.E., Abramovich, R.B., Alfano, J.R., Baldo, A., Buell, C.R., Cartinhour, S.W., Chatterjee, A.K., D'Ascenzo, M., Gwinn, M.L., Lazarowitz, S.G., *et al.* (2002). Genomewide identification of *Pseudomonas syringae* pv. *tomato* DC3000 promotors controlled by the HrpL alternative sigma factor. Proc. Natl. Acad. Sci. U.S.A. *99*, 2275–2280.

Furutani, A., Takaoka, M., Sanada, H., Noguchi, Y., Oku, T., Tsuno, K., Ochiai, H., and Tsuge, S. (2009). Identification of novel type III secretion effectors in *Xanthomonas oryzae* pv. *oryzae*. Mol. Plant Microbe Interact. *22*, 96–106.

Gabriel, D.W., Burges, A., and Lazo, G.R. (1986). Gene-for-gene interactions of five cloned avirulence genes from *Xanthomonas campestris* pv. *malvacearum* with specific resistance genes in cotton. Proc. Natl. Acad. Sci. U.S.A. *83*, 6415–6419.

Gardan, L., Shafik, H., Belouin, S., Broch, R., Grimont, F., and Grimont, P.A.D. (1999). DNA relatedness among the pathovars of *Pseudomonas syringae* and description of *Pseudomonas tremae* sp. nov. and *Pseudomonas cannabina* sp. nov. (*ex* Sutic and Dowson 1959). Int. J. Syst. Bacteriol. *49*.

Genin, S., and Denny, T.P. (2012). Pathogenomics of the *Ralstonia solanacearum* species complex. Annu. Rev. Phytopathol. *50*, 67–89.

Genin, S., Gough, C.L., Zischek, C., and Boucher, C.A. (1992). Evidence that the *hrpB* gene encodes a positive regulator of pathogenicity genes from *Pseudomonas solanacearum*. Mol. Microbiol. *6*, 3065–3076.

Gimenez-Ibanez, S., Hann, D.R., Ntoukakis, V., Petutschnig, E., Lipka, V., and Rathjen, J.P. (2009). AvrPtoB targets the LysM receptor Kinase CERK1 to promote bacterial virulence on plants. Curr. Biol. *19*, 423–429.

Glasner, J.D., Yang, C.H., Reverchon, S., Hugouvieux-Cotte-Pattat, N., Condemine, G., Bohin, J.P., Van Gijsegem, F., Yang, S., Franza, T., Expert, D., *et al.* (2011). Genome sequence of the plant-pathogenic bacterium *Dickeya dadantii* 3937. J. Bacteriol. *193*, 2076–2077.

Gohre, V., Spallek, T., Haweker, H., Mersmann, S., Mentzel, T., Boller, T., de Torres, M., Mansfield, J.W., and Robatzek, S. (2008). Plant pattern-recognition receptor FLS2 is directed for degradation by the bacterial ubiquitin ligase AvrPtoB. Curr. Biol. *18*, 1824–1832.

Gomez-Gomez, L., and Boller, T. (2000). FLS2: an LRR receptor like kinase involved in the perception of the bacterial elicitor flagellin in *Arabidopsis*. Mol. Cell *5*, 1003–1011.

Guidot, A., Prior, P., Schoenfeld, J., Carrere, S., Genin, S., and Boucher, C. (2007). Genomic structure and phylogeny of the plant pathogen *Ralstonia solanacearum* inferred from gene distribution analysis. J. Bacteriol. *189*, 377–387.

Guindon, S., Dufayard, J.F., Lefort, V., Anisimova, M., Hordijk, W., and Gascuel, O. (2010). New algorithms and methods to estimate maximum-likelihood phylogenies: assessing the performance of PhyML 3.0. Syst. Biol. *59*, 307–321.

Guttman, D.S., and Greenberg, J.T. (2001). Functional analysis of the type III effectors AvrRpt2 and AvrRpm1 of *Pseudomonas syringae* with the use of a single-copy genomic integration system. Mol. Plant Microbe Interact. *14*, 145–155.

Guttman, D.S., Vinatzer, B.A., Sarkar, S.F., Ranall, M.V., Kettler, G., and Greenberg, J.T. (2002). A functional screen for the type III (Hrp) secretome of the plant pathogen *Pseudomonas syringae*. Science *295*, 1722–1726.

Hauben, L., Moore, E.R.B., Vauterin, L., Steenackers, M., Mergaert, J., Verdonck, L., and Swings, J. (1998). Phylogenetic position of phytopathogens within the Enterobacteriaceae. Syst. Appl. Microbiol. *21*, 384–397.

He, P., Shan, L., Lin, N.C., Martin, G.B., Kemmerling, B., Nurnberger, T., and Sheen, J. (2006). Specific bacterial suppressors of MAMP signaling upstream of MAPKKK in *Arabidopsis* innate immunity. Cell *125*, 563–575.

Heese, A., Hann, D.R., Gimenez-Ibanez, S., Jones, A.M., He, K., Li, J., Schroeder, J.I., Peck, S.C., and Rathjen, J.P. (2007). The receptor-like kinase SERK3/BAK1 is a central regulator of innate immunity in plants. Proc. Natl. Acad. Sci. U.S.A. *104*, 12217–12222.

Holeva, M.C., Bell, K.S., Hyman, L.J., Avrova, A.O., Whisson, S.C., Birch, P.R., and Toth, I.K. (2004). Use of a pooled transposon mutation grid to demonstrate roles in disease development for *Erwinia carotovora* subsp. *atroseptica* putative type III secreted effector (DspE/A) and helper (HrpN) proteins. Mol. Plant Microbe Interact. *17*, 943–950.

Iizasa, E., Mitsutomi, M., and Nagano, Y. (2010). Direct binding of a plant LysM receptor-like kinase, LysM RLK1/CERK1, to chitin *in vitro*. J. Biol. Chem. *285*, 2996–3004.

Innes, R.W., Bisgrove, S.R., Smith, N.M., Bent, A.F., Staskawicz, B.J., and Liu, Y.C. (1993). Identification of a disease resistance locus in *Arabidopsis* that is functionally homologous to the RPG1 locus of soybean. Plant J. *4*, 813–820.

Innes, R.W., Ameline-Torregrosa, C., Ashfield, T., Cannon, E., Cannon, S.B., Chacko, B., Chen, N.W., Couloux, A., Dalwani, A., Denny, R., *et al.* (2008). Differential accumulation of retroelements and diversification of NB-LRR disease resistance genes in duplicated regions following polyploidy in the ancestor of soybean. Plant Physiol. *148*, 1740–1759.

Jackson, R.W., Athanassopoulos, E., Tsiamis, G., Mansfield, J.W., Sesma, A., Arnold, D.L., Gibbon, M.J., Murillo, J., Taylor, J.D., and Vivian, A. (1999). Identification of a pathogenicity island, which contains genes for virulence and avirulence, on a large native plasmid in the bean pathogen *Pseudomonas syringae* pathovar *phaseolicola*. Proc. Natl. Acad. Sci. U.S.A. *96*, 10875–10880.

Janjusevic, R., Abramovitch, R.B., Martin, G.B., and Stebbins, C.E. (2006). A bacterial inhibitor of host programmed cell death defenses is an E3 ubiquitin ligase. Science *311*, 222–226.

Jeuken, M.J., Zhang, N.W., McHale, L.K., Pelgrom, K., den Boer, E., Lindhout, P., Michelmore, R.W., Visser, R.G., and Niks, R.E. (2009). RIN4 causes hybrid necrosis and race-specific resistance in an interspecific lettuce hybrid. Plant Cell *21*, 3368–3378.

Jones, J.D., and Dangl, J.L. (2006). The plant immune system. Nature *444*, 323–329.

Jones, R.M., and Neish, A.S. (2012). Recognition of bacterial pathogens and mucosal immunity. Cell Microbiol. *13*, 670–676.

Jones, R.M., Wu, H., Wentworth, C., Luo, L., Collier-Hyams, L., and Neish, A.S. (2008). *Salmonella* AvrA coordinates suppression of host immune and apoptotic defenses via JNK pathway blockade. Cell Host Microbe 3, 233–244.

Jones, R.M., Luo, L., and Moberg, K.H. (2012). *Aeromonas salmonicida*-secreted protein AopP is a potent inducer of apoptosis in a mammalian and a *Drosophila* model. Cell Microbiol. *14*, 274–285.

Kadota, Y., and Shirasu, K. (2012). The HSP90 complex of plants. Biochim Biophys Acta *1823*, 689–697.

Katoh, K., Misawa, K., Kuma, K., and Miyata, T. (2002). MAFFT: a novel method for rapid multiple sequence alignment based on fast Fourier transform. Nucleic Acids Res. *30*, 3059–3066.

Kim, M.G., da Cunha, L., McFall, A.J., Belkhadir, Y., DebRoy, S., Dangl, J.L., and Mackey, D. (2005). Two *Pseudomonas syringae* type III effectors inhibit RIN4-regulated basal defense in *Arabidopsis*. Cell *121*, 749–759.

Kim, Y.J., Lin, N.C., and Martin, G.B. (2002). Two distinct *Pseudomonas* effector proteins interact with the Pto kinase and activate plant immunity. Cell *109*, 589–598.

Kinch, L.N., Yarbrough, M.L., Orth, K., and Grishin, N.V. (2009). Fido, a novel AMPylation domain common to fic, doc, and AvrB. PLoS One *4*, e5818.

Kobayashi, D., Tamaki, S., and Keen, N. (1990). Molecular characterization of avirulence gene D from *Pseudomonas syringae* pv. *tomato*. Mol. Plant Microbe Interact. 3, 94–102.

Krachler, A.M., Woolery, A.R., and Orth, K. (2011). Manipulation of kinase signaling by bacterial pathogens. J. Cell Biol. *195*, 1083–1092.

Kramer, R.W., Slagowski, N.L., Eze, N.A., Giddings, K.S., Morrison, M.F., Siggers, K.A., Starnbach, M.N., and Lesser, C.F. (2007). Yeast functional genomic screens lead to identification of a role for a bacterial effector in innate immunity regulation. PLoS Pathog. *3*, e21.

Kube, M., Migdoll, A.M., Muller, I., Kuhl, H., Beck, A., Reinhardt, R., and Geider, K. (2008). The genome of *Erwinia tasmaniensis* strain Et1/99, a non-pathogenic bacterium in the genus *Erwinia*. Environ. Microbiol. *10*, 2211–2222.

Kunkeaw, S., Tan, S., and Coaker, G. (2010). Molecular and evolutionary analyses of *Pseudomonas syringae* pv. *tomato* race 1. Mol. Plant Microbe Interact. *23*, 415–424.

Kvitko, B.H., Park, D.H., Velasquez, A.C., Wei, C.F., Russell, A.B., Martin, G.B., Schneider, D.J., and Collmer, A. (2009). Deletions in the repertoire of *Pseudomonas syringae* pv. *tomato* DC3000 type III secretion effector genes reveal functional overlap among effectors. PLoS Pathog. *5*, e1000388.

Kwon, S.W., Go, S.J., Kang, H.W., Ryu, J.C., and Jo, J.K. (1997). Phylogenetic analysis of *Erwinia* species based on 16S rRNA gene sequences. Int. J. Syst. Bacteriol. *47*, 1061–1067.

Lavie, M., Shillington, E., Eguiluz, C., Grimsley, N., and Boucher, C. (2002). PopP1, a new member of the YopJ/AvrRxv family of type III effector proteins, acts as a host-specificity factor and modulates aggressiveness of *Ralstonia solanacearum*. Mol. Plant Microbe Interact. *15*, 1058–1068.

Lavie, M., Seunes, B., Prior, P., and Boucher, C. (2004). Distribution and sequence analysis of a family of type III-dependent effectors correlate with the phylogeny of *Ralstonia solanacearum* strains. Mol. Plant Microbe Interact. *17*, 931–940.

Lee, A.H., Hurley, B., Felsensteiner, C., Yea, C., Ckurshumova, W., Bartetzko, V., Wang, P.W., Quach, V., Lewis, J.D., Liu, Y.C., *et al.* (2012). A bacterial acetyltransferase destroys plant microtubule networks and blocks secretion. PLoS Pathog. *8*, e1002523.

Lee, C.C., Wood, M.D., Ng, K., Andersen, C.B., Liu, Y., Luginbuhl, P., Spraggon, G., and Katagiri, F. (2004). Crystal structure of the type III effector AvrB from *Pseudomonas syringae*. Structure *12*, 487–494.

Lehti-Shiu, M.D., and Shiu, S.H. (2012). Diversity, classification and function of the plant protein kinase superfamily. Philos. Trans. R. Soc. Lond. B Biol. Sci. *367*, 2619–2639.

Lewis, J.D., Abada, W., Ma, W., Guttman, D.S., and Desveaux, D. (2008). The HopZ family of *Pseudomonas syringae* type III effectors require myristoylation for virulence and avirulence functions in *Arabidopsis thaliana*. J. Bacteriol. *190*, 2880–2891.

Lewis, J.D., Wu, R., Guttman, D.S., and Desveaux, D. (2010). Allele-specific virulence attenuation of the *Pseudomonas syringae* HopZ1a type III effector via the *Arabidopsis* ZAR1 resistance protein. PLoS Genet. *6*, e1000894.

Lewis, J.D., Lee, A., Ma, W., Zhou, H., Guttman, D.S., and Desveaux, D. (2011). The YopJ superfamily in plant-associated bacteria. Mol. Plant Pathol. *12*, 928–937.

Lewis, J.D., Wan, J., Ford, R., Gong, Y., Fung, P., Nahal, H., Wang, P.W., Desveaux, D., and Guttman, D.S. (2012). Quantitative interactor screening with next-generation sequencing (QIS-Seq) identifies *Arabidopsis thaliana* MLO2 as a target of the *Pseudomonas syringae* type III effector HopZ2. BMC Genomics *13*, 8.

Li, H., Xu, H., Zhou, Y., Zhang, J., Long, C., Li, S., Chen, S., Zhou, J.-M., and Shao, F. (2007). The phosphothreonine lyase activity of a bacterial type III effector family. Science *315*, 1000–1003.

Li, J., Wen, J., Lease, K.A., Doke, J.T., Tax, F.E., and Walker, J.C. (2002). BAK1, an *Arabidopsis* LRR receptor-like protein kinase, interacts with BRI1 and modulates brassinosteroid signaling. Cell *110*, 213–222.

Lin, N.C., Abramovitch, R.B., Kim, Y.J., and Martin, G.B. (2006). Diverse AvrPtoB homologs from several *Pseudomonas syringae* pathovars elicit Pto-dependent resistance and have similar virulence activities. Appl. Environ. Microbiol. *72*, 702–712.

Lindeberg, M., Stavrinides, J., Chang, J.H., Alfano, J.R., Collmer, A., Dangl, J.L., Greenberg, J.T., Mansfield, J.W., and Guttman, D.S. (2005). Proposed guidelines for a unified nomenclature and phylogenetic analysis of type III Hop effector proteins in the plant pathogen *Pseudomonas syringae*. Mol. Plant Microbe Interact. *18*, 275–282.

Liu, J., Elmore, J.M., Lin, Z.-J.D., and Coaker, G. (2011). A receptor-like cytoplasmic kinase phosphorylates the host target RIN4, leading to the activation of a plant innate immune receptor. Cell Host Microbe *9*, 137–146.

Ma, W., Dong, F.F., Stavrinides, J., and Guttman, D.S. (2006). Type III effector diversification via both pathoadaptation and horizontal transfer in response to a coevolutionary arms race. PLoS Genet. *2*, e209.

Mackey, D., Holt, B.F., 3rd, Wiig, A., and Dangl, J.L. (2002). RIN4 interacts with *Pseudomonas syringae* type III effector molecules and is required for RPM1-mediated resistance in *Arabidopsis*. Cell *108*, 743–754.

Malnoy, M., Martens, S., Norelli, J.L., Barny, M.-A., Sundin, G.W., Smits, T.H.M., and Duffy, B. (2012). Fire blight: applied genomic insights of the pathogen and host. Annu. Rev. Phytopathol. *50*, 475–494.

Mazurkiewicz, P., Thomas, J., Thompson, J.A., Liu, M., Arbibe, L., Sansonetti, P., and Holden, D.W. (2008). SpvC is a *Salmonella* effector with phosphothreonine lyase activity on host mitogen-activated protein kinases. Mol. Microbiol. *67*, 1371–1383.

Meinzer, U., Barreau, F., Esmiol-Welterlin, S., Jung, C., Villard, C., Leger, T., Ben-Mkaddem, S., Berrebi, D., Dussaillant, M., Alnabhani, Z., *et al.* (2012). *Yersinia pseudotuberculosis* effector YopJ subverts the Nod2/RICK/TAK1 pathway and activates caspase-1 to induce intestinal barrier dysfunction. Cell Host Microbe *11*, 337–351.

Mergaert, J., Verdonck, L., and Kersters, K. (1993). Transfer of *Erwinia ananas* (synonym, *Erwinia uredovora*) and *Erwinia stewartii* to the genus *Pantoea* emend. as *Pantoea ananas* (Serrano 1928) comb. nov. and *Pantoea stewartii* (Smith 1898) comb, nov., respectively, and description of *Pantoea stewartii* subsp. *indologenes* subsp. nov. Int. J. Sys. Bacteriol. *43*, 162–173.

Mittal, R., Peak-Chew, S.Y., and McMahon, H.T. (2006). Acetylation of MEK2 and I kappa B kinase (IKK) activation loop residues by YopJ inhibits signaling. Proc. Natl. Acad. Sci. U.S.A. *103*, 18574–18579.

Mittal, R., Peak-Chew, S.Y., Sade, R.S., Vallis, Y., and McMahon, H.T. (2010). The acetyltransferase activity of the bacterial toxin YopJ of *Yersinia* is activated by eukaryotic host cell inositol hexakisphosphate. J. Biol. Chem. *285*, 19927–19934.

Monack, D.M., Mecsas, J., Ghori, N., and Falkow, S. (1997). *Yersinia* signals macrophages to undergo apoptosis and YopJ is necessary for this cell death. Proc. Natl. Acad. Sci. U.S.A. *94*, 10385–10390.

Mor, H., Manulis, S., Zuck, M., Nizan, R., Coplin, D.L., and Barash, I. (2001). Genetic organization of the *hrp* gene cluster and *dspAE/BF* operon in *Erwinia herbicola* pv. *gypsophilae*. Mol. Plant Microbe Interact. *14*, 431–436.

Morgan, R.L., Zhou, H., Lehto, E., Nguyen, N., Bains, A., Wang, X., and Ma, W. (2010). Catalytic domain of the diversified *Pseudomonas syringae* type III effector HopZ1 determines the allelic specificity in plant hosts. Mol. Microbiol. *76*, 437–455.

Mudgett, M.B., Chesnokova, O., Dahlbeck, D., Clark, E.T., Rossier, O., Bonas, U., and Staskawicz, B.J. (2000). Molecular signals required for type III secretion and translocation of the *Xanthomonas campestris* AvrBs2 protein to pepper plants. Proc. Natl. Acad. Sci. U.S.A. *97*, 13324–13329.

Mukaihara, T., Tamura, N., Murata, Y., and Iwabuchi, M. (2004). Genetic screening of Hrp type III-related pathogenicity genes controlled by the HrpB transcriptional activator in *Ralstonia solanacearum*. Mol. Microbiol. *54*, 863–875.

Mukaihara, T., Tamura, N., and Iwabuchi, M. (2010). Genome-wide identification of a large repertoire of *Ralstonia solanacearum* type III effector proteins by a new functional screen. Mol. Plant Microbe Interact. *23*, 251–262.

Mukherjee, S., Keitany, G., Li, Y., Wang, Y., Ball, H.L., Goldsmith, E.J., and Orth, K. (2006). *Yersinia* YopJ acetylates and inhibits kinase activation by blocking phosphorylation. Science *312*, 1211–1214.

Mukherjee, S., Hao, Y.H., and Orth, K. (2007). A newly discovered post-translational modification – the acetylation of serine and threonine residues. Trends Biochem. Sci. 32, 210–216.

Nam, K.H., and Li, J. (2002). BRI1/BAK1, a receptor kinase pair mediating brassinosteroid signaling. Cell *110*, 203–212.

Narusaka, M., Shirasu, K., Noutoshi, Y., Kubo, Y., Shiraishi, T., Iwabuchi, M., and Narusaka, Y. (2009). RRS1 and RPS4 provide a dual resistance-gene system against fungal and bacterial pathogens. Plant J. *60*, 218–226.

Nguyen, H.P., Yeam, I., Angot, A., and Martin, G.B. (2010). Two virulence determinants of type III effector AvrPto are functionally conserved in diverse *Pseudomonas syringae* pathovars. New Phytol. *187*, 969–982.

Noël, L., Thieme, F., Nennstiel, D., and Bonas, U. (2001). cDNA-AFLP analysis unravels a genome-wide *hrpG*-regulon in the plant pathogen *Xanthomonas campestris* pv. *vesicatoria*. Mol. Microbiol. *41*, 1271–1281.

O'Brien, H.E., Desveaux, D., and Guttman, D.S. (2011a). Next-generation genomics of *Pseudomonas syringae*. Curr. Opin. Microbiol. *14*, 24–30.

O'Brien, H.E., Thakur, S., and Guttman, D.S. (2011b). Evolution of plant pathogenesis in *Pseudomonas syringae*: a genomics perspective. Annu. Rev. Phytopathol. *49*, 269–289.

O'Brien, H.E., Thakur, S., Gong, Y., Fung, P., Zhang, J., Yuan, L., Wang, P.W., Yong, C., Scortichini, M., and Guttman, D.S. (2012). Extensive remodeling of the *Pseudomonas syringae* pv. *avellanae* type III secretome associated with two independent host shifts onto hazelnut. BMC Microbiol. *12*, 141.

Occhialini, A., Cunnac, S., Reymond, N., Genin, S., and Boucher, C. (2005). Genome-wide analysis of gene expression in *Ralstonia solanacearum* reveals that the *hrpB* gene acts as a regulatory switch controlling multiple virulence pathways. Mol. Plant Microbe Interact. *18*, 938–949.

Ong, L.E., and Innes, R.W. (2006). AvrB mutants lose both virulence and avirulence activities on soybean and *Arabidopsis*. Mol. Microbiol. *60*, 951–962.

Orth, K., Palmer, L.E., Bao, Z.Q., Stewart, S., Rudolph, A.E., Bliska, J.B., and Dixon, J.E. (1999). Inhibition of the mitogen-activated protein kinase kinase superfamily by a *Yersinia* effector. Science *285*, 1920–1923.

Orth, K., Xu, Z., Mudgett, M.B., Bao, Z.Q., Palmer, L.E., Bliska, J.B., Mangel, W.F., Staskawicz, B., and Dixon, J.E. (2000). Disruption of signaling by *Yersinia* effector YopJ, a ubiquitin-like protein protease. Science *290*, 1594–1597.

Paquette, N., Conlon, J., Sweet, C., Rus, F., Wilson, L., Pereira, A., Rosadini, C.V., Goutagny, N., Weber, A.N., Lane, W.S., *et al.* (2012). Serine/threonine acetylation of TGFbeta-activated kinase (TAK1) by *Yersinia pestis* YopJ inhibits innate immune signaling. Proc. Natl. Acad. Sci. U.S.A. *109*, 12710–12715.

Parkinson, N., Aritua, V., Heeney, J., Cowie, C., Bew, J., and Stead, D. (2007). Phylogenetic analysis of *Xanthomonas* species by comparison of partial gyrase B gene sequences. Int. J. Syst. Evol. Microbiol. *57*, 2881–2887.

Petersen, M., Brodersen, P., Naested, H., Andreasson, E., Lindhart, U., Johansen, B., Nielsen, H.B., Lacy, M., Austin, M.J., Parker, J.E., *et al.* (2000). *Arabidopsis* MAP kinase 4 negatively regulates systemic acquired resistance. Cell *103*, 1111–1120.

Petnicki-Ocwieja, T., Schneider, D.J., Tam, V.C., Chancey, S.T., Shan, L., Jamir, Y., Schechter, L.M., Janes, M.D., Buell, C.R., Tang, X., *et al.* (2002). Genomewide identification of proteins secreted by the Hrp type III protein secretion system of *Pseudomonas syringae* pv. *tomato* DC3000. Proc. Natl. Acad. Sci. U.S.A. *99*, 7652–7657.

Petutschnig, E.K., Jones, A.M., Serazetdinova, L., Lipka, U., and Lipka, V. (2010). The lysin motif receptor-like kinase (LysM-RLK) CERK1 is a major chitin-binding protein in *Arabidopsis thaliana* and subject to chitin-induced phosphorylation. J. Biol. Chem. *285*, 28902–28911.

Poueymiro, M., and Genin, S. (2009). Secreted proteins from *Ralstonia solanacearum*: a hundred tricks to kill a plant. Curr. Opin. Microbiol. *12*, 44–52.

Prior, P., and Fegan, M. (2005). Recent developments in the phylogeny and classification of *Ralstonia solanacearum*. Acta Hortic *695*, 127–136.

Rademaker, J.L.W., Louws, F.J., Schultz, M.H., Rossbach, U., Vauterin, L., Swings, J., and de Bruijn, F.J. (2005). A comprehensive species to strain taxonomic framework for *Xanthomonas*. Phytopathology *95*, 1098–1111.

Remenant, B., De Cambiaire, J.-C., Cellier, G., Jacobs, J.M., Mangenot, S., Barbe, V., Lajus, A., Vallenet, D., Medigue, C., Fegan, M., *et al.* (2011). *Ralstonia syzygii*, the blood disease bacterium and some Asian *R. solanacearum* strains form a single genomic species despite divergent lifestyles. PLoS One *6*, e24356.

Roden, J., Eardley, L., Hotson, A., Cao, Y., and Mudgett, M.B. (2004a). Characterization of the *Xanthomonas* AvrXv4 effector, a SUMO protease translocated into plant cells. Mol. Plant Microbe Interact. *17*, 633–643.

Roden, J.A., Belt, B., Ross, J.B., Tachibana, T., Vargas, J., and Mudgett, M.B. (2004b). A genetic screen to isolate type III effectors translocated into pepper cells during *Xanthomonas* infection. Proc. Natl. Acad. Sci. U.S.A. *101*, 16624–16629.

Ronald, P.C., Salmeron, J.M., Carland, F.M., and Staskawicz, B.J. (1992). The cloned avirulence gene *avrPto* induces disease resistance in tomato cultivars containing the Pto resistance gene. J. Bacteriol. *174*, 1604–1611.

Rosebrock, T.R., Zeng, L., Brady, J.J., Abramovitch, R.B., Xiao, F., and Martin, G.B. (2007). A bacterial E3 ubiquitin ligase targets a host protein kinase to disrupt plant immunity. Nature *448*, 370–374.

Samson, R., Legendre, J.B., Christen, R., Saux, M.F.-L., Achouak, W., and Gardan, L. (2005). Transfer of *Pectobacterium chrysanthemi* (Burkholder *et al.*, 1953) Brenner *et al.*, 1973 and *Brenneria paradisiaca* to the genus *Dickeya* gen. nov. as *Dickeya chrysanthemi* comb. nov. and *Dickeya paradisiaca* comb. nov. and delineation of four novel species, *Dickeya dadantii* sp. nov., *Dickeya dianthicola* sp. nov., *Dickeya dieffenbachiae* sp. nov. and *Dickeya zeae* sp. nov. Int. J. Syst. Evol. Microbiol. *55*, 1415–1427.

Sarkar, S.F., and Guttman, D.S. (2004). Evolution of the core genome of *Pseudomonas syringae*, a highly clonal, endemic plant pathogen. Appl. Environ. Microbiol. *70*, 1999–2012.

Schaad, N.W., Postnikova, E., Sechler, A., Claflin, L.E., Vidaver, A.K., Jones, J.B., Agarkova, I., Ignatov, A., Dickstein, E., and Ramundo, B.A. (2008). Reclassification of subspecies of *Acidovorax avenae* as *A. Avenae* (Manns 1905) emend., *A. cattleyae* (Pavarino, 1911) comb. nov., *A. citrulli* Schaad *et al.*, 1978) comb. nov., and proposal of *A. oryzae* sp. nov. Syst. Appl. Microbiol. *31*, 434–446.

Schechter, L.M., Roberts, K.A., Jamir, Y., Alfano, J.R., and Collmer, A. (2004). *Pseudomonas syringae* type III secretion system targeting signals and novel effectors studied with a Cya translocation reporter. J. Bacteriol. *186*, 543–555.

Schechter, L.M., Vencato, M., Jordan, K.L., Schneider, S.E., Schneider, D.J., and Collmer, A. (2006). Multiple approaches to a complete inventory of *Pseudomonas syringae* pv. *tomato* DC3000 type III secretion system effector proteins. Mol. Plant Microbe Interact. *19*, 1180–1192.

Schulze, B., Mentzel, T., Jehle, A.K., Mueller, K., Beeler, S., Boller, T., Felix, G., and Chinchilla, D. (2010). Rapid heteromerization and phosphorylation of ligand-activated plant transmembrane receptors and their associated kinase BAK1. J. Biol. Chem. *285*, 9444–9451.

Selote, D., and Kachroo, A. (2010). RPG1-B-derived resistance to AvrB-expressing *Pseudomonas syringae* requires RIN4-like proteins in soybean. Plant Physiol. *153*, 1199–1211.

Selote, D., Robin, G.P., and Kachroo, A. (2012). GmRIN4 protein family members function nonredundantly in soybean race-specific resistance against *Pseudomonas syringae*. New Phytol. *197*, 1225–1235.

Shan, L., Thara, V.K., Martin, G.B., Zhou, J.M., and Tang, X. (2000). The *Pseudomonas* AvrPto protein is differentially recognized by tomato and tobacco and is localized to the plant plasma membrane. Plant Cell *12*, 2323–2338.

Shan, L., He, P., Li, J., Heese, A., Peck, S.C., Nurnberger, T., Martin, G.B., and Sheen, J. (2008). Bacterial effectors target the common signaling partner BAK1 to disrupt multiple MAMP receptor-signaling complexes and impede plant immunity. Cell Host Microbe *4*, 17–27.

Shang, Y., Li, X., Cui, H., He, P., Thilmony, R., Chintamanani, S., Zwiesler-Vollick, J., Gopalan, S., Tang, X., and Zhou, J.-M. (2006). RAR1, a central player in plant immunity, is targeted by *Pseudomonas syringae* effector AvrB. Proc. Natl. Acad. Sci. U.S.A. *103*, 19200–19205.

Singer, A.U., Wu, B., Yee, A., Houliston, S., Xu, X., Cui, H., Skarina, T., Garcia, M., Semesi, A., Arrowsmith, C.H., *et al.* (2012). Structural analysis of HopPmaL reveals the presence of a second adaptor domain common to the HopAB family of *Pseudomonas syringae* type III effectors. Biochemistry *51*, 1–3.

Staskawicz, B., Dahlbeck, D., Keen, N., and Napoli, C. (1987). Molecular characterization of cloned avirulence genes from race 0 and race 1 of *Pseudomonas syringae* pv. *glycinea*. J. Bacteriol. *169*, 5789–5794.

Studholme, D.J., Kemen, E., Maclean, D., Schornack, S., Aritua, V., Thwaites, R., Grant, M., Smith, J., and Jones, J.D.G. (2010). Genome-wide sequencing data reveals virulence factors implicated in banana *Xanthomonas* wilt. FEMS Microbiol. Lett. *310*, 182–192.

Swanson, J., Kearney, B., Dahlbeck, D., and Staskawicz, B. (1988). Cloned avirulence gene of *Xanthomonas campestris* pv. *vesicatoria* complements spontaneous race-change mutants. Mol. Plant Microbe Interact. *1*, 5–9.

Szczesny, R., Buttner, D., Escolar, L., Schulze, S., Seiferth, A., and Bonas, U. (2010). Suppression of the AvrBs1-specific hypersensitive response by the YopJ effector homolog AvrBsT from *Xanthomonas* depends on a SNF1-related kinase. New Phytol. *187*, 1058–1074.

Tamaki, S., Dahlbeck, D., Staskawicz, B., and Keen, N.T. (1988). Characterization and expression of two avirulence genes cloned from *Pseudomonas syringae* pv. *glycinea*. J. Bacteriol. *170*, 4846–4854.

Tasset, C., Bernoux, M., Jauneau, A., Pouzet, C., Briere, C., Kieffer-Jacquinod, S., Rivas, S., Marco, Y., and Deslandes, L. (2010). Autoacetylation of the *Ralstonia solanacearum* effector PopP2 targets a lysine residue essential for RRS1-R-mediated immunity in *Arabidopsis*. PLoS Pathog. *6*, e1001202.

de Torres-Zabala, M., Truman, W., Bennett, M.H., Lafforgue, G., Mansfield, J.W., Rodriguez Egea, P., Bogre, L., and Grant, M. (2007). *Pseudomonas syringae* pv. *tomato* hijacks the *Arabidopsis* abscisic acid signalling pathway to cause disease. EMBO J. *26*, 1434–1443.

Trosky, J.E., Mukherjee, S., Burdette, D.L., Roberts, M., McCarter, L., Siegel, R.M., and Orth, K. (2004). Inhibition of MAPK signaling pathways by VopA from *Vibrio parahaemolyticus*. J. Biol. Chem. *279*, 51953–51957.

Trosky, J.E., Li, Y., Mukherjee, S., Keitany, G., Ball, H., and Orth, K. (2007). VopA inhibits ATP binding by acetylating the catalytic loop of MAPK kinases. J. Biol. Chem. *282*, 34299–34305.

Tsiamis, G., Mansfield, J.W., Hockenhull, R., Jackson, R.W., Sesma, A., Athanassopoulos, E., Bennett, M.A., Stevens, C., Vivian, A., Taylor, J.D., *et al.* (2000). Cultivar-specific avirulence and virulence functions assigned to *avrPphF* in *Pseudomonas syringae* pv. *phaseolicola*, the cause of bean halo-blight disease. EMBO J. *19*, 3204–3214.

Vauterin, L., Hoste, B., Kersters, K., and Swings, J. (1995). Reclassification of *Xanthomonas*. Int. J. Syst. Bacteriol. *45*, 472–489.

Vinatzer, B.A., Teitzel, G.M., Lee, M.W., Jelenska, J., Hotton, S., Fairfax, K., Jenrette, J., and Greenberg, J.T. (2006). The type III effector repertoire of *Pseudomonas syringae* pv. *syringae* B728a and its role in survival and disease on host and non-host plants. Mol. Microbiol. *62*, 26–44.

Wang, Y., Zhang, Q., Sun, M.-A., and Guo, D. (2011). High-accuracy prediction of bacterial type III secreted effectors based on position-specific amino acid composition profiles. Bioinformatics 27, 777–784.

Wanner, L.A., Mittal, S., and Davis, K.R. (1993). Recognition of the avirulence gene *avrB* from *Pseudomonas syringae* pv. *glycinea* by *Arabidopsis thaliana*. Mol. Plant Microbe Interact. *6*, 582–591.

Wengelnik, K., and Bonas, U. (1996). HrpXv, an AraC-type regulator, activates expression of five of the six loci in the *hrp* cluster of *Xanthomonas campestris* pv. *vesicatoria*. J. Bacteriol. *178*, 3462–3469.

Whalen, M.C., Stall, R.E., and Staskawicz, B.J. (1988). Characterization of a gene from a tomato pathogen determining hypersensitive resistance in non-host species and genetic analysis of this resistance in bean. Proc. Natl. Acad. Sci. U.S.A. *85*, 6743–6747.

Whalen, M.C., Wang, J.F., Carland, F.M., Heiskell, M.E., Dahlbeck, D., Minsavage, G.V., Jones, J.B., Scott, J.W., Stall, R.E., and Staskawicz, B.J. (1993). Avirulence gene *avrRxv* from *Xanthomonas campestris* pv. *vesicatoria* specifies resistance on tomato line Hawaii 7998. Mol. Plant Microbe Interact. *6*, 616–627.

Whalen, M.C., Richter, T., Zakhareyvich, K., Yoshikawa, M., Al-Azzeh, D., Adefioye, A., Spicer, G., Mendoza, L.L., Morales, C.Q., Klassen, V., *et al.* (2008). Identification of a host 14-3-3 protein that interacts with *Xanthomonas* effector AvrRxv. Physiol. Mol. Plant Pathol. 72, 46–55.

White, F.F., Potnis, N., Jones, J.B., and Koebnik, R. (2009). The type III effectors of *Xanthomonas*. Mol. Plant Pathol. *10*, 749–766.

Wroblewski, T., Caldwell, K.S., Piskurewicz, U., Cavanaugh, K.A., Xu, H., Kozik, A., Ochoa, O., McHale, L.K., Lahre, K., Jelenska, J., *et al.* (2009). Comparative large-scale analysis of interactions between several crop species and the effector repertoires from multiple pathovars of *Pseudomonas* and *Ralstonia*. Plant Physiol. *150*, 1733–1749.

Wu, H., Jones, R.M., and Neish, A.S. (2010). The *Salmonella* effector AvrA mediates bacterial intracellular survival during infection *in vivo*. Cell Microbiol. *14*, 28–39.

Wulf, J., Pascuzzi, P.E., Fahmy, A., Martin, G.B., and Nicholson, L.K. (2004). The solution structure of type III effector protein AvrPto reveals conformational and dynamic features important for plant pathogenesis. Structure *12*, 1257–1268.

Xiang, T., Zong, N., Zou, Y., Wu, Y., Zhang, J., Xing, W., Li, Y., Tang, X., Zhu, L., Chai, J., *et al.* (2008). *Pseudomonas syringae* effector AvrPto blocks innate immunity by targeting receptor kinases. Curr. Biol. *18*, 74–80.

Xiang, T., Zong, N., Zhang, J., Chen, J., Chen, M., and Zhou, J.M. (2011). BAK1 is not a target of the *Pseudomonas syringae* effector AvrPto. Mol. Plant Microbe Interact. *24*, 100–107.

Xiao, F., He, P., Abramovitch, R.B., Dawson, J.E., Nicholson, L.K., Sheen, J., and Martin, G.B. (2007). The N-terminal region of *Pseudomonas* type III effector AvrPtoB elicits Pto-dependent immunity and has two distinct virulence determinants. Plant J. *52*, 595–614.

Xiao, Y., and Hutcheson, S.W. (1994). A single promotor sequence recognized by a newly identified alternate sigma factor directs expression of pathogenicity and host range determinants in *Pseudomonas syringae*. J. Bacteriol. *176*, 3089–3091.

Xing, W., Zou, Y., Liu, Q., Liu, J., Luo, X., Huang, Q., Chen, S., Zhu, L., Bi, R., Hao, Q., *et al.* (2007). The structural basis for activation of plant immunity by bacterial effector protein AvrPto. Nature *449*, 243–247.

Yang, S., Peng, Q., Zhang, Q., Zou, L., Li, Y., Robert, C., Pritchard, L., Liu, H., Hovey, R., Wang, Q., *et al.* (2010a). Genome-wide identification of HrpL-regulated genes in the necrotrophic phytopathogen *Dickeya dadantii* 3937. PLoS One 5, e13472.

Yang, Y., Zhao, J., Morgan, R.L., Ma, W., and Jiang, T. (2010b). Computational prediction of type III secreted proteins from Gram-negative bacteria. BMC Bioinformatics *11*, S47.

Yeam, I., Nguyen, H.P., and Martin, G.B. (2010). Phosphorylation of the *Pseudomonas syringae* effector AvrPto is required for FLS2/BAK1-independent virulence activity and recognition by tobacco. Plant J. *61*, 16–24.

Yoon, S., Liu, Z., Eyobo, Y., and Orth, K. (2003). *Yersinia* effector YopJ inhibits yeast MAPK signaling pathways by an evolutionarily conserved mechanism. J. Biol. Chem. *278*, 2131–2135.

Young, J., Park, D.C., Shearman, H., and Fargier, E. (2008). A multilocus sequence analysis of the genus *Xanthomonas*. Syst. Appl. Microbiol. *31*, 366–377.

Yucel, I., Slaymaker, D., Boyd, C., Murillo, J., Buzzell, R.I., and Keen, N.T. (1994). Avirulence gene *avrPphC* from *Pseudomonas syringae* pv. *phaseolicola* 3121: a plasmid-borne homologue of *avrC* closely linked to an *avrD* allele. Mol. Plant Microbe Interact. *7*, 677–679.

Zeng, L., Velasquez, A.C., Munkvold, K.R., Zhang, J., and Martin, G.B. (2012). A tomato LysM receptor-like kinase promotes immunity and its kinase activity is inhibited by AvrPtoB. Plant J. *69*, 92–103.

Zhang, J., Shao, F., Li, Y., Cui, H., Chen, L., Li, H., Zou, Y., Long, C., Lan, L., Chai, J., *et al.* (2007). A *Pseudomonas syringae* effector inactivates MAPKs to suppress PAMP-induced immunity in plants. Cell Host Microbe *1*, 175–185.

Zhao, Y., and Qi, M. (2011). Comparative genomics of *Erwinia amylovora* and related *Erwinia* species: what do we learn? Genes 2, 627–639.

Zhao, Y., Blumer, S.E., and Sundin, G.W. (2005). Identification of *Erwinia amylovora* genes induced during infection of immature pear tissue. J. Bacteriol. *187*, 8088–8103.

Zhao, Y., He, S.Y., and Sundin, G.W. (2006). The *Erwinia amylovora avrRpt2EA* gene contributes to virulence on pear and AvrRpt2EA is recognized by *Arabidopsis* RPS2 when expressed in *Pseudomonas syringae*. Mol. Plant Microbe Interact. *19*, 644–654.

Zhou, H., Morgan, R.L., Guttman, D.S., and Ma, W. (2009). Allelic variants of the *Pseudomonas syringae* type III effector HopZ1 are differentially recognized by plant resistance systems. Mol. Plant Microbe Interact. *22*, 176–189.

Zhou, H., Lin, J., Johnson, A., Morgan, R.L., Zhong, W., and Ma, W. (2011). *Pseudomonas syringae* type III effector HopZ1 targets a host enzyme to suppress isoflavone biosynthesis and promote infection in soybean. Cell Host Microbe *9*, 177–186.

Zwiesler-Vollick, J., Plovanich-Jones, A.E., Nomura, K., Bandyopadhyay, S., Joardar, V., Kunkel, B.N., and He, S.Y. (2002). Identification of novel *hrp*-regulated genes through function genomic analysis of the *Pseudomonas syringae* pv. *tomato* DC3000 genome. Mol. Microbiol. *45*, 1207–1218.

Systems Biology of *Pseudomonas Syringae* Type III Secretion Effector Repertoires

2

Magdalen Lindeberg and Alan Collmer

Abstract

Pseudomonas syringae defeats the two-tiered innate immune system of plants primarily with effector proteins that are injected into host cells by the type III secretion system. Although effector repertoires are remarkably diverse among *P. syringae* pathovars and even among strains virulent on the same host, they share several properties suggesting that effectors in a repertoire operate as components of a system. Much progress has been made in understanding the structure, function, and evolution of *P. syringae* effector repertoires. Multiple approaches for identifying effector genes have been applied to several sequenced reference strains and then extended to yield a tentative super-repertoire for the *P. syringae* species pan-genome. Comparative and functional genomic studies have revealed patterns in repertoire composition and mechanisms of effector gene gain, loss, and allelic polymorphism. Substantial progress has been made in understanding the mechanisms by which effectors subvert host immunity-associated proteins. Information on *P. syringae* effectors is thus accumulating at multiple levels, from dynamics of gene distribution in field strains to structural biology of effector interactions with host targets. A challenge for the future is to identify vulnerabilities in the apparent robustness of effector repertoires that can be exploited for more durable crop resistance to the many diseases caused by *P. syringae*.

Introduction

Pseudomonas syringae is representative of many biotrophic and hemibiotrophic plant pathogens that defeat plant immunity with effector proteins that are delivered by various means into the cytosol of host cells. *P. syringae* and other bacterial pathogens deliver such effectors using type III secretion systems (T3SSs) (Block and Alfano, 2011; Tampakaki *et al.*, 2010). Genomics approaches have revealed that bacterial, fungal, and oomycete pathogens that use cytoplasmic effectors to promote (hemi)biotrophic pathogenesis often deploy large and remarkably variable repertoires of such effectors (Baxter *et al.*, 2010; Cunnac *et al.*, 2009; Kay and Bonas, 2009; Poueymiro and Genin, 2009; Raffaele *et al.*, 2010; Spanu *et al.*, 2010). As effector interactions with the innate immune system of plants are a major factor underlying the breeding of resistant crops, understanding the systems-level properties of repertoires that determine their composition has both conceptual and practical value. *P. syringae* has become a model for systems analysis because (i) strains in its various host-specific pathovars cause disease in many important crops and model plants (Hirano and Upper, 2000), (ii) the bacterium is experimentally amenable, (iii) virtually all effector genes in several reference strains and many effector genes in the pan-genome have been identified (Baltrus *et al.*, 2011) and (iv) a functionally complete repertoire of effector genes has been deleted and experimentally reassembled in the model strain *P. syringae* pv. *tomato* DC3000 (*Pto* DC3000) (Cunnac *et al.*, 2011).

Expansion and diversification of effector repertoires in *P. syringae* appears to be driven by bacterial coevolution with the innate immune system of plants. According to the current model for a two-tiered plant immune system (Jones and Dangl, 2006), the primary function of effectors is to suppress pathogen- or microbe-associated molecular pattern (PAMP or MAMP)-triggered immunity (PTI or MTI), which is elicited by bacterial features such as flagellin, lipopolysaccharide (LPS), peptidoglycan, and elongation factor Tu (Boller and Felix, 2009). PAMPs are perceived by pattern recognition receptors at the surface of plant cells. In a second level of immunity, which occurs inside plant cells, effectors or their activity are recognized by resistance (R) proteins, typically nucleotide-binding leucine-rich repeat receptor (NLR) proteins, resulting in effector-triggered immunity (ETI) and typically a localized programmed cell death known as the hypersensitive response (HR) (Maekawa *et al.*, 2011). *R* genes can be introgressed into crops to confer resistance against (hemi)biotrophic pathogens, but such resistance is generally not durable in agricultural use because apparent redundancy renders individual effectors dispensable, and virulent new races of the pathogen lacking betraying effectors repeatedly arise and become predominant in the field (Kunkeaw *et al.*, 2010; McDonald and Linde, 2002).

This review addresses the systems-level properties of effector repertoires, particularly those of the model strain *Pto* DC3000. A biological system characteristically has properties that exceed the sum of its parts, and exploring such emergent properties is enhanced by a complete inventory of the parts. Thus, we will begin by addressing the experimental evidence that the repertoire of active (expressed and delivered at a level sufficient to impact the interaction) true effectors (functioning primarily within plant cells rather than outside as delivery helpers) is complete for reference strains, such as *Pto* DC3000. Then we will address genetic factors underlying the evolutionarily rapid gain and loss of effector genes. The functions of individual effectors within plant cells will be addressed in the organizing framework of the Gene Ontology, which furthers systems analysis by making complex biological information universally accessible and machine readable. We will then turn to a model for the evolution of effector repertoires that is based on combining knowledge of effector function and distribution among diverse *P. syringae* strains. Finally, we will assess resources needed to foster a deeper understanding of *P. syringae* effector repertoires as biological systems.

Establishing the effector repertoires of individual strains and of the *P. syringae* pan-genome

Tools for identifying effector genes

Important early steps towards a systems-level understanding of *P. syringae* type III effectors are the identification of all effectors in the repertoires of reference strains and then identification of all (or at least most) effectors in the pan-genome for the species. Tools for identifying *P. syringae* effectors continue to evolve, but they presently rely on three typical characteristics of effector genes. First, effector genes are activated by the HrpL alternative sigma factor acting on Hrp (for hypersensitive response and pathogenicity) promoters recognizable in genome sequences (Fouts *et al.*, 2002; Guttman *et al.*, 2002; Zwiesler-Vollick *et al.*, 2002). Second, the N-terminal regions of the encoded effectors carry amino acid patterns that are predictive of targeting to the T3SS (Guttman *et al.*, 2002; Petnicki-Ocwieja *et al.*, 2002). Third, gene fusions generating reporter hybrids with full-length effector candidates permit experimental validation that the test proteins can be translocated into plant cells by the T3SS (Schechter *et al.*, 2004). Finally, ongoing discovery of effector activities and phenotypes *in planta* are providing the ultimate validation of effectors and their role (Block and Alfano, 2011; Cunnac *et al.*, 2011; Cunnac *et al.*, 2009; Lindeberg *et al.*, 2012).

Plant interaction phenotypes associated with effector gene gain or loss have been important in the process of effector discovery. An initial subset of effector genes was identified through gain-of-function assays for avirulence in inoculated plants carrying matching resistance (*R*) genes (Keen, 1990). The first example of this was the *avrA* gene in *P. syringae* pv. *glycinea*, whose cloning revealed it to be responsible for the gene-for-gene interactions of certain races of this pathovar with cultivars of the host plant soybean (Staskawicz *et al.*, 1984). Effector gene cloning based on gain-of-function avirulence phenotypes was also shown to work with gene transfers between pathovars rather than between races of a given pathovar (Kobayashi *et al.*, 1989).

The development of transposon mutagenesis tools during the period when *avrA* was cloned provided a method for comprehensively tagging and cloning virulence genes in *P. syringae* and other bacteria. Transposon-based mutant screens led to the discovery of the T3SS machinery genes in *P. syringae* because mutations in these genes produced strong Hrp$^-$ phenotype, that is, mutants lost elicitation of the HR in resistant plants and pathogenicity in host plants (Lindgren *et al.*, 1986). However, transposon mutant screens failed to yield effector genes because such mutants were not detectably reduced in virulence. Two exceptions to this observation are illustrative. First, a screen of 947 *Pto* DC3000 Tn5 mutants for reduced symptoms following a dip inoculation of *Arabidopsis* yielded one mutant with an insertion in *avrE* (Brooks *et al.*, 2004). As we will discuss later, AvrE is one of a few effectors that are particularly important to the virulence of *Pto* DC3000. Second, the loss of virulence accompanying the curing of a 154-kb plasmid from *P. syringae* pv. *phaseolicola* 1449B led to the discovery of effector VirPphA (now known as HopAB1) as a factor masking latent avirulence activity in the effector repertoire of 1449B (Jackson *et al.*, 1999). This observation provided early evidence that complex interplay among effectors in a repertoire could determine infection outcomes. However, the requirements for a matching *R* gene in test plants and the rarity of effector gene mutations with strong phenotypes required a genomics approach for comprehensive identification of effector repertoires (Lindeberg *et al.*, 2006; Schneider and Collmer, 2010).

The genomics approach exploited first the Hrp promoters and then the T3SS targeting-associated patterns of *P. syringae* effectors (Collmer *et al.*, 2002; Greenberg and Vinatzer, 2003). AvrPto has been particularly useful as a reference effector in these studies. The first *avrPto* gene was cloned from *P. syringae* pv. *tomato* JL1065 via the avirulence phenotype it conferred on *P. syringae* test strains in tomato cultivars carrying the cognate *Pto R* gene (Ronald *et al.*, 1992). Like a typical *P. syringae* effector, AvrPto is preceded by a Hrp box promoter and has an N-terminus with an aliphatic amino acid in position 3 or 4, no acidic amino acids in the first 12 residues, and a high percentage of Ser in the first 50 residues (Guttman *et al.*, 2002; Petnicki-Ocwieja *et al.*, 2002; Shen and Keen, 1993). T3SS-dependent translocation can be monitored with C-terminal fusions of the test protein with the avirulence domain of another effector or with a calmodulin-dependent adenylate cyclase (Cya) reporter (Guttman and Greenberg, 2001; Schechter *et al.*, 2004). Cya reporter fusions revealed that the N-terminal 16 amino acids of AvrPto are sufficient for weak T3SS-dependent translocation but the first 100 amino acids are needed for near-native translocation levels (Schechter *et al.*, 2004).

Although the T3SS targeting-associated patterns in the first 50 residues are found in most *P. syringae* effectors, it is important to note that proteins that are clearly not effectors may possess these patterns and an AvrPto protein engineered to lack all three patterns was still capable of T3SS-dependent secretion in culture and translocation *in planta* (Schechter *et al.*, 2012; Vencato *et al.*, 2006). It is also useful to note that a subset of *P. syringae* effector genes are linked with chaperone genes, which provides another aid to identifying effector genes (Guo *et al.*, 2005; Kabisch *et al.*, 2005; van Dijk *et al.*, 2002; Wehling *et al.*, 2004). The current recommendation for the validation of novel effectors is for translocation tests to involve reporter fusions with full-length candidates (Lindeberg *et al.*, 2005).

The model effector AvrPto has provided a useful positive control for genome-enabled searches for type III effector genes in *Pto* DC3000. AvrPto was positively identified in an *in planta* expression technology (IPET) screen for genes induced during infection of Arabidopsis (Boch *et al.*, 2002), in a miniTn*5gus* screen for genes activated by HrpL (Fouts *et al.*, 2002), in genomic searches for genes preceded by Hrp promoters (Fouts *et al.*, 2002; Guttman *et al.*, 2002; Zwiesler-Vollick *et al.*, 2002), in searches for genes predicting type III targeting signals (Guttman *et al.*, 2002; Petnicki-Ocwieja *et al.*, 2002), in a differential fluorescence induction screen for genes activated by HrpL (Chang *et al.*, 2005), and in a whole-genome microarray analysis (Ferreira *et al.*, 2006). However, each of these methods failed to find one or more of the other effectors in the *Pto* DC3000 repertoire and produced different, overlapping sets of candidates.

Challenges in identifying the complete repertoire of genes encoding active, true effectors in *P. syringae* strains

The differences in the sets of candidate effector genes identified in the approaches described above relate to ambiguities in Hrp promoter sequences and type III targeting-associated patterns, as well as differences in observed levels of expression and translocation. Resolving these ambiguities and establishing the complete, active repertoire of effectors for a sequenced *P. syringae* strain involves three challenges. First, the coevolutionary processes that led to expansion and variation in repertoires results in the presence of a significant number of effector pseudogenes, weakly expressed effector genes, or otherwise inactive effector genes (Lindeberg *et al.*, 2008). For example, the effector gene *hopAI1* appears inactive in *Pto* DC3000 (Vinatzer *et al.*, 2005), but demonstrably contributes to virulence in *P. syringae* pv. *tomato* 0288-9 (Zhang *et al.*, 2007). Similarly, *avrPtoB* is transcribed in *P. syringae* pv. *tomato* T1 but confers no avirulence to tomato plants carrying *Pto* resistance (Lin *et al.*, 2006). Thus, it is a challenge to know which effector genes in a given strain are actually functional.

Second, it is possible that there are effector genes in *P. syringae* that evade current detection methods because they are not part of the Hrp regulon and do not carry recognizable targeting-associated patterns. Thus, another challenge is how to find any atypical effectors and how to define criteria that would indicate that the identified repertoire for a reference strain is complete.

Third, there are proteins that travel the T3SS and may be capable of being translocated into plant cells but whose primary role seems to be as extracellular component 'helpers' of the T3SS that contribute to the translocation of true effectors into plant cells. Thus, a final challenge is how to systematically distinguish such T3SS components and true effectors.

The *P. syringae* reference strains with closed genome sequences have supported functional analyses that address these challenging questions. These strains are *Pto* DC3000 (Buell *et al.*, 2003), *P. syringae* pv. *syringae* B728a (Feil *et al.*, 2005), and *P. syringae* pv. *phaseolicola* 1448A (Joardar *et al.*, 2005). The effector repertoire of *Pto* DC3000 has received the most attention, as summarized in previous reviews (Cunnac *et al.*, 2009; Greenberg and Vinatzer, 2003; Lindeberg *et al.*, 2006; Vinatzer and Yan, 2008). Extensive effector mining and validation has also been done with B728a and 1448A (Chang *et al.*, 2005; Vencato *et al.*, 2006; Vinatzer *et al.*, 2005, 2006). The findings with the latter two strains are fundamentally similar to those with *Pto* DC3000 and support the validity of ongoing study of *Pto* DC3000 as a model for all *P. syringae* strains in confronting the three challenges in identifying complete, active effector repertoires. Recent progress with *Pto* DC3000 addresses all three challenges.

First, regarding the question of which effector genes are functional in a repertoire, detectable translocation of a full-length fusion of the test effector with a Δ79AvrRpt2 reporter, expressed from its native promoter, provides one useful criterion (Chang *et al.*, 2005). RNASeq experiments are now yielding a more quantitative picture of the relative expression of these genes (Filiatrault *et al.*, 2010), and assays for subtle reductions in the virulence of *Pto* DC3000 strains lacking one, few, or many effectors are providing evidence for an active role of individual effectors in the interaction (Cunnac *et al.*, 2009, 2011).

Second, regarding the possibility of undiscovered, atypical effectors, no evidence has yet emerged for the existence of functional effectors that are expressed independently of the Hrp regulon. This is important because identification of a HrpL-responsive promoter (bioinformatically or experimentally) is the first filter that is used to identify effectors in a sequenced strain of *P. syringae*. In this regard, it is worth noting that the *P. syringae* effector genes in the founding set were found on the basis of their gain-of-function avirulence activity and only subsequently shown to all be expressed from Hrp promoters. Similarly, an *in vivo* screen of *P. syringae* pv. *maculicola* ES4326 based on a transposon-generated translocation reporter fusions yielded 12 subsequently validated *hop* genes, and all were found to be preceded by Hrp promoters (Guttman *et al.*, 2002).

Third, regarding the challenge of distinguishing T3SS helper proteins from true effectors, there is a growing list of distinguishing properties of helper proteins and growing evidence of the role of helpers in translocating true effectors. For example, translocation into plant cells (but not secretion in culture) of effectors by *Pto* DC3000 is abolished in a deletion polymutant lacking the HrpK1 putative translocator protein and the harpins HrpZ1, HrpW1, HopAK1, and HopP1 (Kvitko *et al.*, 2007; Petnicki-Ocwiega *et al.*, 2005). Harpins are defined by several characteristics, including a low isoelectric point, enrichment for glycine, and a lack of cysteine residues. Harpin–Cya fusions can be translocated into plant cells, but the levels of translocation are typically much lower than that observed with effectors (Chang *et al.*, 2005; Kvitko *et al.*, 2007). In contrast, harpins, HrpK1, and the HrpA pilus protein are secreted far more abundantly in Hrp-inducing minimal media than effectors, and bacteria in such media appear to be stuck in a mode of futilely preparing the translocation machinery (Wei and Collmer, 2012). Also, harpins share the unusual ability to elicit the HR when isolated and infiltrated into leaf intercellular spaces (Kvitko *et al.*, 2007). Restoration of individual helper genes to translocation-deficient polymutants reveals that HrpK1, HrpZ1, HrpW1, and HopAK1 can each partially restore effector translocation phenotypes (with HrpK1 having the major role) (Kvitko *et al.*, 2007). HopP1 differs from the other harpins in carrying a lytic transglycosylase domain and in not restoring effector translocation activity to the translocation-deficient polymutant (Kvitko *et al.*, 2007; Oh *et al.*, 2007). Also, it has recently been discovered that the four translocation-promoting proteins differ from HopP1 and established effector proteins in no longer being secreted in mutants lacking the HrpJ T3SS assembly regulator (Crabill *et al.*, 2012). These observations further distinguish the four translocators from effectors, but they leave the primary function of HopP1 unresolved. The T3SSs of plant and animal pathogens typically have a co-regulated lytic transglycosylase, for example, HrpH in *Pto* DC3000. HopP1 has some functional overlap with HrpH, which provides support for a primary role as a helper (Oh *et al.*, 2007).

When considering the function of the complete repertoire of effectors produced by *Pto* DC3000, it is important to consider that some of the T3SS extracellular components may have additional functions that are similar to those of effectors. The *Pto* DC3000 T3SS secretes 12 such substrate/components (Wei and Collmer, 2012). The four translocators (HrpK1, HrpZ1, HrpW1, and HrpAK1) and the two lytic transglycosylases (HrpH and HopP1) were discussed above. In addition there are assembly regulators (HrpJ and HrpP), the pilus protein (HrpA), the putative inner rod (HrpB), a candidate sorting platform protein (HrpD), and a strongly secreted protein that is needed for full *hrp/hrc* gene expression (HrpF). HrpJ has recently been shown to suppress innate immune responses when transgenically expressed in Arabidopsis (Crabill *et al.*, 2012). The stronger contribution of HrpH to *Pto* DC3000 translocation of effectors *in planta*, rather than secretion of effectors in culture, suggests a potential role in immune suppression (Oh *et al.*, 2007). Furthermore, the membrane pore-forming ability of harpins could benefit pathogens independently of effector translocation (Lee *et al.*, 2001), and the high level of variability among the HrpP proteins of different *P. syringae* strains suggests interaction with host factors (Morello and Collmer, 2009).

Based on the multiple, independent analyses described above, *Pto* DC3000 appears to actively deploy 28 type III effectors in addition to the 12 substrate/component/helper proteins. The ultimate test of whether this effector inventory is complete is to delete the genes encoding all 28 effectors and determine if the mutant is functionally effectorless. To this end, the deletion polymutant *Pto* DC3000D28E was constructed and then shown to meet the primary criteria for such a strain based on its poor growth *in planta* but wild-type ability to translocate a test effector (Cunnac *et al.*, 2011). Furthermore, as discussed below, restoration of a minimal functional repertoire of 8 effector genes enabled DC3000D28E to grow in the model plant *Nicotiana benthamiana* to near wild-type levels and produce disease symptoms. However, it is important to note that DC3000D28E appears to have weak ETI elicitation activity because it grows worse in *N. benthamiana* than a DC3000D28E T3SS$^-$ derivative and it retains a residual ability to elicit cell death at very high levels of inoculum (Cunnac *et al.*, 2011). It remains to be determined whether this phenotype is due to the activity of one or more T3SS component/substrates, or to any of the effector genes that were considered too weakly expressed to warrant deletion, or to an undiscovered effector.

Repertoires comprised of a mix of core and variable effectors revealed by comparative genomics

Analysis of the effector repertoires of the three reference strains (*Pto* DC3000, *P. syringae* pv. *syringae* B728a, and *P. syringae* pv. *phaseolicola* 1448A) suggested that functional repertoires are comprised of a few, near-universal effectors and a larger collection of variable effectors (Lindeberg *et al.*, 2006). Next-generation methods have enabled the sequencing of additional strains, which has yielded insights into pathogenesis on important hosts such as tobacco, rice, and olive (Reinhardt *et al.*, 2009; Rodriguez-Palenzuela *et al.*, 2010; Studholme *et al.*, 2009), as well as a comprehensive picture of effector repertoire composition among 19 strains that represent the phylogenetic diversity within *P. syringae* and its many pathovars (Baltrus *et al.*, 2011). This diversity, largely defined by multilocus sequence typing (MLST), provides an important framework for exploring the composition of different effector repertories in the context of evolution and host specificity.

A basic question that is addressable with a set of 19 diverse effector repertoires is the size of the *P. syringae* pan-genome super-repertoire. Since the 14 new genome sequences yielded only nine new effector families and most of these were present in multiple strains, it appears that the super-repertoire is nearly complete with 57 effector families. The larger collection of repertoires reveals that only four effectors can be considered core members (AvrE, HopM, HopAA, and HopI) (Baltrus *et al.*, 2011). The first three of these are encoded by the conserved effector locus (CEL) in the Hrp pathogenicity island, whose acquisition along with the *hrp*/*hrc* T3SS machinery genes by horizontal gene transfer preceded the radiation of *P. syringae* into its pathovars (Alfano *et al.*, 2000). In contrast, the set of variable effectors is much larger, and most of the effectors found on the basis of their avirulence phenotypes are among the rarer of the variable effectors (Lindeberg *et al.*, 2006). Phylogenetic analyses of the variable effectors suggest that several have been acquired, lost, and reacquired multiple times in diverse pathovars (Baltrus *et al.*, 2011). It is also interesting that the sizes of effector repertoires vary, with *Pto* DC3000 having the largest and strains such as *P. syringae* pv. *syringae* B728a in MLST group II having the smallest. Those strains having the smallest effector repertoires produce highly destructive toxins (syringomycin, syringopeptin, and syringolin), which may reduce the number of effectors needed to suppress plant immunity (Bender *et al.*, 1999; Groll *et al.*, 2008). A final, notable feature of *P. syringae* effector repertoires is that divergent repertoires can be found in strains that are pathogenic on the same host, for example *P. syringae* pv. *tomato* strains DC3000 and T1 on tomato or *Pto* DC3000 (if lacking *hopQ1-1*), *P. syringae* pv. *syringae* B728a, and *P. syringae* pv. *tabaci* 11528 on *N. benthamiana* (Almeida *et al.*, 2009; Baltrus *et al.*, 2011; Lindeberg *et al.*, 2006; O'Brien *et al.*, 2011; Wei *et al.*, 2007). Sequencing of three strains of *P. syringae* pv. *avellanae*, a pathogen of hazelnut, further highlights the variability of effector repertoires among even highly related strains (O'Brien *et al.*, 2011).

Potential systems properties of effector repertoires revealed by genetic disassembly and reassembly

What factors determine the composition of *P. syringae* effector repertoires? To address this question it is useful to recall three general behaviours of *P. syringae* effectors in the context of complete repertoires. First, any single effector in a repertoire can elicit ETI and therefore confer avirulence in a test plant with a cognate R protein. Second, individual effectors in a repertoire typically make only a small contribution to virulence. Third, some effectors are able to suppress ETI that would otherwise be elicited by one or more other effectors in the same repertoire (discussed further below). These properties highlight the overriding importance of ETI in determining repertoire composition, and they suggest extensive redundancy and interplay among effectors in a functional repertoire. One approach to understanding how effectors might operate as components of rapidly changing but robust systems has been to disassemble the repertoire of a model strain, such as *Pto* DC3000, and then reassemble a minimal functional repertoire. As discussed above, DC3000 appears to defeat plants with 28 effectors, and this repertoire represents half of the effector families in the *P. syringae* pan-genome.

The first lesson gained from the disassembly of the *Pto* DC3000 effector repertoire was that a single effector can limit host range. That is, deleting the *hopQ1-1* effector gene enables DC3000 to cause disease on *N. benthamiana* (Wei *et al.*, 2007). This finding is consistent with the observation that *P. syringae* strains typically carry multiple effectors that can elicit ETI-like responses in non-host plants (Baltrus *et al.*, 2011; Lindeberg *et al.*, 2009; Wroblewski *et al.*, 2009), and with increasing evidence that disrupting recognition of these avirulence determinants can enable compatible growth in non-hosts. For example, compromising tomato ETI detection of AvrPto and AvrPtoB (or other HopAB family members) enables several *P. syringae* pathovars that are adapted to other plant species to cause disease in tomato (Lin and Martin, 2007). Similarly, *P. syringae* pv. *tomato* T1 mutant lacking HopAS1 and AvrRpt2, another effector recognized by Arabidopsis Col-0, grows significantly better in non-host Arabidopsis (Sohn *et al.*, 2011). Interestingly, *hopQ1-1* is also found in the sequenced strain *P. syringae* pv. *phaseolicola* 1448A but is lacking from seven strains of *P. syringae* pv. *tabaci*, a pathogen of tobacco and *N. benthamiana*, whose virulence on the latter plant is disrupted by expression of the 1448A *hopQ1-1* gene (Ferrante *et al.*, 2009).

The second lesson gained from the disassembly of the *Pto* DC3000 effector repertoire is that two redundant effector groups (REGs) are particularly important for virulence. Mutants containing progressive deletions in the 28 well-expressed effector genes were constructed and analysed for changes in growth in *N. benthamiana* (Cunnac *et al.*, 2011; Kvitko *et al.*, 2009; Wei *et al.*, 2007). Whereas only slight reductions in pathogen growth resulted from deleting 15 of the effector genes (Kvitko *et al.*, 2009), strong reductions in growth accompanied loss of two REGs: AvrPto/AvrPtoB and AvrE/HopM1/HopR1. These REGs contain internally redundant effectors that appear to disrupt two distinct phases in PTI, PAMP perception and vesicle trafficking of antimicrobials, respectively (Block and Alfano, 2011; Kvitko *et al.*, 2009). Consistent with this concept, deleting *fliC*, which encodes flagellin, the major *P. syringae* PAMP, rescued growth of an *avrPto/avrPtoB* mutant but not of an *avrE/hopM1/hopR1* mutant (Kvitko *et al.*, 2009). According to this model, AvrPto and AvrPtoB functionally overlap in blocking PAMP perception and signal transduction. Such structured redundancy would promote system robustness. That is, mutation of both *avrPto* and *avrPtoB* would be needed to fully unblock PAMP signal processing for robust PTI development in infected plants.

A minimal set of *Pto* DC3000 effectors that restored substantial virulence in *N. benthamiana* was identified by first constructing polymutant DC3000D28E and then inserting concatenated subsets of effector genes into the genome (Cunnac *et al.*, 2011). DC3000D28E lacks all 28 of the well-expressed effector genes and grows poorly in *N. benthamiana*. The minimal functional repertoire was constructed in three steps. First, members of the two known REGs were introduced, which revealed that *avrPto* or *avrPtoB* could promote significant growth, but *hopM1* or *avrE* could not unless either *avrPto* or *avrPtoB* was present. Second, a DC3000D28E derivative carrying *avrPto* and *hopM1*, growing almost 2 logs better than DC3000D28E, was used as a recipient for shuffled, concatenated sets of 3 or 5 effector genes from a collection of 17 effectors. The concatenated gene sets were integrated into the exchangeable effector locus in the Hrp pathogenicity island of DC3000D28E using the programmable or random *in vivo* assembly shuttle (PRIVAS) system (Cunnac *et al.*, 2011). Analysis of the resulting strains that grew best revealed that the presence of *hopE1*, *hopG1*, and *hopAM1* was correlated with increased virulence. Third, insights from the previous observations enabled construction of a minimal functional set of 8 effectors (AvrPtoB, HopE1, HopG1, HopAM1, AvrE, HopM1, HopAA1, HopN1) that promoted growth of DC3000D28E in *N. benthamiana* to near wild-type levels along with the production of necrotic/chlorotic disease lesions.

The following statements summarize systems-level properties of *P. syringae* effector repertoires (Cunnac *et al.*, 2011): (i) effectors collectively are essential for virulence; (ii) but no single effector is essential; (iii) no single effector is sufficient for significant virulence; (iv) some effectors can be assigned to REGs that redundantly target distinct processes in plant defence, (v) effector genes reassembled in small groups can be observed to contribute in a hierarchical fashion to growth and symptom production; (vi) a minimal functional repertoire appears to require members of at least two REGs and several other effectors, which represent a mix of core and variable effectors; (vii) variability in the repertoires of pathogens of the same host indicates that a given plant may be defeated by many alternative combinations of effectors; (viii) heterologous expression and delivery of effectors from other strains, or even from oomycetes (Sohn *et al.*, 2007), can increase the virulence of wild-type strains, which suggests that the virulence of wild-type *P. syringae* strains is not maximized.

The latter two statements warrant further comment. It is noteworthy that the screen for random combinations of effectors that could promote growth of DC3000D28E expressing just AvrPto and HopM1 yielded effectors other than HopE1, HopG1, and HopAM1 (Cunnac *et al.*, 2011). Furthermore, *P. syringae* pv. *syringae* B728a and *P. syringae* pv. *tabaci* 11528 lack these three effectors, but they are able to cause disease in *N. benthamiana* (Studholme *et al.*, 2009; Vinatzer *et al.*, 2006). It is also noteworthy that mutations involving single or multiple effector genes can produce significantly different phenotypes in alternative hosts of a given strain (Crabill *et al.*, 2010; Vinatzer *et al.*, 2006; Wei *et al.*, 2007). It is not yet clear whether this results from differences in the abilities of effectors to evade weak ETI or suppress PTI or weak ETI in different plants.

Mechanisms by which effector genes are acquired

The diversity in effector repertoires discussed above appears to be the product of an ongoing evolutionary arms race between pathogens and their hosts, where bacteria possess the key advantage of being able to rapidly adjust to selection pressure through mutagenesis and acquisition of novel genetic material. Horizontal gene transfer events shaping the effector repertoires of diverse strains and pathovars have been mediated by various mobile genetic elements (MGEs) including plasmids, phage, integrative and conjugative elements (ICElands), and transposons. Some MGEs integrate at very specific sites in the chromosome, such as the phage integrases which preferentially integrate adjacent to tRNAs, but many have less stringent specificity (Boyd *et al.*, 2009). Once integrated, MGEs can function as repetitive sequences, further contributing to genome rearrangement through intra-genome homologous recombination (Phadnis and Berg, 1985). Mechanisms of effector gene acquisition are further described below.

Role of horizontal gene transfer

The impact of these processes on bacterial genome structure is evident from examination of whole genome sequences. With the exception of highly reduced genomes, most bacterial genome sequences exhibit a mosaic structure composed of core regions conserved across strains in both gene content and gene synteny relative, and variable regions characterized by significant differences (Hacker and Kaper, 2000; Welch *et al.*, 2002). Variable regions can be identified by compositional deviation relative to the core genome regions, typically manifested by differences in GC content, enrichment for MGEs as well as by lack of syntenic gene alignment and conservation with related genomes (Gal-Mor and Finlay, 2006). Variable regions are frequently referred to as pathogenicity islands when containing virulence genes.

Genes encoding the T3SS in plant pathogens appear to have been acquired by horizontal transfer. The T3SS is composed of eight highly conserved injectisome proteins evolutionarily related to flagellar components, plus 10 or more accessory proteins with varying degrees of conservation. While it is still debated whether the T3SS is evolutionarily derived from the flagellum or from a common ancestor of the two, comparison of the core injectisome proteins reveals seven different families among bacteria (Cornelis, 2006; McCann and Guttman, 2008) with HrpI injectisomes characteristic of *P. syringae*. *P. syringae* pv. *phaseolicola* 1448A (Pph 1448A) encodes a second cluster of T3SS genes, more closely related to T3SS families found in *Yersinia* and *Rhizobium* but of as yet unknown biological significance (Joardar *et al.*, 2005). The overall lack of correspondence between injectisome phylogeny and that of their host bacteria represents the most significant evidence supporting initial acquisition by horizontal transfer (Naum *et al.*, 2009), though in cases where acquisition occurred prior to strain or pathovar diversification, such as among *P. syringae* pathovars, T3SS structural genes are highly conserved.

While selected effector genes such as those in the *P. syringae* conserved effector locus have a long association with the lineage, most were acquired more recently, are found in variable genome regions, and are frequently associated with MGEs, such as phage, ICElands, plasmids, and insertion sequences (Guidot *et al.*, 2007; Lima *et al.*, 2008; Lindeberg *et al.*, 2008; Salzberg *et al.*, 2008). Also supporting the prevalence of horizontal transfer is the lack of congruity between effector repertoires and strain phylogeny, ever more apparent as sequences for closely related strains have become available.

Relative timing of gene acquisition can be estimated using several approaches, including analysis of representation among strains, assessment of gene location, and differences in sequence composition relative to the core genome. Compositional deviation of horizontally transferred regions is believed to ameliorate with increasing time of association (Lawrence and Ochman, 1997). The use of this as a metric for estimating relative time of gene acquisition suggested that among the variable effector families, HopD, HopI, HopX and HopZ are the most ancient (Rohmer *et al.*, 2004). Sequencing of over 30 P. syringae strains confirms that HopI and HopX are among the most widely distributed (Baltrus *et al.*, 2011).

Role of recombination and eukaryotic domain acquisition

Horizontal transfer, together with recombination, serves as an important means for generation of new effector families. Alignments of effector sequences reveal that selected *P. syringae* effectors share regulatory and translocation modules of common origin, supporting the hypothesis that reassortment of N-terminal regions has played an important role in the generation of novel effectors and recruitment of novel host-interaction domains to the effector repertoire (Stavrinides *et al.*, 2006). In addition, many effectors have evolved eukaryotic host targeting signals by which they are directed to various sites within the host cell. These include HopI which is targeted to the chloroplast, and multiple Hops with confirmed myristoylation motifs including AvrRpm1, AvrRpt2, AvrB, AvrPto, HopF2, HopO1-1, HopAR1, and members of the HopZ family. Modified with the myristate fatty acid, these motifs serve to localize effectors to host plasma membrane. In summary, horizontal gene transfer and recombination provide *P. syringae* with an efficient means for acquisition of new functional properties and targeting information to deploy in the ongoing battle to suppress diverse elements of the plant defence machinery.

Mechanisms by which effector genes are inactivated or lost

Probably the most significant force driving effector repertoire diversification is the need to evade ETI. Analysis of effector genes and pseudogenes in several sequenced strains has revealed a variety of mechanisms by which betraying effector genes can be inactivated.

Pathoadaptation

Pathoadaptation refers to evasion of host detection through introduction of point mutations (Ma *et al.*, 2006), evidence for which can be found among several effector families. Examples include the HopX1 (AvrPphE) family for which loss of recognition by the *R2* resistance gene is linked to selected point mutations among the HopX1 family members present in multiple strains (Stevens *et al.*, 1998). The effector AvrPto represents a particularly illuminating example given the extent to which its structural interactions with the resistance protein Pto have been characterized. *P. syringae* pv. *tomato* strains isolated from tomato carrying the *Pto* resistance gene were found to have *avrPto* alleles with mutations in the GINP loop required for the AvrPto–Pto interaction (Kunkeaw *et al.*, 2010).

The HopZ1 subfamily has also been found to contain variation among alleles suggestive of selective pressure to evade host detection; specifically, the differential plant response to the HopZ1a and HopZ1b alleles was found to be caused by a single residue change (Morgan *et al.*, 2010; Zhou *et al.*, 2009). Subsequent characterization of the ZAR1 resistance gene confirmed differential physical interaction by the two proteins (Lewis *et al.*, 2010).

The extent of pathoadaptation can be predicted from sequences of otherwise uncharacterized effectors by calculating ratios of non-synonymous to synonymous mutations. In this way proteins or domains under positive or diversifying selection can be identified, pointing to gene products under different degrees of selective pressure. Diversifying selection has been shown for the T3SS pilus protein HrpA and more recently for several effector subfamilies (Baltrus *et al.*, 2011), with nucleotide-by-nucleotide analyses such as those applied to HopZ1 and HopM1 pointing to regions subject to the most intensive selection pressure (Baltrus *et al.*, 2011; Ma *et al.*, 2006).

Effector gene loss

It is apparent from genome sequence analysis that evasion of detection caused by internal frameshifts, premature stop codons, and insertion sequences has occurred repeatedly in *P. syringae* evolution. Even among those effectors where virulence and host-detection phenotypes are genetically separable, complete disruption of the reading frame apparently confers a strong selective advantage (Stevens *et al.*, 1998). An alternative strategy for effector inactivation is exhibited by AvrPtoB. T1-like strains of *P. syringae* pv. *tomato* evade detection by the Pto R protein by post-transcriptional reduction of AvrPtoB protein levels (Lin *et al.*, 2006). Interestingly, the expression level appears sufficient to confer a virulence benefit as T1-like strains containing down-regulated AvrPtoB are more virulent than those in which *avrPtoB* has been deleted (Kunkeaw *et al.*, 2010).

In some cases, effector gene loss is achieved by deletion of a larger chromosomal region. *P. syringae* pv. *phaseolicola* strains lacking HopF1 have been shown to lack a 9.5-kb region bordered by a chimeric mobile element that is present in HopF1-containing strains (Rivas *et al.*, 2005). Conservation of the deletion site suggests that HopF1 loss occurred by a single deletion event in an ancestor of the six strains analysed. MGE-mediated effector loss and gain has been more thoroughly documented for HopAR1 (AvrPphB). Genomic island PPHGI-1 in *Pph* 1302A encodes an ICEland capable of excising from *Pph* 1302A and generating a circular episome in response to selection pressure from R3-based resistance in bean (Pitman *et al.*, 2005). This episome can be transformed into alternate bacterial strains *in planta* in a process activated by plant defence (Lovell *et al.*, 2009). Remnants of a similar ICEland associated with varied suites of effector genes have been found in other *P. syringae* genomes (Lindeberg *et al.*, 2008).

These examples illustrate the many paths by which effectors evade detection through changes in their coding sequences; however, host detection can also be blocked by acquisition of new functional modules. In perhaps the best characterized example of this phenomenon, AvrPtoB appears to have acquired a C-terminal E3 ubiquitin ligase domain that mediates the destruction of the tomato Fen kinase, which targets the AvrPtoB N-terminal domain (Rosebrock *et al.*, 2007). Sequence analysis of AvrRpm2 from *P. syringae* pv. *actinidae* reveals an N-terminal region with similarity to HopF and HopO, a central region similar to AvrRpm1, and a C-terminal region lacking similarity to any previously characterized effector, raising the possibility that sequential acquisition of functional domains as a means of evading detection may not be unique to the HopAB family. In summary, the functional diversity of effector repertoires among *P. syringae* strains is likely to be vastly greater when analysis is extended beyond the presence or absence of effector genes to polymorphisms among alleles, and knowing the functions of effectors is essential to understanding the significance of these polymorphisms.

Effector functions, targets, and sites of action in plants, as summarized in universal Gene Ontology terms

Effectors are collectively essential for *P. syringae* virulence and powerful probes for exploring the plant immune system. However, the complex and varied nature of their interference with host biology, as characterized over years of research, can impede coalescence of larger themes. Gene Ontology (GO) provides a useful tool for addressing these challenges (Ashburner *et al.*, 2000). By virtue of its three component ontologies, data concerning the cellular locations, molecular functions, and biological processes of individual gene products can be systematically translated into a universal format that includes the identity of the host organism, the nature of the supporting evidence, and the source publication. GO terms are arranged in a tree-like structure, with more specific terms positioned as children of more general parent terms, such that GO annotations can be assigned with different levels of specificity according to the degree to which the given gene product has been characterized. To address the need for terms that systematically encompass processes occurring during inter-organismal interactions the Plant and Animal-associated Microbe Gene Ontology Consortium (PAMGO) was formed and the biological process term 'GO:0044403 symbiosis, encompassing mutualism through parasitism' was created to serve as a parent term for all processes occurring during interactions between organisms (Torto-Alalibo *et al.*, 2010). Consistent with the goal of using terminology that is meaningful to researchers in diverse fields and the need to accommodate the range of inter-organismal relationships to which these processes can contribute, the word 'symbiont' is used rather than 'pathogen' in the newly created terms.

Over 650 annotations have been generated for *P. syringae* effectors and are available at the Pseudomonas–Plant Interaction website (http://www.pseudomonas-syringae.org) where they can be sorted by a variety of fields, including effector name, GO term, and plant host. Fig. 2.1 provides a broad overview of *P. syringae* effector actions in plants in the context of host protein targets and immunity related biological processes that are disrupted. In the next section, we will present a selection of representative effectors and associated higher level GO biological process terms as a guide for discussing unifying themes in effector function.

Negative regulation by symbiont of PAMP-induced host innate immunity (GO:0052034)

PTI is a multistage process encompassing recognition of PAMPS by receptor-like kinases followed by signalling via a kinase cascade, transcriptome re-programming, and transport of defence compounds to the site of PAMP-detection. Detection of PAMPs is achieved through their interaction with pattern recognition receptors typically linked to kinase domains critical to initiation of signalling cascades that communicate the detection event to the cell's regulatory machinery. One of the best characterized of these receptor-like kinases (RLKs) is FLS2, which recognizes components of bacterial flagellin and has homologues in *Arabidopsis, N. benthamiana,* and tomato (Boller and Felix, 2009). The EFR receptor, which recognizes components of elongation factor Tu is limited to the Brassicaceae (Zipfel *et al.*, 2006), while CERK1, in combination with Lym1 and Lym3, plays a role in perception of bacterial PGN by *Arabidopsis* (Willmann *et al.*, 2011).

PTI signalling in *Arabidopsis* is best characterized in relation to the response triggered by FLS2 and EFR upon binding to their respective PAMPS. Both RLKs bind and phosphorylate the signalling adapter and amplifier BAK1, required for downstream signal transmission (Chinchilla *et al.*, 2007). Activation of FLS2 results in phosphorylation of additional *Arabidopsis* proteins, including BIK1 and PBL1, mutations of which diminish the PTI response to multiple PAMPS, suggesting that they integrate immune signalling from multiple receptors (Zhang *et al.*, 2010). MAP kinase signalling cascades play a critical role in bridging early events in RLK signalling to the transcription of defence compounds. While elucidation of signalling has been complicated by MAP kinase involvement in multiple biological processes and lethality of individual MAP kinase mutants, there appear to be at least

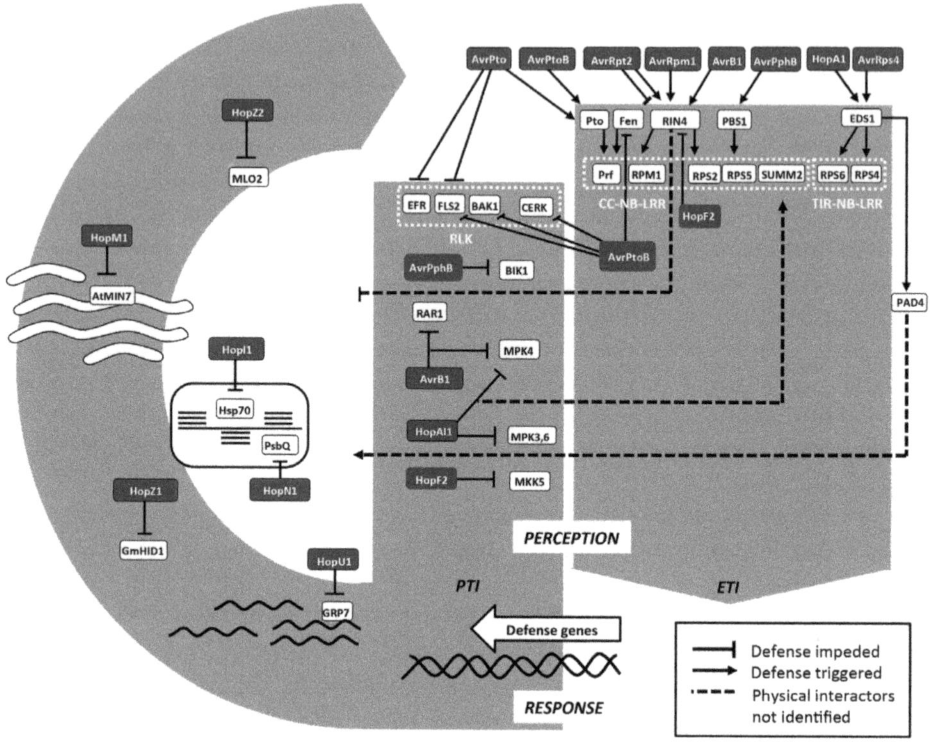

Figure 2.1 *Pseudomonas syringae* effector targets in the context of PAMP and effector perception and host responses. Specific effector–target interactions are superimposed on a schematic diagram of ETI and PTI. Interactions shown emphasize those for which physical association has been demonstrated (described in detail in the main text and Table 2.2). Three instances of effector impact on host defence are included (shown with dashed lines) despite lacking confirmed physical bases, to illustrate interplay between the ETI and PTI pathways. Two subclasses of NLR proteins are distinguished: CC-NB-LRR and TIR-NB-LRR (coiled-coil or Toll-interleukin-1 receptor-nucleotide binding-leucine-rich).

two MAPK cascades involved in *Arabidopsis* PAMP signalling, with MPK3/6 and MKK5 involved in positive regulation of PAMP responses and MPK4 in negative regulation (Bittel and Robatzek, 2007).

Type III effectors interfere with various stages of PTI signalling and have proven to be valuable probes for identification of the individual players in the host. Effectors experimentally linked to PTI suppression are united by the GO term 'Negative regulation by symbiont of PAMP-induced host innate immunity' (GO:0052034) in the Biological Process ontology. Suppression of the early stages of PTI signalling is best characterized for effectors AvrPto and AvrPtoB. AvrPtoB structurally mimics a eukaryotic E3 ubiquitin ligase, targeting host proteins for proteasome-dependent degradation (Abramovitch *et al.*, 2006). AvrPtoB has been reported to bind and ubiquitinate FLS2, resulting in loss of detectable FLS2 protein and reduction in flg22-dependent downstream signalling events (Gohre *et al.*, 2008). Ubiquitination by AvrPtoB and associated degradation has also been shown for CERK1 (Gimenez-Ibanez *et al.*, 2009). Interestingly, PTI suppression by AvrPtoB via binding to

Bti9, a CERK homologue in tomato for which the PAMP inducer remains uncharacterized, occurs in an E3-ligase independent manner (Zeng *et al.*, 2011).

AvrPto in contrast, inhibits the activity of diverse kinase proteins by occupying the P+1 loop, originally characterized in the host protein Pto and suggested by structural modelling for EFR and FLS2 (Xiang *et al.*, 2008). AvrPto was shown to interact specifically with EFR and FLS2 in a manner dependent on Y89, a residue required for its interaction with Pto, thereby inhibiting autophosphorylation of EFR and FLS2 and inhibiting signalling associated with the PTI response to flg22 (Xiang *et al.*, 2008). It has been reported that disruption of PTI upon flg22-FLS2 binding results from the interaction of AvrPto or AvrPtoB with the downstream signalling partner, BAK1 (Shan *et al.*, 2008), and low levels of AvrPtoB –dependent BAK1 ubiquitination have been observed (Gohre *et al.*, 2008). However, a requirement for BAK1-effector binding has not been shown in other studies (Gimenez-Ibanez *et al.*, 2009; Xiang *et al.*, 2008, 2011).

Interference with downstream kinase signalling has been demonstrated for multiple effectors, including HopAR1 (AvrPphB), AvrB1, HopF2, and HopAI1. HopAR1 acts at an early stage in the signalling cascade, proteolytically cleaving BIK1 and other proteins belonging to a subfamily of cytoplasmic receptor-like protein kinases (Zhang *et al.*, 2010). HopAI1 is a phosphothreonine lyase that inactivates *Arabidopsis* MPK3 and MPK6 (Zhang *et al.*, 2007). HopF2 similarly inactivates components of the MAP kinase cascade that positively regulate PTI. HopF2 binds to several MAP kinases and ADP-ribosylates MKK5, inhibiting PAMP-induced MPK activation (Wang *et al.*, 2010). AvrB1 suppresses PTI by binding to MPK4 in a RAR1-dependent manner, resulting in perturbations in hormone signalling mediated by enhanced MPK4 phosphorylation and increased kinase activity (Cui *et al.*, 2010).

PAMP detection and kinase signalling ultimately lead to transcriptional induction of defence genes and the production and delivery of antimicrobial factors to the site of PAMP detection. These processes represent obvious targets for effector-dependent disruption of PTI. Evidence for disruption of the defence-related transcriptional response is best supported by characterization of HopU1 activity and target proteins. One of several effectors with mono-ADP-ribosyltransferase activity, HopU1 was shown to ADP-ribosylate the RNA binding protein, GRP7. The enzymatic activity is required for PTI-suppression, and loss of GRP7 enhances host susceptibility (Fu *et al.*, 2007; Jeong *et al.*, 2011). Although the RNAs bound by GRP7 remain undefined, GRP7 has been shown to play a role in response to abiotic stress, facilitating movement of mRNAs from the nucleus to the cytoplasm under stress conditions, and implicated in stress-induced stomatal opening and closure (Kim *et al.*, 2008). Overexpression of GRP7 is also linked to up-regulation of proteins associated with SA-mediated defence including *PR1*, *PR2* and *PR5* (Streitner *et al.*, 2010). Given the importance of stomatal closure and SA-induction in plant defence, it is speculated that HopU1 may exert its PTI suppressing activity by interfering with GRP7-dependent post-transcriptional regulation of transcripts required for PTI-induced stomatal closure. There is also evidence suggesting that GRP7 suppresses PTI through negative regulation of FLS2 (J.R. Alfano, personal communication).

Chloroplasts play a central role in defence signalling. Not only are they the point of synthesis for salicylic acid, jasmonic acid, nitric oxide, and reactive oxygen species (Padmanabhan and Dinesh-Kumar, 2010), but the chloroplast protein CAS, required for both PTI and ETI, has recently been shown to mediate control of flg22-induced SA biosynthesis (Nomura *et al.*, 2012). HopI1 is one of the most highly conserved effectors across strains and is targeted to the chloroplast where it binds Hsp70 (Jelenska *et al.*, 2007, 2010) silencing of which enhances host susceptibility to T3SS^{-} *P. syringae* strains, implicating this protein in PTI. While the targets of Hsp70 remain uncharacterized it is hypothesized that it exerts its effect by impacting the protein folding/complex assembly integral to a chloroplast-based defence response. It should also be noted that in addition to binding glycine-rich RNA-binding proteins, GRP7 and GRP8, HopU1 was also shown to complex with three chloroplast RNA-binding proteins (Fu *et al.*, 2007).

Among the defence compounds induced by bacterial infection are antimicrobial secondary metabolites known as phytoalexins. A search for host targets of HopZ1a identified the soybean protein GmHID1, which is required for biosynthesis of the soybean isoflavone daidzein; HopZ1 binds to and promotes degradation of GmHID1, reducing the level of daidzein production and enhancing susceptibility to T3SS^{-} *P. syringae* strains (Zhou *et al.*, 2011).

The host endomembrane system plays an important role in transporting defence compounds to the sites of PAMP detection. Disruption of vesicle trafficking is best characterized for HopM1, which is encoded by the conserved effector locus and was shown to interact with and mediate proteasome-dependent degradation of AtMIN7, an ADP ribosylation factor (ARF) guanine nucleotide exchange factor (GEF) protein. ARF GEF proteins are known to be key components of the vesicle trafficking system in eukaryotic cells. AtMIN7 is post-transcriptionally regulated by PTI and localized to the trans–Golgi network/early endosome compartment, where it facilitates callose deposition (Nomura *et al.*, 2006, 2011). MLO2 is a regulator of vesicular trafficking plasma membrane regions associated with pathogen entry (Meyer *et al.*, 2009). Research by Lewis *et al.* (2012) has shown that MLO2 binding by HopZ2 is required for its role in virulence, suggesting that this effector may suppress secretion of virulence compounds by interfering with normal MLO2 function.

Many additional effector families, including HopA, HopC, HopS, HopT, HopAA, HopAF, hopAM, HopAO, HopAR, AvrE, AvrRpms, AvrRps4, and AvrRpt2, have been shown capable of PTI suppression, typically demonstrated by reduced expression of PTI associated phenotypes or gene expression upon challenge with a non-pathogen or flg22 (Block and Alfano, 2011; Cunnac *et al.*, 2009). Further exploration of their mechanisms of action is expected to yield new insight into the host pathways involved in PTI.

Positive regulation by symbiont of host resistance gene-dependent defence response (GO:0052527)

ETI induction by *P. syringae* effectors appears to typically involve indirect recognition by NLR proteins mediated by interactions between the effectors and initial target proteins, referred to alternately as 'guardees' or 'decoys' according to the guard hypothesis and decoy model (Zhou and Chai, 2008). Modification of the guardee/decoy proteins is detected by the associated NLR protein to trigger ETI and typically the HR but also a range of other defence responses including callose deposition. The best characterized interactions between effectors and decoy/guardees are those of AvrPto and AvrPtoB with Pto, leading to Prf-mediated ETI, and the interactions of AvrRpt2, AvrRpm1, and AvrB1 with RIN4 resulting in ETI mediated by RPM1 and RPS2.

Attempts to identify Arabidopsis components of ETI signalling pathways downstream of the NLR proteins have met with more limited success than similar investigations of PTI signalling. There appears to be significant overlap in genes involved in PAMP and effector signalling (Navarro *et al.*, 2004; Tsuda *et al.*, 2009), suggesting that ETI is a more recently evolved phenomena that may have adapted pre-existing PTI signalling pathways. It has also been proposed that the significant differences in PTI and ETI signalling are less in the components than in duration and the greater degree of compensatory relationships among ETI signalling pathways that results in lower sensitivity to perturbation (Tsuda and Katagiri, 2010; Tsuda *et al.*, 2009). In several instances it appears that the decoy/guardees themselves may function as important components in PTI signalling. For example, EDS1, linked to ETI induction upon effector-driven dissociation from NLR proteins RPS4 and RPS6 (Bhattacharjee *et al.*, 2011), appears to be required for PTI signalling in its PAD4-associated configuration (Rietz *et al.*, 2011). RIN4, well established as a decoy/guardee for RPM1 and RPS2, regulates PTI responses by regulating stomatal aperture in concert with guard cell plasma membrane ATPases (Liu *et al.*, 2009). Finally, MPK4, integral to PTI signalling, triggers a SUMM2-dependent ETI response when targeted by HopAI1 (Zhang *et al.*, 2012).

Several MAP kinases have been linked to ETI signalling in tomato and *N. benthamiana*. A mitogen-activated protein kinase kinase kinase α (MAPKKKα) has been identified as a positive regulator of Pto-mediated ETI (del Pozo *et al.*, 2004), which activates MPK2 and MPK3 (Pedley and Martin, 2004). In addition, MAPKKKα interacts with a 14-3-3 protein, TFT7, which appears to mediate ETI responses induced by Pto as well as other *R* gene/effector gene pairs (Oh *et al.*, 2010).

Negative regulation by symbiont of host defence-related PCD (GO:0034054)

In contrast to PTI where effector interference has been described for multiple stages in the process including detection, signalling, and defence-compound delivery, effector interference with ETI thus far identified is chiefly aimed at modification of the decoy/guardee proteins. AvrPtoB, one of the best studied examples of ETI interference, is a multidomain protein with its N-terminal region being detected in tomato by Fen kinase which in turn triggers ETI via the NLR protein, Prf. The C-terminal domain of AvrPtoB is an E3-ubiquitin ligase that facilitates ubiquitination of Fen, targeting it for destruction (Rosebrock *et al.*, 2007). HopF2 likewise inhibits ETI by targeting a decoy/guardee, in this instance interacting with RIN4 to inhibit AvrRpt2-dependent degradation and the resulting suppression of *P. syringae* growth (Wilton *et al.*, 2010). ETI suppressing capabilities have been shown for up to 15 additional Hop families (Jamir, 2004; Guo, 2009; Fu, 2007), though their mechanisms of action remain uncharacterized.

Capturing information about effectors with the Molecular Function and Cellular Component ontologies of GO

In addition to capturing information on the impact of individual proteins on biological processes, GO can be used to describe the molecular functions and host cellular locations of effectors. Most effectors have not been localized more precisely than the host cell, but in several cases locations have been more thoroughly characterized. A summary of Cellular Component annotations is shown in Table 2.1. Molecular Function Ontology assignments can be used to highlight those effectors for which 'protein binding' (GO:0005515) or 'catalytic activity' (GO:000384) has been demonstrated. 'Chaperone binding' (GO:0051087), a child term of the former, can be used to capture the numerous effectors that bind T3SS chaperone proteins in the bacterium. Binding to host proteins can be captured with other

Table 2.1 Selected Gene Ontology terms in the Cellular Component Ontology assigned to *Pseudomonas syringae* type III effectors

Hop[1]	GO term	GO number	Reference (PMID)[2]	Host[3]
AvrB1 (Pgy race 0)	Host cell plasma membrane	GO:0020002	10830163	A thal
AvrPto (Pto DC3000)	Host cell plasma membrane	GO:0020002	11148281	N tab
AvrRpm1 (Pma M2)	Host cell plasma membrane	GO:0020002	10830163	A thal
HopF2 (Pto DC3000)	Host cell plasma membrane	GO:0020002	16529375	N tab, A thal
HopG1 (Pto DC3000)	Host cell mitochondrion	GO:0033650	19863557	N tab
HopI1 (Pma ES4326)	Host cell chloroplast	GO:0033652	17350264	A thal, N tab, N benth
HopM1 (Pto DC3000)	Host cell endomembrane system	GO:0033645	16840699	A thal
HopN1 (Pto DC3000)	Host cell chloroplast thylakoid membrane	GO:0033654	22233353	N benth
HopZ1a (Psy A2)	Host cell plasma membrane	GO:0020002	18263728	A thal
HopZ2 (Ppi 895A)	Host cell plasma membrane	GO:0020002	18263728	N tab
HopAA1–1 (Pto DC3000)	Host cell mitochondrion	GO:0033650	18321194	Yeast
HopAR1 (AvrPphB) (Pph race 3)	Host cell plasma membrane	GO:0020002	19346252	B cam nap

[1]Pathovar abbreviations can be found on the PPI website (http://www.pseudomonas-syringae.org).
[2]All annotations in this table are inferred from direct evidence.
[3]A thal, *Arabidopsis thaliana*; N tab, *Nicotiana tabacum*; N benth, *Nicotiana benthamiana*; B cam nap, *Brassica campestris napus*.

child terms such as 'Protein kinase binding' (GO:0019901). Selected Molecular Function annotations for *P. syringae* effectors are shown in Table 2.2. Tables 2.1 and 2.2 illustrate the many fields of information included in GO annotations. In addition to the GO term itself, each annotation includes the source of the supporting information, the host organism, and the evidence code. The major categories of supporting evidence for the annotations shown are 'Inferred from Physical Interaction' (IPI), 'Inferred from Direct Assay' (IDA), 'Inferred from Mutant Phenotype' (IMP), and 'Inferred from Sequence Similarity' (ISS). GO guidelines typically contain a 'with' field to specify the identity of bound proteins. For easier interpretation, this field is renamed here as 'interacting host protein' and the identities of host proteins are specified not only for binding interactions but also for substrates of enzymatic activity on the part of the effector. The source strain of the characterized effector is specified in the 'Hop' field.

A model for the evolution of *P. syringae* effector repertoires based on integrating knowledge of effector functions and distribution among diverse strains

The relative importance of effectors in the two defined REGs (HopM1/AvrE/HopR1 and AvrPto/AvrPtoB) was an intriguing finding from the disassembly and reassembly of the *Pto* DC3000 effector repertoire, as discussed above. Notably, this work revealed that restoring

Table 2.2 Selected Gene Ontology terms in the Molecular Function Ontology assigned to *Pseudomonas syringae* type III effectors

Hop[1]	GO term	GO number	Reference (PMID)	Evidence[2]	Host[3]	Interacting host protein
AvrB1 (Pgy race 0)	Protein binding	GO:0005515	11955429	IPI	A thal	Binds RIN4
AvrB1 (Pgy race 0)	Protein binding	GO:0005515	19671880	IPI	S lyco	Binds RIN4
AvrB1 (Pgy race 0)	Protein binding	GO:0005515	17148606	IDA	A thal	Binds RAR1
AvrB1 (Pgy race 0)	Protein kinase binding	GO:0019901	20159621	IPI	A thal	Binds MPK4
AvrPto1 (Pto DC3000)	Protein kinase binding	GO:0019901	18158241	IPI	A thal	Binds FLS2 and EFR
AvrPto1 (Pto DC3000)	Protein kinase binding	GO:0019901	18158241	IPI	S lyco	Binds LeFLS2
AvrPto1 (Pto DC3000)	Protein kinase binding	GO:0019901	18621007	IPI	A thal	Binds BAK1
AvrPto1 (Pto DC3000)	Protein kinase binding	GO:0019901	20923364	IPI	A thal	Binds BSK3 and CDG1
AvrPto1 (Pto DC3000)	Protein binding	GO:0005515	19671880	IPI	S lyco	Binds RIN4
HopA1 (Pto DC3000)	Protein binding	GO:0005515	22158819	IPI	N benth	Binds EDS1
HopC1 (Pto DC3000)	Cysteine-type peptidase activity	GO:0008234	12062101	IMP, ISS	–	–
AvrPtoB (Pto DC3000)	Protein kinase binding	GO:0019901	12062102	IPI	S lyco	Binds Pto
AvrPtoB (Pto DC3000)	Protein kinase binding	GO:0019901	17637671	IPI	S pimp	Binds Pto and Fen
AvrPtoB (Pto DC3000)	Protein kinase binding	GO:0019901	18621007	IPI	A thal	Binds BAK1
AvrPtoB (Pto DC3000)	Protein kinase binding	GO:0019901	19062288	IPI	A thal	Binds FLS2
AvrPtoB (Pto DC3000)	Protein kinase binding	GO:0019901	21880077	IPI	S lyco	Binds Bti9 and SlLyk13
AvrPtoB (Pto DC3000)	Protein kinase binding	GO:0019901	19249211	IPI	A thal	Binds CERK1
AvrPtoB (Pto DC3000)	Protein binding	GO:0005515	17637671	IPI	S pimp	Binds PtoC
AvrPtoB (Pto DC3000)	Protein binding	GO:0005515	19671880	IPI	S lyco	Binds RIN4
AvrPtoB (Pto DC3000)	Ubiquitin–protein ligase activity	GO:0004842	16373536	IDA	–	–
AvrRpm1 (Pma M2)	Protein binding	GO:0005515	11955429	IPI	–	RIN4
AvrRps4 (Ppi 151)	Protein binding	GO:0005515	22158819	IPI	N benth	Binds EDS1

Table 2.2 Continued

Hop[1]	GO term	GO number	Reference (PMID)	Evidence[2]	Host[3]	Interacting host protein
AvrRpt2 (Pto JL1065)	Cysteine-type peptidase activity	GO:0008234	12950919	ISS	–	–
AvrRpt2 (Pto JL1065)	Cysteine-type peptidase activity	GO:0008234	12950919	IMP	–	–
AvrRpt2 (Pto JL1065)	cysteine-type peptidase Activity	GO:0008234	15684089	IDA	–	–
HopF2 (Pto DC3000)	Protein binding	GO:0005515	20133879	IDA	A thal	Binds RIN4
HopF2 (Pto DC3000)	Protein kinase binding	GO:0019901	20571112	IPI	A thal	Binds MKK3, MKK4, MKK5, MKK6, MKK10
HopF2 (Pto DC3000)	NAD(P)+-protein-arginine ADP-ribosyltransferase activity	GO:0003956	20571112	IDA	A thal	ADP-ribosylation of MKK5 and RIN4
HopI1 (Pma ES4326)	Protein binding	GO:0005515	20615948	ISS	A thal	Hsp70
HopM1 (Pto DC3000)	Small GTPase binding (covers ARF and GEF)	GO:0031267	16840699	IPI	A thal	Binds AtMIN7
HopM1 (Pto DC3000)	Protein binding	GO:0005515	16840699	IPI	yeast	Binds AtMIN2–7, AtMIN9–11
HopN1 (Pto DC3000)	Cysteine-type peptidase activity	GO:0008234	15469508	IDA	–	–
HopN1 (Pto DC3000)	Protein binding	GO:0005515	22233353	IPI	S lyco	Binds PsbQ
HopO1–2 (Pto DC3000)	NAD(P)+-protein–arginine ADP-ribosyltransferase activity	GO:0003956	17450127	ISS	–	–
HopU1 (Pto DC3000)	NAD(P)+-protein–arginine ADP-ribosyltransferase activity	GO:0003956	17450127	IDA	A thal	ADP-ribosylates cp29-RPB, cp31-RPB, RPB31, GRP7, GRP8
HopW1–1 (Pma ES4326)	Protein binding	GO:0005515	18266921	IPI	A thal	Binds WIN1 WIN2 WIN3
HopX1 (Pto DC3000)	Cysteine-type peptidase activity	GO:0008234	17427805	IMP	–	–
HopZ1a (Psy A2)	Cysteine-type peptidase activity	GO:0008234	17194219	IDA		–
HopZ1a (Psy A2)	Protein binding	GO:0005515	21402357	IPI	G max	Binds GmHID1
HopZ1a (Psy A2)	Protein binding	GO:0005515	22319451	IPI	A thal	Binds tubulin

Table 2.2 Continued

Hop[1]	GO term	GO number	Reference (PMID)	Evidence[2]	Host[3]	Interacting host protein
HopZ2 (Ppi 895A)	Cysteine-type peptidase activity	GO:0008234	17194219	IDA		–
HopZ2 (Ppi 895A)	Protein binding	GO:0020002	22230763	IPI	A thal	Binds MLO2
HopZ3 (Psy B728a)	Cysteine-type peptidase activity	GO:0008234	17194219	IDA		–
HopAI1 (Pto DC3000)	Phosphothreonine lyase activity	GO:0034598	18005697	IDA	–	Reduces phosphorylation of MPK3 MPK6
HopAI1 (Pto DC3000)	Protein kinase binding	GO:0019901	18005697	IPI	A thal	Binds MPK3 MPK6
HopAO1 (Pto DC3000)	Protein tyrosine phosphatase activity	GO:0004725	12828636	IDA	–	–
HopAO1 (Pto DC3000)	Protein tyrosine phosphatase activity	GO:0004725	12828637	IMP	–	–
HopAR1 (AvrPphB) (Pph race 3)	Cysteine-type peptidase activity	GO:0008234	12062101	IDA	A thal	–

[1]Pathovar abbreviations can be found on the PPI website (http://www.pseudomonas-syringae.org).
[2]Gene ontology evidence categories: IDA, inferred from direct evidence; IPI, inferred from physical interaction; IMP, inferred from mutant phenotype; ISS, inferred from sequence similarity.
[3]A thal, *Arabidopsis thaliana*; N benth, *Nicotiana benthamiana*; S lyco, *Solanum lycopersicum*; S pimp, *Solanum pimpinellifolium*; G max, *Gycine max*.

production of HopM1 and either AvrPto or AvrPtoB to DC3000D28E significantly promoted growth in *N. benthamiana* (Cunnac *et al.*, 2011). As also discussed above, HopM1 and AvrE are encoded in the conserved effector locus and are core components of diverse effector repertoires, whereas AvrPto and AvrPtoB are members of the variable portion of *P. syringae* effector repertoires. Also, the biological process targeted by HopM1 is quite different from that targeted by AvrPto and AvrPtoB. As depicted in Fig. 2.1, HopM1 disrupts the antimicrobial deployment phase of PTI, whereas AvrPto and AvrPtoB disrupt the pathogen perception phase. The target of AvrE is unknown, but the effector carries a WxxxE motif and a putative C-terminal endoplasmic reticulum membrane retention/retrieval signal, which are required for all AvrE biological activities and suggest an activity, like that of HopM1, with G-protein signalling and vesicle trafficking (Ham *et al.*, 2009). Furthermore, genetic evidence suggests that AvrE and HopM1 disrupt vesicle trafficking (Nomura *et al.*, 2006), and both effectors have a general ability to elicit cell death in plants in the absence of any known matching *R* genes, which contrasts with AvrPto and AvrPtoB (Badel *et al.*, 2006; Lindeberg *et al.*, 2012).

These observations have suggested a 'protected core' model for the evolution of *P. syringae* effector repertoires (Lindeberg *et al.*, 2012), which is an extension of the general concept that a necessary function of some effectors in a functional repertoire is the suppression of ETI that would otherwise be elicited by other effectors in the same repertoire (Abramovitch *et al.*, 2003; Jackson *et al.*, 1999). The protected core model postulates that ETI is triggered by HopM1 damage to a cellular process (vesicle trafficking) in combination with a second signal generated by PAMP perception. This model further suggests a scenario for the evolution of effector repertoires in which (i) ancestral effectors targeted vesicle trafficking, thereby blocking PTI, (ii) plants evolved an ancestral ETI based on the two signals of vesicle trafficking disruption and PAMP perception, (iii) pathogens evolved effectors such as AvrPto that targeted PAMP perception, and (iv) subsequent evolution of the ETI system and redundancies in plant PTI signalling led to expansion and diversification of pathogen effector repertoires (Lindeberg *et al.*, 2012).

A toolkit for systems-level study of *P. syringae* effector biology and future challenges

Our overriding hypothesis is that *P. syringae* effectors function as components of robust systems that involve redundancies and interplay with other effectors and with host targets and that these targets, in turn, function as components of host immune systems that have similarly evolved to promote robustness (Pritchard and Birch, 2011; Schneider and Collmer, 2010; Tsuda *et al.*, 2009). Understanding effector activity in such complex systems requires the development of two types of knowledge bases (Schneider and Collmer, 2010). The first is a reductionist, factor-centric, components-biology knowledge base that is universally accessible and linked to genomic loci through GO annotation, as we have described above. The second knowledge base is specific to each pathosystem and is aimed at understanding how effectors operate in the context of other virulence factors and the overall ecology of the pathosystem. The first knowledge base will facilitate comparisons between the translocated effectors of diverse pathogens. One question that could be efficiently addressed with this knowledge base is whether fungal or oomycete pathogens also have cytoplasmic effector repertoires with core effectors that target vesicle trafficking. The second knowledge base will facilitate, for example, analysis of how *Pto* DC3000 effectors operate in coordination with the phytotoxin coronatine in suppressing salicylic acid-mediated immune signalling.

Many resources are making *P. syringae* effector repertoires increasingly amenable for systems-level study. A growing MLST database enables newly sequenced strains to be phylogenetically placed with greater accuracy (Almeida *et al.*, 2010). Genome sequences of environmental strains and strains isolated from the same crop over time afford insights into effector repertoire dynamics in a global and temporal context (Cai *et al.*, 2011; Morris *et al.*, 2008; O'Brien *et al.*, 2011). Ongoing study and annotation of reference genomes (particularly that of *Pto* DC3000) enhance understanding of the operation of effector repertoires in the context of multiple regulatory systems and deep transcriptomics (Bronstein *et al.*, 2008; Filiatrault *et al.*, 2010). The development of a web-based community nomenclature system for *P. syringae* effectors, coupled with ongoing GO annotation derived from journal articles, facilitates comparative studies and hypothesis generation. A massive yeast-two-hybrid screen for Arabidopsis proteins that interact with a large collection of *P. syringae* effectors provides many leads for seeking biological targets of effectors (Mukhtar *et al.*, 2011). Also a large collection of *Pto* DC3000 combinatorial mutants and of DC3000D28E derivatives with customized repertoires facilitates testing postulated roles of effectors (Cunnac *et al.*, 2011; Kvitko *et al.*, 2009).

If effector repertoires have evolved to maximize system robustness in interactions with plants in the field, then a practical goal for the future is to identify vulnerabilities in these systems that can be exploited for more durable crop protection. One approach could be the deployment of combinations of *R* genes designed to defeat internal redundancies in effector repertoires or to target effectors needed to suppress ETI elicited by other effectors. Another approach would be to exploit advances in understanding the interrelated phenomena of host specialization in pathogens and non-host resistance in plants that underlie the host specificity that *P. syringae* pathovars have for various crop species. In that regard, it is likely that more information on variations in crop immune systems will provide another source of clues for explaining the composition of functional effector repertoires. Thus, the interactions of *P. syringae* with crop plants provide both knowledge and a focus to systems-level study of *P. syringae* effector repertoires.

Acknowledgements

This work was supported by NSF grant IOS-1025642.

Web resources

Pseudomonas–Plant Interaction site: http://www.pseudomonas-syringae.org/.html.
PAMDB: http://genome.ppws.vt.edu/cgi-bin/MLST/home.pl.
Pseudomonas Genome Database: http://www.pseudomonas.com/
The Gene Ontology: http://www.geneontology.org/

References

Abramovitch, R.B., Kim, Y.J., Chen, S., Dickman, M.B., and Martin, G.B. (2003). *Pseudomonas* type III effector AvrPtoB induces plant disease susceptibility by inhibition of host programmed cell death. EMBO J. 22, 60–69.

Abramovitch, R.B., Janjusevic, R., Stebbins, C.E., and Martin, G.B. (2006). Type III effector AvrPtoB requires intrinsic E3 ubiquitin ligase activity to suppress plant cell death and immunity. Proc. Natl. Acad. Sci. U.S.A. *103*, 2851–2856.

Alfano, J.R., Charkowski, A.O., Deng, W.-L., Badel, J.L., Petnicki-Ocwieja, T., van Dijk, K., and Collmer, A. (2000). The *Pseudomonas syringae* Hrp pathogenicity island has a tripartite mosaic structure composed of a cluster of type III secretion genes bounded by exchangeable effector and conserved effector loci that contribute to parasitic fitness and pathogenicity in plants. Proc. Natl. Acad. Sci. U.S.A. *97*, 4856–4861.

Almeida, N.F., Yan, S., Lindeberg, M., Studholme, D.J., Schneider, D.J., Condon, B., Liu, H., Viana, C.J., Warren, A., Evans, C., *et al.* (2009). A draft genome sequence of *Pseudomonas syringae* pv. *tomato* strain T1 reveals a repertoire of type III related genes significantly divergent from that of *Pseudomonas syringae* pv. *tomato* strain DC3000. Mol. Plant Microbe Interact. *22*, 52–62.

Almeida, N.F., Yan, S., Cai, R., Clarke, C.R., Morris, C.E., Schaad, N.W., Schuenzel, E.L., Lacy, G.H., Sun, X., Jones, J.B., *et al.* (2010). PAMDB, a multilocus sequence typing and analysis database and website for plant-associated microbes. Phytopathology *100*, 208–215.

Ashburner, M., Ball, C.A., Blake, J.A., Botstein, D., Butler, H., Cherry, J.M., Davis, A.P., Dolinski, K., Dwight, S.S., Eppig, J.T., *et al.* (2000). Gene ontology: tool for the unification of biology. Nat. Genet. *25*, 25–29.

Badel, J.L., Shimizu, R., Oh, H.-S., and Collmer, A. (2006). A *Pseudomonas syringae* pv. *tomato avrE1/hopM1* mutant is severely reduced in growth and lesion formation in tomato. Mol. Plant Microbe Interact. *19*, 99–111.

Baltrus, D.A., Nishimura, M.T., Romanchuk, A., Chang, J.H., Mukhtar, M.S., Cherkis, K., Roach, J., Grant, S.R., Jones, C.D., and Dangl, J.L. (2011). Dynamic evolution of pathogenicity revealed by sequencing and comparative genomics of 19 *Pseudomonas syringae* isolates. PLoS Pathog. *7*, e1002132.

Baxter, L., Tripathy, S., Ishaque, N., Boot, N., Cabral, A., Kemen, E., Thines, M., Ah-Fong, A., Anderson, R., Badejoko, W., *et al.* (2010). Signatures of adaptation to obligate biotrophy in the *Hyaloperonospora arabidopsidis* genome. Science *330*, 1549–1551.

Bender, C.L., Alarcon-Chaidez, F., and Gross, D.C. (1999). *Pseudomonas syringae* phytotoxins: mode of action, regulation, and biosynthesis by peptide and polyketide synthetases. Microbiol. Mol. Biol. Rev. *63*, 266–292.

Bhattacharjee, S., Halane, M.K., Kim, S.H., and Gassmann, W. (2011). Pathogen effectors target *Arabidopsis* EDS1 and alter its interactions with immune regulators. Science *334*, 1405–1408.

Bittel, P., and Robatzek, S. (2007). Microbe-associated molecular patterns (MAMPs) probe plant immunity. Curr. Opin. Plant Biol. *10*, 335–341.

Block, A., and Alfano, J.R. (2011). Plant targets for *Pseudomonas syringae* type III effectors: virulence targets or guarded decoys? Curr. Opin. Microbiol. *14*, 39–46.

Boch, J., Joardar, V., Gao, L., Robertson, T.L., Lim, M., and Kunkel, B.N. (2002). Identification of *Pseudomonas syringae* genes induced during infection of *Arabidopsis thaliana*. Mol. Microbiol. *44*, 73–88.

Boller, T., and Felix, G. (2009). A renaissance of elicitors: perception of microbe-associated molecular patterns and danger signals by pattern-recognition receptors. Annu. Rev. Plant Biol. *60*, 379–406.

Boyd, E.F., Almagro-Moreno, S., and Parent, M.A. (2009). Genomic islands are dynamic, ancient integrative elements in bacterial evolution. Trends Microbiol. *17*, 47–53.

Bronstein, P.A., Filiatrault, M.J., Myers, C.R., Rutzke, M., Schneider, D.J., and Cartinhour, S.W. (2008). Global transcriptional responses of *Pseudomonas syringae* DC3000 to changes in iron bioavailability *in vitro*. BMC Microbiol. *8*, 209.

Brooks, D., Hernandez-Guzman, G., Koek, A.P., Alarcon-Chaidez, F., Sreedharan, A., Rangaswarmy, V., Penaloza-Vasquez, A., Bender, C.L., and Kunkel, B.N. (2004). Identification and characterization of a well-defined series of coronatine biosynthetic mutants of *Pseudomonas syringae* pv. *tomato* DC3000. Mol. Plant Microbe Interact. *17*, 162–174.

Buell, C.R., Joardar, V., Lindeberg, M., Selengut, J., Paulsen, I.T., Gwinn, M.L., Dodson, R.J., Deboy, R.T., Durkin, A.S., Kolonay, J.F., *et al.* (2003). The complete sequence of the *Arabidopsis* and tomato pathogen *Pseudomonas syringae* pv. *tomato* DC3000. Proc. Natl. Acad. Sci. U.S.A. *100*, 10181–10186.

Cai, R., Lewis, J., Yan, S., Liu, H., Clarke, C.R., Campanile, F., Almeida, N.F., Studholme, D.J., Lindeberg, M., Schneider, D., *et al.* (2011). The plant pathogen *Pseudomonas syringae* pv. *tomato* is genetically monomorphic and under strong selection to evade tomato immunity. PLoS Pathog. *7*, e1002130.

Chang, J.H., Urbach, J.M., Law, T.F., Arnold, L.W., Hu, A., Gombar, S., Grant, S.R., Ausubel, F.M., and Dangl, J.L. (2005). A high-throughput, near-saturating screen for type III effector genes from *Pseudomonas syringae*. Proc. Natl. Acad. Sci. U.S.A. *102*, 2549–2554.

Chinchilla, D., Zipfel, C., Robatzek, S., Kemmerling, B., Nurnberger, T., Jones, J.D., Felix, G., and Boller, T. (2007). A flagellin-induced complex of the receptor FLS2 and BAK1 initiates plant defence. Nature *448*, 497–500.

Collmer, A., Lindeberg, M., Petnicki-Ocwieja, T., Schneider, D.J., and Alfano, J.R. (2002). Genomic mining type III secretion system effectors in *Pseudomonas syringae* yields new picks for all TTSS prospectors. Trends Microbiol. *10*, 462–469.

Cornelis, G.R. (2006). The type III secretion injectisome. Nat. Rev. Microbiol. *4*, 811–825.

Crabill, E., Joe, A., Block, A., Van Rooyen, J.M., and Alfano, J.R. (2010). Plant immunity directly or indirectly restricts the injection of type III effectors by the *Pseudomonas syringae* type III secretion system. Plant Physiol. 154, 233–244.

Crabill, E., Karpisek, A., and Alfano, J.R. (2012). The *Pseudomonas syringae* HrpJ protein controls the secretion of type III translocator proteins and has a virulence role inside plant cells. Mol. Microbiol. *85*, 225–238.

Cui, H., Wang, Y., Xue, L., Chu, J., Yan, C., Fu, J., Chen, M., Innes, R.W., and Zhou, J.M. (2010). *Pseudomonas syringae* effector protein AvrB perturbs *Arabidopsis* hormone signaling by activating MAP kinase 4. Cell Host Microbe 7, 164–175.

Cunnac, S., Lindeberg, M., and Collmer, A. (2009). *Pseudomonas syringae* type III secretion system effectors: repertoires in search of functions. Curr. Opin. Microbiol. *12*, 53–60.

Cunnac, S., Chakravarthy, S., Kvitko, B.H., Russell, A.B., Martin, G.B., and Collmer, A. (2011). Genetic disassembly and combinatorial reassembly identify a minimal functional repertoire of type III effectors in *Pseudomonas syringae*. Proc. Natl. Acad. Sci. U.S.A. *108*, 2975–2980.

van Dijk, K., Tam, V.C., Records, A.R., Petnicki-Ocwieja, T., and Alfano, J.R. (2002). The ShcA protein is a molecular chaperone that assists in the secretion of the HopPsyA effector from the type III (Hrp) protein secretion system of *Pseudomonas syringae*. Mol. Microbiol. *44*, 1469–1481.

Feil, H., Feil, W.S., Chain, P., Larimer, F., Dibartolo, G., Copeland, A., Lykidis, A., Trong, S., Nolan, M., Goltsman, E., *et al.* (2005). Comparison of the complete genome sequences of *Pseudomonas syringae* pv. *syringae* B728a and pv. *tomato* DC3000. Proc. Natl. Acad. Sci. U.S.A. *102*, 11064–11069.

Ferrante, P., Clarke, C.R., Cavanaugh, K.A., Michelmore, R.W., Buonaurio, R., and Vinatzer, B.A. (2009). Contributions of the effector gene *hopQ1-1* to differences in host range between *Pseudomonas syringae* pv. *phaseolicola* and *P. syringae* pv. *tabaci*. Mol. Plant Pathol. *10*, 837–842.

Ferreira, A.O., Myers, C.R., Gordon, J.S., Martin, G.B., Vencato, M., Collmer, A., Wehling, M.D., Alfano, J.R., Moreno-Hagelsieb, G., Lamboy, W.F., *et al.* (2006). Whole-genome expression profiling defines the HrpL regulon of *Pseudomonas syringae* pv. *tomato* DC3000, allows *de novo* reconstruction of the Hrp *cis* element, and identifies novel co-regulated gene. Mol. Plant Microbe Interact. *19*, 1167–1179.

Filiatrault, M.J., Stodghill, P.V., Bronstein, P.A., Moll, S., Lindeberg, M., Grills, G., Schweitzer, P., Wang, W., Schroth, G.P., Luo, S., *et al.* (2010). Transcriptome analysis of *Pseudomonas syringae* identifies new genes, ncRNAs, and antisense activity. J. Bacteriol. *192*, 2359–2372.

Fouts, D.E., Abramovitch, R.B., Alfano, J.R., Baldo, A.M., Buell, C.R., Cartinhour, S., Chatterjee, A.K., D'Ascenzo, M., Gwinn, M.L., Lazarowitz, S.G., *et al.* (2002). Genomewide identification of *Pseudomonas syringae* pv. *tomato* DC3000 promoters controlled by the HrpL alternative sigma factor. Proc. Natl. Acad. Sci. U.S.A. *99*, 2275–2280.

Fu, Z.Q., Guo, M., Jeong, B.R., Tian, F., Elthon, T.E., Cerny, R.L., Staiger, D., and Alfano, J.R. (2007). A type III effector ADP-ribosylates RNA-binding proteins and quells plant immunity. Nature *447*, 284–288.

Gal-Mor, O., and Finlay, B.B. (2006). Pathogenicity islands: a molecular toolbox for bacterial virulence. Cell Microbiol. *8*, 1707–1719.

Gimenez-Ibanez, S., Hann, D.R., Ntoukakis, V., Petutschnig, E., Lipka, V., and Rathjen, J.P. (2009). AvrPtoB targets the LysM receptor kinase CERK1 to promote bacterial virulence on plants. Curr. Biol. *19*, 423–429.

Gohre, V., Spallek, T., Haweker, H., Mersmann, S., Mentzel, T., Boller, T., de Torres, M., Mansfield, J.W., and Robatzek, S. (2008). Plant pattern-recognition receptor FLS2 is directed for degradation by the bacterial ubiquitin ligase AvrPtoB. Curr. Biol. *18*, 1824–1832.

Greenberg, J.T., and Vinatzer, B.A. (2003). Identifying type III effectors of plant pathogens and analyzing their interaction with plant cells. Curr. Opin. Microbiol. *6*, 20–28.

Groll, M., Schellenberg, B., Bachmann, A.S., Archer, C.R., Huber, R., Powell, T.K., Lindow, S., Kaiser, M., and Dudler, R. (2008). A plant pathogen virulence factor inhibits the eukaryotic proteasome by a novel mechanism. Nature *452*, 755–758.

Guidot, A., Prior, P., Schoenfeld, J., Carrere, S., Genin, S., and Boucher, C. (2007). Genomic structure and phylogeny of the plant pathogen *Ralstonia solanacearum* inferred from gene distribution analysis. J. Bacteriol. *189*, 377–387.

Guo, M., Chancey, S.T., Tian, F., Ge, Z., Jamir, Y., and Alfano, J.R. (2005). *Pseudomonas syringae* type III chaperones ShcO1, ShcS1, and ShcS2 facilitate translocation of their cognate effectors and can substitute for each other in the secretion of HopO1-1. J. Bacteriol. *187*, 4257–4269.

Guo, M., Tian, F., Wamboldt, Y., and Alfano, J.R. (2009). The majority of the type III effector inventory of *Pseudomonas syringae* pv. *tomato* DC3000 can suppress plant immunity. Mol. Plant Microbe Interact. *22*, 1069–1080.

Guttman, D.S., and Greenberg, J.T. (2001). Functional analysis of the type III effectors AvrRpt2 and AvrRpm1 of *Pseudomonas syringae* with the use of a single-copy genomic integration system. Mol. Plant Microbe Interact. *14*, 145–155.

Guttman, D.S., Vinatzer, B.A., Sarkar, S.F., Ranall, M.V., Kettler, G., and Greenberg, J.T. (2002). A functional screen for the Type III (Hrp) secretome of the plant pathogen *Pseudomonas syringae*. Science *295*, 1722–1726.

Hacker, J., and Kaper, J.B. (2000). Pathogenicity islands and the evolution of microbes. Annu. Rev. Microbiol. *54*, 641–679.

Ham, J.H., Majerczak, D.R., Nomura, K., Mecey, C., Uribe, F., He, S.Y., Mackey, D., and Coplin, D.L. (2009). Multiple activities of the plant pathogen type III effector proteins WtsE and AvrE require WxxxE motifs. Mol. Plant Microbe Interact. *22*, 703–712.

Hirano, S.S., and Upper, C.D. (2000). Bacteria in the leaf ecosystem with emphasis on *Pseudomonas syringae* – a pathogen, ice nucleus, and epiphyte. Microbiol. Mol. Biol. Rev. *64*, 624–653.

Jackson, R.W., Athanassopoulos, E., Tsiamis, G., Mansfield, J.W., Sesma, A., Arnold, D.L., Gibbon, M.J., Murillo, J., Taylor, J.D., and Vivian, A. (1999). Identification of a pathogenicity island, which contains genes for virulence and avirulence, on a large native plasmid in the bean pathogen *Pseudomonas syringae* pathovar phaseolicola. Proc. Natl. Acad. Sci. U.S.A. *96*, 10875–10880.

Jamir, Y., Guo, M., Oh, H.-S., Petnicki-Ocwieja, T., Chen, S., Tang, X., Dickman, M.B., Collmer, A., and Alfano, J.R. (2004). Identification of *Pseudomonas syringae* type III secreted effectors that suppress programmed cell death in plants and yeast. Plant J. *37*, 554–565.

Jelenska, J., Yao, N., Vinatzer, B.A., Wright, C.M., Brodsky, J.L., and Greenberg, J.T. (2007). A J domain virulence effector of *Pseudomonas syringae* remodels host chloroplasts and suppresses defenses. Curr. Biol. *17*, 499–508.

Jelenska, J., van Hal, J.A., and Greenberg, J.T. (2010). *Pseudomonas syringae* hijacks plant stress chaperone machinery for virulence. Proc. Natl. Acad. Sci. U.S.A. *107*, 13177–13182.

Jeong, B.R., Lin, Y., Joe, A., Guo, M., Korneli, C., Yang, H., Wang, P., Yu, M., Cerny, R.L., Staiger, D., *et al.* (2011). Structure function analysis of an ADP-ribosyltransferase type III effector and its RNA-binding target in plant immunity. J. Biol. Chem. *286*, 43272–43281.

Joardar, V., Lindeberg, M., Jackson, R.W., Selengut, J., Dodson, R., Brinkac, L.M., Daugherty, S.C., DeBoy, R., Durkin, A.S., Giglio, M.G., *et al.* (2005). Whole genome sequence analysis of *Pseudomonas syringae* pv. phaseolicola 1448A reveals sequence divergence among pathovars in genes involved in virulence and transposition. J. Bacteriol. *187*, 6488–6498.

Jones, J.D., and Dangl, J.L. (2006). The plant immune system. Nature *444*, 323–329.

Kabisch, U., Landgraf, A., Krause, J., Bonas, U., and Boch, J. (2005). Type III secretion chaperones ShcS1 and ShcO1 from *Pseudomonas syringae* pv. *tomato* DC3000 bind more than one effector. Microbiology *151*, 269–280.

Kay, S., and Bonas, U. (2009). How *Xanthomonas* type III effectors manipulate the host plant. Curr. Opin. Microbiol. *12*, 37–43.

Keen, N.T. (1990). Gene-for-gene complementarity in plant–pathogen interactions. Annu. Rev. Genet. *24*, 447–463.

Kim, J.S., Jung, H.J., Lee, H.J., Kim, K.A., Goh, C.H., Woo, Y., Oh, S.H., Han, Y.S., and Kang, H. (2008). Glycine-rich RNA-binding protein 7 affects abiotic stress responses by regulating stomata opening and closing in *Arabidopsis thaliana*. Plant J. *55*, 455–466.

Kobayashi, D.Y., Tamaki, S.J., and Keen, N.T. (1989). Cloned avirulence genes from the tomato pathogen *Pseudomonas syringae* pv. *tomato* confer cultivar specificity on soybean. Proc. Natl. Acad. Sci. U.S.A. *86*, 157–161.

Kunkeaw, S., Tan, S., and Coaker, G. (2010). Molecular and evolutionary analyses of *Pseudomonas syringae* pv. *tomato* race 1. Mol. Plant Microbe Interact. *23*, 415–424.

Kvitko, B.H., Ramos, A.R., Morello, J.E., Oh, H.-S., and Collmer, A. (2007). Identification of harpins in *Pseudomonas syringae* pv. tomato DC3000, which are functionally similar to HrpK1 in promoting translocation of type III secretion system effectors. J. Bacteriol. *189*, 8059–8072.

Kvitko, B.H., Park, D.H., Velásquez, A.C., Wei, C.-F., Russell, A.B., Martin, G.B., Schneider, D.J., and Collmer, A. (2009). Deletions in the repertoire of *Pseudomonas syringae* pv. *tomato* DC3000 type III secretion effector genes reveal functional overlap among effectors. PLoS Pathog. *5*, e1000388.

Lawrence, J.G., and Ochman, H. (1997). Amelioration of bacterial genomes: rates of change and exchange. J. Mol. Evol. *44*, 383–397.

Lee, J., Klusener, B., Tsiamis, G., Stevens, C., Neyt, C., Tampakaki, A.P., Panopoulos, N.J., Noller, J., Weiler, E.W., Cornelis, G.R., *et al.* (2001). $HrpZ_{Psph}$ from the plant pathogen *Pseudomonas syringae* pv. *phaseolicola* binds to lipid bilayers and forms an ion-conducting pore *in vitro*. Proc. Natl. Acad. Sci. U.S.A. *98*, 289–294.

Lewis, J.D., Wu, R., Guttman, D.S., and Desveaux, D. (2010). Allele-specific virulence attenuation of the *Pseudomonas syringae* HopZ1a type III effector via the Arabidopsis ZAR1 resistance protein. PLoS Genet. *6*, e1000894.

Lewis, J.D., Wan, J., Ford, R., Gong, Y., Fung, P., Nahal, H., Wang, P.W., Desveaux, D., and Guttman, D.S. (2012). Quantitative interactor screening with next-generation sequencing (QIS-Seq) identifies *Arabidopsis thaliana* MLO2 as a target of the *Pseudomonas syringae* type III effector HopZ2. BMC Genomics *13*, 8.

Lima, W.C., Paquola, A.C., Varani, A.M., Van Sluys, M.A., and Menck, C.F. (2008). Laterally transferred genomic islands in Xanthomonadales related to pathogenicity and primary metabolism. FEMS Microbiol. Lett. *281*, 87–97.

Lin, N.-C., and Martin, G.B. (2007). Pto/Prf-mediated recognition of AvrPto and AvrPtoB restricts the ability of diverse *Pseudomonas syringae* pathovars to infect tomato. Mol. Plant Microbe Interact. *20*, 806–815.

Lin, N.C., Abramovitch, R.B., Kim, Y.J., and Martin, G.B. (2006). Diverse AvrPtoB homologs from several *Pseudomonas syringae* pathovars elicit Pto-dependent resistance and have similar virulence activities. Appl. Environ. Microbiol. 72, 702–712.

Lindeberg, M., Stavrinides, J., Chang, J.H., Alfano, J.R., Collmer, A., Dangl, J.L., Greenberg, J.T., Mansfield, J.W., and Guttman, D.S. (2005). Proposed guidelines for a unified nomenclature and phylogenetic analysis of type III Hop effector proteins in the plant pathogen *Pseudomonas syringae*. Mol. Plant Microbe Interact. *18*, 275–282.

Lindeberg, M., Cartinhour, S., Myers, C.R., Schechter, L.M., Schneider, D.J., and Collmer, A. (2006). Closing the circle on the discovery of genes encoding Hrp regulon members and type III secretion system effectors in the genomes of three model *Pseudomonas syringae* strains. Mol. Plant Microbe Interact. *19*, 1151–1158.

Lindeberg, M., Myers, C.R., Collmer, A., and Schneider, D.J. (2008). Roadmap to new virulence determinants in *Pseudomonas syringae*: insights from comparative genomics and genome organization. Mol. Plant Microbe Interact. *21*, 685–700.

Lindeberg, M., Cunnac, S., and Collmer, A. (2009). The evolution of *Pseudomonas syringae* host specificity and type III effector repertoires. Mol. Plant Pathol. *10*, 767–775.

Lindeberg, M., Cunnac, S., and Collmer, A. (2012). *Pseudomonas syringae* type III effector repertoires: last words in endless arguments. Trends Microbiol. *20*, 199–208.

Lindgren, P.B., Peet, R.C., and Panopoulos, N.J. (1986). Gene cluster of *Pseudomonas syringae* pv. '*phaseolicola*' controls pathogenicity of bean plants and hypersensitivity on nonhost plants. J. Bacteriol. *168*, 512–522.

Liu, J., Elmore, J.M., Fuglsang, A.T., Palmgren, M.G., Staskawicz, B.J., and Coaker, G. (2009). RIN4 functions with plasma membrane H+-ATPases to regulate stomatal apertures during pathogen attack. PLoS Biol. *7*, e1000139.

Lovell, H.C., Mansfield, J.W., Godfrey, S.A., Jackson, R.W., Hancock, J.T., and Arnold, D.L. (2009). Bacterial evolution by genomic island transfer occurs via DNA transformation *in planta*. Curr. Biol. *19*, 1586–1590.

Ma, W., Dong, F.F., Stavrinides, J., and Guttman, D.S. (2006). Type III effector diversification via both pathoadaptation and horizontal transfer in response to a coevolutionary arms race. PLoS Genet. *2*, e209.

McCann, H.C., and Guttman, D.S. (2008). Evolution of the type III secretion system and its effectors in plant–microbe interactions. New Phytologist *177*, 33–47.

McDonald, B.A., and Linde, C. (2002). Pathogen population genetics, evolutionary potential, and durable resistance. Ann. Rev. Phytopathol. *40*, 349–379.

Maekawa, T., Kufer, T.A., and Schulze-Lefert, P. (2011). NLR functions in plant and animal immune systems: so far and yet so close. Nat. Immunol. *12*, 817–826.

Meyer, D., Pajonk, S., Micali, C., O'Connell, R., and Schulze-Lefert, P. (2009). Extracellular transport and integration of plant secretory proteins into pathogen-induced cell wall compartments. Plant J. *57*, 986–999.

Morello, J.E., and Collmer, A. (2009). *Pseudomonas syringae* HrpP is a type III secretion substrate specificity switch domain protein that is translocated into plant cells but functions atypically for a substrate switching protein. J. Bacteriol. *191*, 3120–3131.

Morgan, R.L., Zhou, H., Lehto, E., Nguyen, N., Bains, A., Wang, X., and Ma, W. (2010). Catalytic domain of the diversified *Pseudomonas syringae* type III effector HopZ1 determines the allelic specificity in plant hosts. Mol. Microbiol. *76*, 437–455.

Morris, C.E., Sands, D.C., Vinatzer, B.A., Glaux, C., Guilbaud, C., Buffiere, A., Yan, S., Dominguez, H., and Thompson, B.M. (2008). The life history of the plant pathogen *Pseudomonas syringae* is linked to the water cycle. ISME J. *2*, 321–334.

Mukhtar, M.S., Carvunis, A.R., Dreze, M., Epple, P., Steinbrenner, J., Moore, J., Tasan, M., Galli, M., Hao, T., Nishimura, M.T., *et al.* (2011). Independently evolved virulence effectors converge onto hubs in a plant immune system network. Science *333*, 596–601.

Naum, M., Brown, E.W., and Mason-Gamer, R.J. (2009). Phylogenetic evidence for extensive horizontal gene transfer of type III secretion system genes among enterobacterial plant pathogens. Microbiology *155*, 3187–3199.

Navarro, L., Zipfel, C., Rowland, O., Keller, I., Robatzek, S., Boller, T., and Jones, J.D. (2004). The transcriptional innate immune response to flg22. Interplay and overlap with Avr gene-dependent defense responses and bacterial pathogenesis. Plant Physiol. *135*, 1113–1128.

Nomura, H., Komori, T., Uemura, S., Kanda, Y., Shimotani, K., Nakai, K., Furuichi, T., Takebayashi, K., Sugimoto, T., Sano, S., *et al.* (2012). Chloroplast-mediated activation of plant immune signalling in *Arabidopsis*. Nat. Commun. *3*, 926.

Nomura, K., Debroy, S., Lee, Y.H., Pumplin, N., Jones, J., and He, S.Y. (2006). A bacterial virulence protein suppresses host innate immunity to cause plant disease. Science *313*, 220–223.

Nomura, K., Mecey, C., Lee, Y.N., Imboden, L.A., Chang, J.H., and He, S.Y. (2011). Effector-triggered immunity blocks pathogen degradation of an immunity-associated vesicle traffic regulator in *Arabidopsis*. Proc. Natl. Acad. Sci. U.S.A. *108*, 10774–10779.

O'Brien, H.E., Thakur, S., and Guttman, D.S. (2011). Evolution of plant pathogenesis in *Pseudomonas syringae*: a genomics perspective. Annu. Rev. Phytopathol. *49*, 269–289.

Oh, C.S., Pedley, K.F., and Martin, G.B. (2010). Tomato 14–3–3 protein 7 positively regulates immunity-associated programmed cell death by enhancing protein abundance and signaling ability of MAPKKKα. Plant Cell *22*, 260–272.

Oh, H.-S., Kvitko, B.H., Morello, J.E., and Collmer, A. (2007). *Pseudomonas syringae* lytic transglycosylases co-regulated with the type III secretion system contribute to the translocation of effector proteins into plant cells. J. Bacteriol. *189*, 8277–8289.

Padmanabhan, M.S., and Dinesh-Kumar, S.P. (2010). All hands on deck-the role of chloroplasts, endoplasmic reticulum, and the nucleus in driving plant innate immunity. Mol. Plant Microbe Interact. *23*, 1368–1380.

Pedley, K.F., and Martin, G.B. (2004). Identification of MAPKs and their possible MAPKK activators involved in the Pto-mediated defense response of tomato. J. Biol. Chem. *279*, 49229–49235.

Petnicki-Ocwieja, T., Schneider, D.J., Tam, V.C., Chancey, S.T., Shan, L., Jamir, Y., Schechter, L.M., Buell, C.R., Tang, X., Collmer, A., and Alfano, J.R. (2002). Genomewide identification of proteins secreted by the Hrp type III protein secretion system of *Pseudomonas syringae* pv. *tomato* DC3000. Proc. Natl. Acad. Sci. U.S.A. *99*, 7652–7657.

Petnicki-Ocwiega, T., van Dijk, K., and Alfano, J.R. (2005). The *hrpK* operon of *Pseudomonas syringae* pv. tomato DC3000 encodes two proteins secreted by the type III (Hrp) protein secretion system: HopB1 and HrpK, a putative type III translocator. J. Bacteriol. *187*, 649–663.

Phadnis, S.H., and Berg, D.E. (1985). *recA*-independent recombination between repeated IS*50* elements is not caused by an IS*50*-encoded function. J. Bacteriol. *161*, 928–932.

Pitman, A.R., Jackson, R.W., Mansfield, J.W., Kaitell, V., Thwaites, R., and Arnold, D.L. (2005). Exposure to host resistance mechanisms drives evolution of bacterial virulence in plants. Curr. Biol. *15*, 2230–2235.

Poueymiro, M., and Genin, S. (2009). Secreted proteins from *Ralstonia solanacearum*: a hundred tricks to kill a plant. Curr. Opin. Microbiol. *12*, 44–52.

del Pozo, O., Pedley, K.F., and Martin, G.B. (2004). MAPKKKα is a positive regulator of cell death associated with both plant immunity and disease. EMBO J. *23*, 3072–3082.

Pritchard, L., and Birch, P. (2011). A systems biology perspective on plant–microbe interactions: biochemical and structural targets of pathogen effectors. Plant Sci. *180*, 584–603.

Raffaele, S., Farrer, R.A., Cano, L.M., Studholme, D.J., MacLean, D., Thines, M., Jiang, R.H., Zody, M.C., Kunjeti, S.G., Donofrio, N.M., *et al.* (2010). Genome evolution following host jumps in the Irish potato famine pathogen lineage. Science *330*, 1540–1543.

Reinhardt, J.A., Baltrus, D.A., Nishimura, M.T., Jeck, W.R., Jones, C.D., and Dangl, J.L. (2009). *De novo* assembly using low-coverage short read sequence data from the rice pathogen *Pseudomonas syringae* pv. *oryzae*. Genome Res. *19*, 294–305.

Rietz, S., Stamm, A., Malonek, S., Wagner, S., Becker, D., Medina-Escobar, N., Vlot, A.C., Feys, B.J., Niefind, K., and Parker, J.E. (2011). Different roles of Enhanced Disease Susceptibility1 (EDS1) bound to and dissociated from Phytoalexin Deficient4 (PAD4) in *Arabidopsis* immunity. New Phytol. *191*, 107–119.

Rivas, L.A., Mansfield, J., Tsiamis, G., Jackson, R.W., and Murillo, J. (2005). Changes in race-specific virulence in *Pseudomonas syringae* pv. phaseolicola are associated with a chimeric transposable element and rare deletion events in a plasmid-borne pathogenicity island. Appl. Environ. Microbiol. *71*, 3778–3785.

Rodriguez-Palenzuela, P., Matas, I.M., Murillo, J., Lopez-Solanilla, E., Bardaji, L., Perez-Martinez, I., Rodriguez-Moskera, M.E., Penyalver, R., Lopez, M.M., Quesada, J.M., *et al.* (2010). Annotation and overview of the *Pseudomonas savastanoi* pv. savastanoi NCPPB 3335 draft genome reveals the virulence gene complement of a tumour-inducing pathogen of woody hosts. Environ. Microbiol. *12*, 1604–1620.

Rohmer, L., Guttman, D.S., and Dangl, J.L. (2004). Diverse evolutionary mechanisms shape the type III effector virulence factor repertoire in the plant pathogen *Pseudomonas syringae*. Genetics *167*, 1341–1360.

Ronald, P.C., Salmeron, J.M., Carland, F.M., and Staskawicz, B.J. (1992). The cloned avirulence gene *avrPto* induces disease resistance in tomato cultivars containing the *Pto* resistance gene. J. Bacteriol. *174*, 1604–1611.

Rosebrock, T.R., Zeng, L., Brady, J.J., Abramovitch, R.B., Xiao, F., and Martin, G.B. (2007). A bacterial E3 ubiquitin ligase targets a host protein kinase to disrupt plant immunity. Nature *448*, 370–374.

Salzberg, S.L., Sommer, D.D., Schatz, M.C., Phillippy, A.M., Rabinowicz, P.D., Tsuge, S., Furutani, A., Ochiai, H., Delcher, A.L., Kelley, D., *et al.* (2008). Genome sequence and rapid evolution of the rice pathogen *Xanthomonas oryzae* pv. oryzae PXO99A. BMC Genomics 9, 204.

Schechter, L.M., Roberts, K.A., Jamir, Y., Alfano, J.R., and Collmer, A. (2004). *Pseudomonas syringae* type III secretion system targeting signals and novel effectors studied with a Cya translocation reporter. J. Bacteriol. *186*, 543–555.

Schechter, L.M., Valenta, J.C., Schneider, D.J., Collmer, A., and Sakk, E. (2012). Functional and computational analysis of amino acid patterns predictive of type III secretion system substrates in *Pseudomonas syringae*. PLoS One 7, e36038.

Schneider, D.J., and Collmer, A. (2010). Studying plant–pathogen interactions in the genomics era: beyond molecular Koch's postulates to systems biology. Annu. Rev. Phytopathol. *48*, 457–479.

Shan, L., He, P., Li, J., Heese, A., Peck, S.C., Nurnberger, T., Martin, G.B., and Sheen, J. (2008). Bacterial effectors target the common signaling partner BAK1 to disrupt multiple MAMP receptor–signaling complexes and impede plant immunity. Cell Host Microbe 4, 17–27.

Shen, H., and Keen, N.T. (1993). Characterization of the promoter of avirulence gene D from *Pseudomonas syringae* pv. tomato. J. Bacteriol. *175*, 5916–5924.

Sohn, K.H., Lei, R., Nemri, A., and Jones, J.D. (2007). The downy mildew effector proteins ATR1 and ATR13 promote disease susceptibility in *Arabidopsis thaliana*. Plant Cell *19*, 4077–4090.

Sohn, K.H., Saucet, S.B., Clarke, C.R., Vinatzer, B.A., O'Brien, H.E., Guttman, D.S., and Jones, J.D.G. (2011). HopAS1 recognition significantly contributes to *Arabidopsis* nonhost resistance to *Pseudomonas syringae* pathogens. New Phytol. *193*, 58–66.

Spanu, P.D., Abbott, J.C., Amselem, J., Burgis, T.A., Soanes, D.M., Stuber, K., Ver Loren van Themaat, E., Brown, J.K., Butcher, S.A., Gurr, S.J., *et al.* (2010). Genome expansion and gene loss in powdery mildew fungi reveal tradeoffs in extreme parasitism. Science *330*, 1543–1546.

Staskawicz, B.J., Dahlbeck, D., and Keen, N.T. (1984). Cloned avirulence gene of *Pseudomonas syringae* pv. *glycinea* determines race specific incompatibility on *Glycine max* (L.) Merr. Proc. Natl. Acad. Sci. U.S.A. *81*, 6024–6028.

Stavrinides, J., Ma, W., and Guttman, D.S. (2006). Terminal reassortment drives the quantum evolution of type III effectors in bacterial pathogens. PLoS Pathog. *2*, e104.

Stevens, C., Bennett, M.A., Athanassopoulos, E., Tsiamis, G., Taylor, J.D., and Mansfield, J.W. (1998). Sequence variations in alleles of the avirulence gene *avrPphE.R2* from *Pseudomonas syringae* pv. *phaseolicola* lead to loss of recognition of the AvrPphE protein within bean cells and a gain in cultivar-specific virulence. Mol. Microbiol. *29*, 165–1677.

Streitner, C., Hennig, L., Korneli, C., and Staiger, D. (2010). Global transcript profiling of transgenic plants constitutively overexpressing the RNA-binding protein AtGRP7. BMC Plant Biol. *10*, 221.

Studholme, D.J., Gimenez Ibanez, S., Maclean, D., Dangl, J.L., Chang, J.H., and Rathjen, J.P. (2009). A draft genome sequence and functional screen reveals the repertoire of type III secreted proteins of *Pseudomonas syringae* pathovar *tabaci* 11528. BMC Genomics *10*, 395.

Tampakaki, A.P., Skandalis, N., Gazi, A.D., Bastaki, M.N., Sarris, P.F., Charova, S.N., Kokkinidis, M., and Panopoulos, N.J. (2010). Playing the 'Harp': evolution of our understanding of *hrp/hrc* genes. Annu. Rev. Phytopathol. *48*, 347–370.

Torto-Alalibo, T., Collmer, C.W., Gwinn-Giglio, M., Lindeberg, M., Meng, S., Chibucos, M.C., Tseng, T.-T., Lomax, J., Biehl, B., Ireland, A., *et al.* (2010). Unifying themes in microbial associations with animal and plant hosts described using the Gene Ontology. Microbiol. Mol. Biol. Rev. *74*, 479–503.

Tsuda, K., and Katagiri, F. (2010). Comparing signaling mechanisms engaged in pattern-triggered and effector-triggered immunity. Curr. Opin. Plant Biol. *13*, 459–465.

Tsuda, K., Sato, M., Stoddard, T., Glazebrook, J., and Katagiri, F. (2009). Network properties of robust immunity in plants. PLoS Genet. *5*, e1000772.

Vencato, M., Tian, T., Alfano, J.R., Buell, C.R., Cartinhour, S., DeClerck, G.A., Guttman, D.S., Stavrinides, J., Joardar, V., Lindeberg, M., *et al.* (2006). Bioinformatics-enabled identification of the HrpL regulon and type III secretion system effector proteins of *Pseudomonas syringae* pv. *phaseolicola* 1448A. Mol. Plant Microbe Interact. *19*, 1193–1206.

Vinatzer, B.A., and Yan, S. (2008). Mining the genomes of plant-pathogenic bacteria: how not to drown in gigabases of sequence. Mol. Plant Pathol. *9*, 105–118.

Vinatzer, B.A., Jelenska, J., and Greenberg, J.T. (2005). Bioinformatics correctly identifies many type III secretion substrates in the plant pathogen *Pseudomonas syringae* and the biocontrol isolate *P. fluorescens* SBW25. Mol. Plant Microbe Interact. *18*, 877–888.

Vinatzer, B.A., Teitzel, G.M., Lee, M.W., Jelenska, J., Hotton, S., Fairfax, K., Jenrette, J., and Greenberg, J.T. (2006). The type III effector repertoire of *Pseudomonas syringae* pv. *syringae* B728a and its role in survival and disease on host and non-host plants. Mol. Microbiol. *62*, 26–44.

Wang, Y., Li, J., Hou, S., Wang, X., Li, Y., Ren, D., Chen, S., Tang, X., and Zhou, J.M. (2010). A *Pseudomonas syringae* ADP-ribosyltransferase inhibits *Arabidopsis* mitogen-activated protein kinase kinases. Plant Cell *22*, 2033–2044.

Wehling, M.D., Guo, M., Fu, Z.Q., and Alfano, J.R. (2004). The *Pseudomonas syringae* HopPtoV protein is secreted in culture and translocated into plant cells via the type III protein secretion system in a manner dependent on the ShcV type III chaperone. J. Bacteriol. *186*, 3621–3630.

Wei, C.-F., Kvitko, B.H., Shimizu, R., Crabill, E., Alfano, J.R., Lin, N.-C., Martin, G.B., Huang, H.-C., and Collmer, A. (2007). A *Pseudomonas syringae* pv. *tomato* DC3000 mutant lacking the type III effector HopQ1-1 is able to cause disease in the model plant *Nicotiana benthamiana*. Plant J. *51*, 32–46.

Wei, H.-L., and Collmer, A. (2012). Multiple lessons from the multiple functions of a regulator of type III secretion system assembly in the plant pathogen *Pseudomonas syringae* Mol. Microbiol. *85*, 195–200.

Welch, R.A., Burland, V., Plunkett, G., 3rd, Redford, P., Roesch, P., Rasko, D., Buckles, E.L., Liou, S.R., Boutin, A., Hackett, J., *et al.* (2002). Extensive mosaic structure revealed by the complete genome sequence of uropathogenic *Escherichia coli*. Proc. Natl. Acad. Sci. U.S.A. *99*, 17020–17024.

Willmann, R., Lajunen, H.M., Erbs, G., Newman, M.A., Kolb, D., Tsuda, K., Katagiri, F., Fliegmann, J., Bono, J.J., Cullimore, J.V., *et al.* (2011). *Arabidopsis* lysin-motif proteins LYM1 LYM3 CERK1 mediate bacterial peptidoglycan sensing and immunity to bacterial infection. Proc. Natl. Acad. Sci. U.S.A. *108*, 19824–19829.

Wilton, M., Subramaniam, R., Elmore, J., Felsensteiner, C., Coaker, G., and Desveaux, D. (2010). The type III effector $HopF2_{Pto}$ targets *Arabidopsis* RIN4 protein to promote *Pseudomonas syringae* virulence. Proc. Natl. Acad. Sci. U.S.A. *107*, 2349–2354.

Wroblewski, T., Caldwell, K.S., Piskurewicz, U., Cavanaugh, K.A., Xu, H., Kozik, A., Ochoa, O., McHale, L.K., Lahre, K., Jelenska, J., *et al.* (2009). Comparative large-scale analysis of interactions between several crop species and the effector repertoires from multiple pathovars of *Pseudomonas* and *Ralstonia*. Plant Physiol. *150*, 1733–1749.

Xiang, T., Zong, N., Zou, Y., Wu, Y., Zhang, J., Xing, W., Li, Y., Tang, X., Zhu, L., Chai, J., and Zhou, J.M. (2008). *Pseudomonas syringae* effector AvrPto blocks innate immunity by targeting receptor kinases. Curr. Biol. *18*, 74–80.

Xiang, T., Zong, N., Zhang, J., Chen, J., Chen, M., and Zhou, J.M. (2011). BAK1 is not a target of the *Pseudomonas syringae* effector AvrPto. Mol. Plant Microbe Interact. *24*, 100–107.

Zeng, L., Velasquez, A.C., Munkvold, K.R., Zhang, J., and Martin, G.B. (2011). A tomato LysM receptor-like kinase promotes immunity and its kinase activity is inhibited by AvrPtoB. Plant J. *69*, 92–103.

Zhang, J., Shao, F., Li, Y., Cui, H., Chen, L., Li, H., Zou, Y., Long, C., Lan, L., Chai, J., *et al.* (2007). A *Pseudomonas syringae* effector inactivates MAPKs to suppress PAMP-induced immunity in plants. Cell Host Microbe *1*, 175–185.

Zhang, J., Li, W., Xiang, T., Liu, Z., Laluk, K., Ding, X., Zou, Y., Gao, M., Zhang, X., Chen, S., *et al.* (2010). Receptor-like cytoplasmic kinases integrate signaling from multiple plant immune receptors and are targeted by a *Pseudomonas syringae* effector. Cell Host Microbe *7*, 290–301.

Zhang, Z., Wu, Y., Gao, M., Zhang, J., Kong, Q., Liu, Y., Ba, H., Zhou, J., and Zhang, Y. (2012). Disruption of PAMP-Induced MAP kinase cascade by a *Pseudomonas syringae* effector activates plant Immunity mediated by the NB-LRR protein SUMM2. Cell Host Microbe *11*, 253–263.

Zhou, H., Morgan, R.L., Guttman, D.S., and Ma, W. (2009). Allelic variants of the *Pseudomonas syringae* type III effector HopZ1 are differentially recognized by plant resistance systems. Mol. Plant Microbe Interact. *22*, 176–189.

Zhou, H., Lin, J., Johnson, A., Morgan, R.L., Zhong, W., and Ma, W. (2011). *Pseudomonas syringae* type III effector HopZ1 targets a host enzyme to suppress isoflavone biosynthesis and promote infection in soybean. Cell Host Microbe *9*, 177–186.

Zhou, J.M., and Chai, J. (2008). Plant-pathogenic bacterial type III effectors subdue host responses. Curr. Opin. Microbiol. *11*, 179–185.

Zipfel, C., Kunze, G., Chinchilla, D., Caniard, A., Jones, J.D., Boller, T., and Felix, G. (2006). Perception of the bacterial PAMP EF-Tu by the receptor EFR restricts *Agrobacterium*-mediated transformation. Cell *125*, 749–760.

Zwiesler-Vollick, J., Plovanich-Jones, A.E., Nomura, K., Brandyopadhyay, S., Joardar, V., Kunkel, B.N., and He, S.Y. (2002). Identification of novel *hrp*-regulated genes through functional genomic analysis of the *Pseudomonas syringae* pv. *tomato* DC3000 genome. Mol. Microbiol. *45*, 1207–1218.

Towards Understanding Fire Blight: Virulence Mechanisms and their Regulation in *Erwinia amylovora*

3

R. Ryan McNally, Youfu Zhao and George W. Sundin

Abstract

Erwinia amylovora is the causal agent of fire blight, a destructive disease of rosaceous species such as apple and pear. For more than 30 years, research regarding the molecular mechanisms underlying fire blight disease development has identified many genetic determinants required for virulence by *E. amylovora*. These included type III secretion and exopolysaccharide biosynthesis as well as other factors that contribute to pathogenicity. A host of regulatory mechanisms coordinate the activity of all virulence factors in *E. amylovora* and are integral to our understanding of the pathogenesis process. As the emergence and spread of antibiotic resistant *E. amylovora* populations threatens our ability to successfully manage fire blight severity, a more complete understanding of *E. amylovora* biology will be of vital importance in developing robust management strategies for the future.

Introduction

Fire blight is a highly destructive disease of rosaceous tree species caused by the Gram-negative enteric bacterium *Erwinia amylovora*. *E. amylovora* is distinguished as the first bacterium identified as a causal agent of a plant disease (Denning, 1794; Winslow *et al.*, 1920). Since the first reported observation of fire blight in 1794 in the northeastern United States, *E. amylovora* has spread globally with fire blight currently occurring in Europe, the Middle East, New Zealand and Japan (Bonn and van der Zwet, 2000).

E. amylovora infects a broad range of species within the Rosaceae including the subfamily Spiraeoideae as well as the genus *Rubus* (Mann *et al.*, 2013). These hosts include economically important crops such as apple and pear in addition to quince, hawthorn, cotoneaster and *Rubus* species blackberry and raspberry.

E. amylovora cells are disseminated by water, wind, insects and human activity (van der Zwet and Beer, 1999). Flowers represent the primary infection court and bacterial populations can exceed 10e6 cells on the stigma surface (Wilson and Lindow, 1992). *E. amylovora* invades hosts via nectarthodes and moves systemically within the vascular system and cortical parenchyma (Sjulin and Beer, 1978; Koczan *et al.*, 2009). Colonization of xylem elements occludes water transportation resulting in the most characteristic signs and symptoms of fire blight: desiccation, the shepherd's crook and bacterial ooze. Ooze is rich in polysaccharides, sap and bacterial cells and serves as an important source of inoculum (Geider, 2000). While young tissues exhibit increased susceptibility to infection, *E. amylovora* is capable of infecting many host tissues including flowers, shoots, leaves, fruits as well as rootstocks (Pusey and Smith, 2008). Infections of rootstocks can kill host trees and, when environmental conditions are conducive, fire blight epidemics can result in major economic losses (Longstroth, 2001).

In the United States, estimated annual costs attributable to fire blight infection are approximately US$100 million and, consequently, successful management of *E. amylovora* populations is imperative (Norelli *et al.*, 2003). Currently, the antibiotic streptomycin serves as the most effective and commonly utilized chemical control of fire blight. However, owing to widespread use, streptomycin resistance in *E. amylovora* has developed in the United States and other countries including Canada, Israel and New Zealand (McManus *et al.*, 2002; McGhee *et al.*, 2011). While less effective controls are available, such as oxytetracycline and copper sulfate, development of new *E. amylovora* control strategies is increasingly urgent.

Since transposon mutagenesis experiments conducted by Steinberger and Beer (1988) revealed the first genetic determinants required for fire blight disease development, the identification and characterization of *E. amylovora* virulence determinants has progressed rapidly. This chapter summarizes our current understanding of the molecular basis for fire blight development and emphasizes non-canonical virulence factors that play a vital, but under-described role in pathogenesis. It is divided into two main segments: virulence mechanisms and regulation.

Virulence mechanisms

Fire blight development of *E. amylovora* is characterized by the function of two main pathogenicity factors: the exopolysaccharide amylovoran and the type III secretion system (T3SS) (Oh and Beer, 2005; Khan *et al.*, 2012; Zhao and Qi, 2011). Many other virulence determinants collectively contribute to fire blight disease development in *E. amylovora*. All of the major virulence determinants are listed in Table 3.1 and summarized below.

Structure and function of the type three secretion system

The T3SS is a macromolecular protein complex consisting of three structural components: the basal body, the pilus and the translocon (Buttner and He, 2009). The basal body spans the Gram-negative bacterial cell wall and contains an ATPase to provide energy for the translocation process (Kubori *et al.*, 1998). Anchored to the basal body is a hollow pilus that extends into extracellular space and is distally capped by a pore-forming protein complex called the translocon (Buttner and Bonas, 2002). The translocon associates with host membranes and allows pathogen effectors to translocate across extracellular space directly into the host cytoplasm in a single step (Buttner and Bonas, 2002; He and Jin, 2003). Post-translocation, pathogen effectors reprogram host cells to create an environment conducive to bacterial survival and proliferation (Alfano and Collmer, 2004; Deslandes and Rivas, 2012).

In *E. amylovora*, the T3SS is encoded on a pathogenicity island consisting of approximately 62 kilobases of genomic DNA and 60 genes (Oh *et al.*, 2005). The T3SS was discovered via transposon mutagenesis that resulted in *E. amylovora* strains unable to elicit hypersensitive response (HR) and pathogenicity (*hrp*) in non-host plants and hosts

Table 3.1 Genes identified as virulence determinants in *Erwinia amylovora*

Gene	Accession	Description	Function	Reference
Movement				
eae	EAM_0301	Invasin	Attachment	Koczan *et al.* (2011)
hofC	EAM_0729	Type IV pilus assembly	Attachment	Koczan *et al.* (2011)
crl	EAM_0898	Curlin activator	Attachment	Koczan *et al.* (2011)
fliH	EAM_1494	Flagellar assembly	Motility	Cesbron *et al.* (2006)
fimD	EAM_2951	Type I pilus usher	Attachment	Koczan *et al.* (2011)
Metabolism				
foxR	EAM_0358	TonB-dependent siderophore receptor	Iron acquisition	Dellagi *et al.* (1998)
dfoA	EAM_0360	Desferrioxamine synthesis	Iron acquisition, oxidative burst survival	Dellagi *et al.* (1998)
srlE	EAM_0522	PTS sorbitol-specific transporter	Sorbitol metabolism	Aldridge *et al.* (1997)
srlD	EAM_0524	Sorbitol-6-phosphate dehydrogenase	Sorbitol metabolism	Aldridge *et al.* (1997)
scrA	EAM_1596	PTS system sucrose-specific transporter	Sucrose metabolism	Bogs and Geider (2000)
scrY	EAM_1597	Sucrose porin	Sucrose metabolism	Bogs and Geider (2000)
galE	EAM_2161	UDP-glucose 4-epimerase	Galactose metabolism	Metzger *et al.* (1994)
Growth				
EAM_0344	EAM_0344	Translation factor	Translation	Wang and Beer (2006)
carB	EAM_0661	Carbamoylphosphate synthase	Pyrimidine and arginine synthesis	Wang and Beer (2006)
pyrD	EAM_1366	Dihydroorotate dehydrogenase	Pyrimidine synthesis	Wang and Beer (2006)
mltE	EAM_1530	Lytic murein transglycosylase	Cell elongation and division	Zhao *et al.* (2005)
cysB	EAM_1869	HTH-type transcriptional regulator	Cysteine synthesis	Wang and Beer (2006)
purM	EAM_2441	Phosphoribosylaminoimidazole synthetase	Purine synthesis	Wang and Beer (2006)
guaB	EAM_2465	Inosine-5′-monophosphate dehydrogenase	Purine synthesis	Eastgate *et al.* (1997)
tyrA	EAM_2620	Bifunctional chorismate mutase	Prephenate synthesis	Wang and Beer (2006)
pyrB	EAM_3054	Aspartate carbamoyltransferase	Pyrimidine synthesis	Wang and Beer (2006)

Table 3.1 Continued

Gene	Accession	Description	Function	Reference
Exopolysaccharides				
amsE	EAM_2167	Glycosyltransferase	Amylovoran synthesis	Bellemann and Geider (1992)
amsD	EAM_2168	Glycosyltransferase	Amylovoran synthesis	Bellemann and Geider (1992)
amsC	EAM_2169	Oligosaccharide repeat unit polymerase	Amylovoran synthesis	Bellemann and Geider (1992)
amsB	EAM_2170	Glycosyltransferase	Amylovoran synthesis	Bellemann and Geider (1992)
amsA	EAM_2171	Tyrosine-protein kinase	Amylovoran synthesis	Bellemann and Geider (1992)
amsI	EAM_2172	Tyrosine phosphatase	Amylovoran synthesis	Bugert and Geider (1995)
amsH	EAM_2173	Amylovoran export	Amylovoran synthesis	Bugert and Geider (1995)
amsG	EAM_2174	UDP-galactose-lipid carrier transferase	Amylovoran synthesis	Bugert and Geider (1995)
lscC	EAM_3468	Levansucrase	Levan biosynthesis	Geier *et al.* (1993)
Host interactions				
waaL	EAM_0086	Lipid A core – O-antigen ligase	Lipopolysaccharide synthesis	Berry *et al.* (2009)
avrRpt2	EAM_0423	Type III cysteine protease effector	Host interactions	Zhao *et al.* (2006)
acrB	EAM_1016	Multidrug efflux transporter	*In planta* survival	Burse *et al.* (2004)
hopX1	EAM_2190	Type III effector	Avirulence	Bocsanczy *et al.* (2012)
dspF	EAM_2871	Type III effector chaperone	Type III secretion	Gaudriault *et al.* (2002)
dspE	EAM_2872	Type III avrE-like effector	Host interactions, pathogenicity	Barny *et al.* (1990)
eop1	EAM_2875	Type III peptidase effector	Avirulence	Asselin *et al.* (2011)
hrpN	EAM_2877	Type III secretion harpin/ translocator	Type III secretion	Wei *et al.* (1992a)
hrpT	EAM_2879	Type III secretion lipoprotein	Type III secretion	Kim *et al.* (1997)
hrcC	EAM_2880	Type III secretion outer membrane pore	Type III secretion	Kim *et al.* (1997)
hrpE	EAM_2883	Type III secretion apparatus protein	Type III secretion	Kim *et al.* (1997)

Table 3.1 Continued

Gene	Accession	Description	Function	Reference
hrcJ	EAM_2885	Type III secretion inner-membrane protein	Type III secretion	Kim *et al.* (1997)
hrpB	EAM_2886	Type III secretion protein	Type III secretion	Kim *et al.* (1997)
hrpA	EAM_2887	Type III pilus protein	Type III secretion	Kim *et al.* (1997)
hrpJ	EAM_2895	Type III secretion regulator	Type III secretion	Nissinen *et al.* (2007)
hrcV	EAM_2896	Type III secretion inner membrane protein	Type III secretion	Wei and Beer (1993)
hrcN	EAM_2898	Type III secretion system ATPase	Type III secretion	Wei and Beer (1993)
hrcT	EAM_2904	Type III secretion apparatus protein	Type III secretion	Bogdanove *et al.* (1996)
hsvC	EAM_2908	Biotin carboxylase	Putative toxin synthesis	Oh *et al.* (2005
hsvB	EAM_2909	Biotin carboxylase	Putative toxin synthesis	Oh *et al.* (2005)
hsvA	EAM_2910	Amidinostransferase	Putative toxin synthesis	Oh *et al.* (2005)
tolC	EAM_2988	Outer membrane efflux	*In planta* survival	Al-Karablieh *et al.* (2009)
prtD	EAM_3366	Type I secretion system ATPase	Type I secretion	Zhang *et al.* (1999)
Unknown				
ydcN	EAM_1248	Transcriptional regulator	Unknown	McNally *et al.* (2012)
EAM_2937	EAM_2937	Hypothetical protein	Unknown	Wang and Beer (2006)
EAM_2938	EAM_2938	Hypothetical protein	Unknown	McNally *et al.* (2012)
nlpI	EAM_3066	Tetratricopeptide repeat lipoprotein	Unknown	McNally *et al.* (2012)

respectively (Steinberger and Beer, 1988). The T3SS basal body includes nine proteins that are highly conserved between *E. amylovora* and other bacterial pathogens of animals and plants (Oh and Beer, 2005). These proteins include the ATPase HrcN and HrcU, a protein known to regulate secretion hierarchy in other type III systems, including flagella (Pozidis *et al.*, 2003; Lorenz and Buttner, 2011). The *hrp* pilus is constructed from numerous HrpA subunits and the translocon, while less characterized than counterparts in animal bacterial pathogens, is likely composed of the harpins HrpN and HrpW as well as HrpK (Jin and He, 2001; Bocsanczy *et al.*, 2008). The *hrp* T3SS in *E. amylovora* also requires multiple helper proteins such as HrpJ which is required for HR elicitation in tobacco as well as for disease in host species and functions in regulating the secretion of both HrpN and HrpW (Nissinen *et al.*, 2007).

The role of type III secretion in *E. amylovora* is to enable the translocation of pathogen effectors directly into host cells. While effector repertories in bacterial pathogens such as *Pseudomonas syringae* pv. *phaseolicola* 1448A have been reported to include as many as 27 candidates, *E. amylovora* only encodes five known effector genes (Vencato *et al.*, 2006; Nissinen *et al.*, 2007; McNally *et al.*, 2012). By far, DspE (also known as DspA) represents the most important effector of *E. amylovora* as null mutants are non-pathogenic in immature pear and apple shoots (Barny *et al.*, 1990; Bogdanove *et al.*, 1998a). Interestingly, while required for pathogen growth and disease development, *dspE* mutants maintain population numbers *in planta*. Conversely, *E. amylovora* strains generally defective in type III secretion are unable to persist in compatible host species and populations quickly decline (Zhao *et al.*, 2005). This suggests that additional constituents of the T3SS secretome in *E. amylovora* are important in mediating successful host–microbe interactions.

Analyses of T3SS effector repertoire

Amongst the small repertoire of *E. amylovora hrp* type III effectors, DspE is the most characterized virulence factor mediating *E. amylovora*–host interactions. DspE is part of the AvrE-family of effectors with homologues in many phytobacterial pathogens (Lorang and Keen, 1995; Bogdanove *et al.*, 1998a; Ham *et al.*, 2006). While AvrE-class effectors in pathogens including *P. syringae* pv. *tomato* DC3000 are believed to function redundantly with other effectors including HopM1, DspE in *E. amylovora* is absolutely required for pathogenicity (Bogdanove *et al.*, 1998; Kvitko *et al.*, 2009).

DspE is secreted by the T3SS (Bogdanove *et al.*, 1998b; Nissinen *et al.*, 2007). While direct efforts to localize DspE *in planta* using co-immunofluorescence and gold labelling confocal microscopy have failed to identify DspE in the host cytoplasm (Boureau *et al.*, 2006), indirect measurements of protein translocation, including fusions with adenylate cyclase (CyaA), suggest that DspE is translocated into host cells (Bocsanczy *et al.*, 2008; Triplett *et al.*, 2009; Oh *et al.*, 2010).

Translocation of DspE depends on a number of factors. Along with a functional T3SS and N-terminal amino acid residues within DspE, a class of proteins known as chaperones play an important role in DspE localization and function (Triplett *et al.*, 2009; Oh *et al.*, 2010). Type III secretion chaperones are small proteins that facilitate the secretion of some T3SS substrates. *E. amylovora* produces at least three such chaperones including Esc1, Esc3 and DspF, the cognate chaperone of DspE (Oh and Beer, 2005). In *E. amylovora*, *dspF* mutants are severely attenuated in virulence and display reduced levels of DspE translocation (Gaudriault *et al.*, 2002; Triplett *et al.*, 2009). Factors responsible for residual DspE translocation in *dspF* mutants remain undescribed. Interestingly, DspE exhibits several chaperone binding sites, and multiple chaperones, including DspF and Esc1, interact with DspE via yeast two-hybrid screening (Triplett *et al.*, 2009, 2010; Oh *et al.*, 2010). The interaction between DspE and multiple type III secretion chaperones highlights a new model for T3SS effector regulation, and accumulating evidence suggests that the pool of T3SS chaperones, most importantly DspF, collectively play an important role in regulating type III secretion hierarchy in pathogens like *E. amylovora*.

The role of DspE in mediating disease development has been explored in a number of host and non-host pathosystems. Primary host responses attributed to DspE include suppression of callose deposition and cell death. Elicitation of callose deposition during host–microbe interactions is a characteristic of plant innate immunity and is typically associated with salicylic acid (SA)-mediated defence responses (Boller and He, 2009). During compatible interactions, DspE mutants generate more callose deposition than wild-type *E. amylovora* implicating DspE as a suppressor of innate immunity (Debroy *et al.*, 2004; Boureau *et al.*, 2011). Conversely, interactions with non-host *Arabidopsis thaliana* are not affected by DspE and elicitation of callose deposition only requires a functional T3SS (Degrave *et al.*, 2008). Interestingly, while DspE does not modulate callose in *A. thaliana*, inducible expression of DspE in transgenic *A. thaliana* can suppress callose deposition in response to *P. syringae* pv. *tomato* lacking a functional T3SS (Degrave *et al.*, 2013).

DspE is also attributed with inducing cell death, however the nature of DspE-mediated cell death is subject to debate. In compatible host interactions, DspE is a pathogenicity factor required for necrotic disease development (Bogdanove *et al.*, 1998). *Agrobacterium*-mediated transient expression of DspE in apple also causes necrotic lesion formation (Boureau *et al.*, 2006). In comparable transient expression assays, DspE also triggers cell death in multiple non-host species including yeast, tobacco and *A. thaliana* (Boureau *et al.*, 2006; Oh *et al.*, 2007; Degrave *et al.*, 2008). Notably, Degrave *et al.*, (2013) found that while production of DspE in transgenic *A. thaliana* induced marker gene expression indicative of SA signalling activity, SA itself was dispensable for DspE-mediated cell death. While this information suggests that DspE may function in a toxin-like capacity, in *Nicotiana benthamiana* DspE-induced cell death was shown to require SGT1, a host factor necessary for programmed cell death (Oh *et al.*, 2007). During infection of SGT1-silenced *N. benthamiana*, *E. amylovora* exhibited higher population numbers compared to wild-type tobacco (Oh *et al.*, 2007). This indicates that host cell death does not promote *E. amylovora* growth and argues against DspE promoting virulence via inducing cell death (Oh *et al.*, 2007).

DspE may function as a toxin or as an elicitor of HR, simultaneously or separately in a host-specific manner. Elucidating the molecular mechanisms underpinning host responses to DspE may have important implications for future fire blight control strategies.

The function of DspE and AvrE-family effectors is a major area of research in the molecular plant–microbe interactions community. To date, four apple proteins have been shown to interact with DspE via yeast two-hybrid assays and *in vitro* western blotting (Meng *et al.*, 2006). These proteins have been termed DIPMs (DspE-interacting proteins from *Malus*) and represent a class of plant signal transducers termed leucine-rich repeat (LRR) receptor-like kinases (RLK). LRR-RLKs transmit signals from the plant apoplast via the ligand-binding potential of extracellular LRR domains and reprogram host cells accordingly by intracellular RLK-initiated phosphorylation signal cascades (Gómez-Gómez and Boller, 2000; Nurnberger and Kemmerling, 2006). In support of DspE translocation and intracellular function, during the host–microbe interaction DspE binds DIPM proteins through RLK domains only and full DspE–DIPM interactions require RLK kinase activity in yeast two-hybrid assays (Meng *et al.*, 2006). Unfortunately, DIPMs appear to be constitutively expressed in *E. amylovora* host species regardless of susceptibility (Meng *et al.*, 2006) and the significance of DspE–DIPM interactions remains to be determined.

Besides DspE, *E. amylovora* is known to produce a suite of additional effectors including Eop1, HopX1$_{Ea}$, HopPtoC$_{Ea}$, and AvrRpt2$_{Ea}$ (Zhou *et al.*, 2005; Nissinen *et al.*, 2007; McNally *et al.*, 2012). Of these, most notable is AvrRpt2$_{Ea}$. Analyses of *avrRpt2*$_{Ea}$ mutants with a range of apple varieties revealed that *avrRpt2*$_{Ea}$ exhibits the characteristics of an avirulence gene during interactions with *Malus* X *robusta* 5 (Mr5) (Vogt *et al.*, 2013). A large screen exploring compatibility between different *E. amylovora* strains and Mr5 found that the degree *E. amylovora* strains could necrotize Mr5 could be traced to a single nucleotide polymorphism in $avrRpt2_{Ea}$ (Voigt *et al.*, 2013). In contrast, during interactions with immature pear fruit, $AvrRpt2_{Ea}$ appears to function as a virulence factor (Zhou *et al.*, 2006). These results suggest a role for $AvrRpt2_{Ea}$ in contributing to virulence in one plant–microbe interaction while functioning as an avirulence factor in a second.

While DspE and $AvrRpt2_{Ea}$ contribute to *E. amylovora* pathogenicity and virulence, Eop1 and $HopX1_{Ea}$ are implicated as host range limiting factors (Asselin *et al.*, 2011; Bocsanczy *et al.*, 2012). As described above, the host range of *E. amylovora* spans a number of rosaceous genera. In 2011, Asselin *et al.*, observed that Eop1, a YopJ-family effector, exhibits significant genetic variation between *E. amylovora* CFBP 1430 and *E. amylovora* ATCC BAA-2158, a *Rubus*-infecting isolate. Though Eop1 has no role in virulence of pear or HR elicitation by *E. amylovora* CFBP 1430, Eop1 from *Rubus*-infecting *E. amylovora* ATCC BAA-2158 restricted disease development in immature pear when expressed in both *E. amylovora* strains (Asselin *et al.*, 2011). Conversely, the Eop1 allele native to *E. amylovora* CFBP 1430 increased virulence on pear when expressed in the *Rubus* isolate *E. amylovora* ATCC BAA-2158 (Asselin *et al.*, 2011).

Like Eop1, $HopX1_{Ea}$, an AvrPphE homologue of Pseudomonas syringae, is not required for virulence (Bocsanczy *et al.*, 2012). However, overexpression of $HopX1_{Ea}$ in *E. amylovora* CFBP 1430 caused delayed disease development in pear fruits and apple shoots while a chromosomal deletion of $hopX1_{Ea}$ retarded HR development in the non-host *Nicotiana tabacum* (Bocsanczy *et al.*, 2012). This indicates that $HopX1_{Ea}$ quantitatively contributes to avirulence and may limit host range as exemplified by Eop1. Interestingly, inoculation of *E. amylovora* strains overexpression $hopX1_{Ea}$ in *N. benthamiana* suppresses HR development (Bocsanczy *et al.*, 2012). Continued dissection of the differential responses of non-host *Nicotiana* species to $HopX1_{Ea}$ may help elucidate the avirulence role of $HopX1_{Ea}$ during compatible interactions with apple and pear.

Harpins and translocation

Harpins represent a class of glycine-rich, heat-stable, cysteine-lacking proteins unique to the T3SS of plant-pathogenic bacteria (Wei *et al.*, 1992a). *E. amylovora* produces two harpins; HrpN and HrpW (Wei *et al.*, 1992a; Kim and Beer, 1998). HrpN is distinguished as a pathogenicity factor and the first protein identified as a cell-free elicitor of the hypersensitive response (Wei *et al.*, 1992a). *In vitro* and *in planta* expression analyses suggest that *hrpN* is transcribed at early stages of the host–bacterium interaction (McNally *et al.*, 2012; Sarowar *et al.*, 2011; Pester *et al.*, 2012) and that the HrpN protein is secreted via the T3SS into the host apoplast (Perino *et al.*, 1999; Nissinen *et al.*, 2007). Fusions of the HrpN protein to the translocation reporter CyaA also indicate low levels of HrpN translocation into tobacco and apple cells (Bocsanczy *et al.*, 2008; Boureau at al., 2011). The significance of HrpN translocation remains to be determined.

During disease development, HrpN is hypothesized to function as a T3SS translocator. Translocators associate with the distal tip of the T3SS pilus, exhibit pore-formation abilities in host membranes, and facilitate the delivery of pathogen effectors into the host cytoplasm (Buttner and He, 2009). To date, no T3SS translocators for plant pathogens have been demonstrated to directly associate with T3SS pili, but indirect evidence indicates that

HrpN does interact with host and non-host membranes causing changes in ion flux (Wei and Beer, 1992; El-Maarouf *et al.*, 2001; Reboutier *et al.*, 2007a,b). Interestingly, ion fluxes triggered by purified HrpN are species specific and opposite in host apple and non-host *Arabidopsis thaliana* (Reboutier *et al.*, 2007a). This species specificity may play an important role is successful disease development versus HR elicitation. In addition, the role of HrpN as a translocator is suggested by its requirement for wild-type levels of DspE translocation into *N. tabacum* (Bocsanczy *et al.*, 2008). Low levels of HrpN translocation also require functional *hrpN* (Boureau *et al.*, 2011).

As plant membrane reactions to HrpN are species specific, host and non-host downstream responses exhibit equal variability. Non–host interactions with HrpN are characterized by the HR and induction of systemic acquired resistance (SAR) (Wei *et al.*, 1992a; Peng *et al.*, 2003). HR elicitation in tobacco requires functional *hrpN* and purified HrpN produces dose-dependent cell death in *A. thaliana* cell cultures (Barny, 1995; Sinn *et al.*, 2008). Interestingly, HrpN-induced HR can be suppressed in tobacco by AvrPtoB, an effector of *P. syringae* known to block the HR (Oh and Beer, 2005). Sinn *et al.* (2008) conducted research that revealed multiple intriguing details regarding HR elicitation in tobacco and virulence in immature pears. Sinn and colleagues found that high concentrations of *E. amylovora* (1×10^8 CFU/ml) cells do not require functional HrpN for HR induction in tobacco indicating that additional elicitors in *E. amylovora* may play a larger role in plant–microbe interactions. The group also created *hrpN* mutant alleles that are required to generate HR and virulence when expressed in *E. amylovora hrpN* null mutants. Conversely, these same *hrpN* mutant alleles were not required for HR elicitation when produced in *Escherichia coli* and infiltrated into tobacco as a cell-free suspension. This suggests that additional factors contribute to HrpN-mediated HR elicitation by *E. amylovora* and that HrpN virulence and avirulence functions may be inseparable during compatible host–microbe interactions. In addition to HR, HrpN also triggers salicylic acid-dependent SAR in non-host and host plants. Transient expression of HrpN in tobacco increased resistance to the necrotrophic fungus *Botrytis cinerea* while exogenous application of HrpN to apple trees has been shown to induce resistance to blue mould and *Venturia* spp. scab diseases (de Capdeville *et al.*, 2003; Barbosa-Mendes *et al.*, 2009; Percival *et al.*, 2009).

The molecular response of host and non-host species to HrpN is characterized by production of reactive oxygen species, callose deposition, mitogen-activated protein kinase signalling, and defence marker gene expression (Baker *et al.*, 1993; Adam *et al.*,, 1999; Dong *et al.*, 1999; Peng *et al.*, 2003; Boureau *et al.*,, 2011). Recent analyses suggest that HrpN is not required for elicitation of the oxidative burst in apple (Boureau *et al.*, 2011). *E. amylovora hrpN* mutants and purified HrpN are not involved in callose deposition in *A. thaliana* (Degrave *et al.*, 2008). Conversely, investigations found that HrpN does elicit callose deposition in apple and that *in trans* production of *hrpN* was capable of triggering callose deposition in a *dspA/E* and *hrpN* double-mutant (Boureau *et al.*, 2011). Together with information implicating DspE as a callose suppressor, a model is emerging that HrpN facilitates DspA/E translocation while DspE inhibits HrpN-mediated callose deposition.

HrpW is a second harpin produced by *E. amylovora* representing a larger group of plant-pathogenic translocators exhibiting C-terminal pectate lyase domains (Kim and Beer, 1998). Like HrpN, HrpW is secreted by the T3SS, elicits HR on non-host species, and purified HrpW triggers dose-dependent cell death of *A. thaliana* cell culture (Kim and Beer, 1998; Nissinen *et al.*, 2007; Reboutier *et al.*, 2007b). Unlike HrpN, *hrpW* mutants do not exhibit a virulence phenotype. HrpW does not induce callose deposition in apple and, while HrpW contributes to DspE translocation rate, DspE-CyaA translocation is not statistically affected by *hrpW* mutations (Bocsanczy *et al.*, 2008; Boureau *et al.*, 2011). Intriguingly, *hrpW* mutants induce more electrolyte leakage than wild-type *E. amylovora* in *A. thaliana* and *N. tabacum* and data from Reboutier *et al.*, (2007b) suggests that the primary function of HrpW is to ameliorate HrpN-induced host defence responses as mixing purified HrpW with purified HrpN decreases HrpN-triggered cell death, oxidative burst, and host membrane changes in conductivity during treatment of *A. thaliana* cell suspensions (Gaudriault *et al.*, 1998; Degrave *et al.*, 2008). Collectively, research into the functions of HrpW, HrpN and DspE suggests that mediating HrpN-induced host defence responses may play a large role in disease development by *E. amylovora*.

In addition to harpin translocators like HrpN and HrpW, *E. amylovora* also produces a third putative translocator HrpK. Little is known about HrpK but, like HrpW, it is not required for disease development (Oh *et al.*, 2005). Likewise, HrpK is also secreted via the T3SS and null mutants exhibit slower DspE-CyaA translocation than wild-type *E. amylovora* (Nissinen *et al.*, 2007; Bocsanczy *et al.*, 2008). Despite overall slower DspE translocation in *hrpK* mutants, total DspE secretion is frequently greater than wild-type *E. amylovora* suggesting that HrpK may function as a negative regulator of translocation (Bocsanczy *et al.*, 2008). It remains to be determined if HrpK function regulates HrpN-induced host defence responses as do HrpW and DspE.

Additional virulence factors affecting fire blight pathogenesis

For successful disease development by *E. amylovora* during compatible host interactions, *E. amylovora* synthesizes the exopolysaccharides levan and amylovoran (Sjulin and Beer, 1978; Geier *et al.*, 1993). While levan mutants are attenuated in virulence and biofilm production, amylovoran is a pathogenicity factor (Bellemann and Geider, 1992; Koczan *et al.*, 2009; Zhao *et al.*, 2009). Originally considered a toxin, amylovoran is currently hypothesized to facilitate fire blight development via the occlusion of water-conducting xylem elements during *E. amylovora* biofilm formation (Sjulin and Beer, 1978). This hypothesis is supported by observations that decreasing amylovoran viscosity decreases xylem blockage (Sijam *et al.*, 1985). Amylovoran also is an important component of the *E. amylovora* biofilm and as such aids in mediating survival to environmental stresses like copper and presumably antimicrobial compounds produced during infection (Ordax *et al.*, 2010). The dual importance of amylovoran production and type III secretion is highlighted by the observation that co-inoculation of immature pears with amylovoran and T3SS *E. amylovora* mutant strains can rescue the loss-of-pathogenicity phenotype exhibited by either strain inoculated independently (Zhao *et al.*, 2009a). Notably, co-inoculation of either mutants with *Rubus*-adapted *E. amylovora* strain MR1 failed to complement their respective phenotypes suggesting fundamentally unsolved issues surrounding *E. amylovora* host-specificity (Zhao *et al.*, 2009a).

Rubus-adapted *E. amylovora* strains are also distinguished from Spiraeoideae-infecting genotypes by lipopolysaccharide (LPS) biosynthesis (Rezzonico *et al.*, 2012). LPS are large molecules composed of lipids and polysaccharides and serve as important components of the outer membrane of Gram-negative bacteria. LPS protects bacteria from antimicrobial compounds while also serving as important elicitors of host defence (Ingle *et al.*, 2006). The LPS ligase WaaL represents an important difference between LPS biosynthetic gene clusters in the two *E. amylovora* pathovars as their respective WaaL proteins share modest to very low similarity at the amino acid level (Rezzonico *et al.*, 2012). In Spiraeoideae-infecting *E. amylovora* Ea1189, WaaL was characterized as an important virulence factor required for tolerance to antimicrobial compounds and twitching motility (Berry *et al.*, 2009).

Erwinia tasmaniensis Et1/99 is a non-pathogenic relative of *E. amylovora* and consequently a useful source for identifying putative virulence factors via genome wide comparisons of the two *Erwinia* species. Exemplifying this concept is the lack of an amylovoran biosynthetic gene cluster in the genome of *E. tasmaniensis* Et1/99 (Kube *et al.*, 2008). In addition to the amylovoran discrepancy, *E. amylovora* produces a number of verified virulence factors not found encoded in the *E. tasmaniensis* genome. These include the *srl* sorbitol metabolism operon, a putative toxin encoded by the *hsv* gene cluster, as well as the protease PrtA (Kube *et al.*, 2008).

In rosaceous species, sorbitol is utilized for carbohydrate transportation. The first gene of the *srl* operon, *srlA*, is co-expressed with the T3SS in minimal media and mutations affecting the *srl* operon caused reduced virulence or loss-of-pathogenicity after inoculation into apple seedlings (Aldridge *et al.*, 1997; McNally *et al.*, 2012). The gene cluster *hsvABC* shares sequence homology with genes involved in the peptide biosynthesis of phaseolotoxin in *P. syringae* pv. *phaseolicola* (Oh *et al.*, 2005). While no phaseolotoxin activity nor small toxin-like peptide has been discovered in *E. amylovora* to date, mutations in *hsvABC* display attenuated virulence in apple shoots (Oh *et al.*, 2005). PrtA is a metalloprotease secreted via the type I secretion system (Zhang *et al.*, 1999). While PrtA is required for full virulence in apple seedlings, the mechanism underlying PrtA expression remains unknown (Zhang *et al.*, 1999).

Iron is an important cofactor for many proteins and consequently of vital nutritional importance. *E. amylovora* synthesizes the siderophore desferrioxamine E (DFO) as well as the TonB-dependent ferrioxamine receptor FoxR (Kachadourian *et al.*, 1996; Dellagi *et al.*, 1998). DFO is also important for protection against oxidative stress (Venisse *et al.*, 2003). In *E. amylovora, dfoE* and *foxR* mutants exhibit severe inhibition of flower infection highlighting iron scarcity *in planta* during host interactions (Dellagi *et al.*, 1998).

Another important nutritional factor is the vitamin thiamine. In *E. amylovora,* the ubiquitous plasmid pEA29 encodes thiamine biosynthesis genes *thiOSGF* (McGhee and Jones, 2000). *E. amylovora* strains cured of pEA29 are less virulent than wild-type *E. amylovora* and exhibited decreased amylovoran production (Falkenstein *et al.*, 1989; McGhee and Jones, 2000).

Survival *in planta* requires that *E. amylovora* not only meet nutritional demands but also persist in and colonize host extracellular spaces characterized by the presence of preformed phytoanticipins and induced antimicrobial compounds. Along with protective biofilm matrices and the LPS ligase WaaL, *E. amylovora* also utilizes a tripartite system consisting of the multidrug efflux pump AcrAB and the outer membrane secretin TolC (Burse *et al.*, 2004; Al-Karablieh *et al.*, 2009). In *E. amylovora, acrB* and *tolC* mutants are non-pathogenic or acutely reduced in shoot virulence and both strains are unable to persist *in planta* (Burse *et al.*, 2004; Al-Karablieh *et al.*, 2009). When challenged with apple leaf extracts, the growth of the *acrB* mutant was inhibited in agar diffusion assays and both AcrB and TolC are required for survival in the presence of a wide range of plant phytoalexins (Burse *et al.*, 2004; Al-Karablieh *et al.*, 2009). Taken together, these data suggest that survival *in planta* is an active process and that AcrAB and TolC function collaboratively to mediate *E. amylovora* survival via the export of host defence compounds.

Regulation of pathogenesis in *E. amylovora*

Environmental and host signals

No generic conclusions can be drawn in terms of virulence gene regulation in different bacterial pathosystems (Pallen *et al.*, 2005). However, as a trend, optimal expression of T3SS genes in mammalian pathogens occurs at 37°C, whereas in plant-pathogenic bacteria, the expression of T3SS genes occurs at lower temperature. Other environmental conditions such as osmolarity, the availability of oxygen, temperature, pH, iron concentration, nutrients and host signals also play important roles in regulating T3SS gene expression (Venecia and Young, 2005). Interestingly, gene regulation of different T3SS in one cell may also differ dramatically, mostly reflecting the niche in which the specific T3SS may function. For example, in *Salmonella* spp., the optimal expression of SPI-1 T3SS genes occurs in conditions such as low oxygen, high osmolarity, slight alkalinity (pH 8), and nutrient-rich media, which reflect extracellular conditions in the lumen of the small intestine; whereas the optimal conditions for expression of SPI-2 T3SS genes are acidic, low osmolarity, low Ca^{2+} and minimal media, which represent the intracellular lifestyle of *Salmonella* inside the host cells (Lober *et al.*, 2006). On the other hand, secretion of Ysc T3SS effector proteins Yops (for *Yersinia* outer proteins) in *Y. enterocolitica* are induced by growth at 37°C in a low calcium medium; whereas secretion of Ysps (for *Yersinia* secreted proteins) by the Ysa T3SS occurs *in vitro* when grown at 26°C in a high salt medium (Pallen *et al.*, 2005).

In *E. amylovora*, three T3SSs coexist and two of them, PAI2 and PAI3, belong to the same T3SS group (Inv/Mxi/Spa group) as SPI-1 and Ysa (Zhao *et al.*, 2011). Generally speaking, early studies have indicated that regulatory genes of *E. amylovora hrp*-T3SS (PAI1) are expressed under conditions such as low nutrients (minimum *hrp*-inducing medium), low pH, and relatively low temperature (18 to 20°C); are induced *in planta*, and repressed in rich media *in vitro* (Wei *et al.*, 1992b). Recent evidence suggested that PAI2 and PAI3 T3SS genes in *E. amylovora* are expressed at higher temperature (28°C) and in rich LB medium (Nakka *et al.*, 2010; Zhao, unpublished data). In addition, the findings also indicated that *E. amylovora* strains produced more amylovoran at high temperature (32°C) or neutral pH (7.0) than at low temperature (22 or 28°C) or acidic pH (5.5) (Wang *et al.*, 2011b). We reasoned that, in *in vitro* condition, acidic pH and temperature may be the environmental signals sensed by two-component systems such as GrrSA, which negatively regulate amylovoran production. Transcriptomic profiling also suggests that host contact may be one of the main signals sensed by the Rcs phosphorelay system, which leads to membrane protein-induced cell surface remodelling and thus positively regulates amylovoran production (Wang *et al.*, 2012a). However, the precise host signals for activating both T3SS and amylovoran remain to be determined.

Genetic regulation and regulatory network

In prokaryotes, gene expression is regulated primarily at the level of transcription initiation. A simplified model of virulence factor regulation in *E. amylovora* is presented in Fig. 3.1. Cell–cell communication systems, two-component systems, sigma factors, AraC-type transcriptional regulators, and other global regulators such as H-NS constitute the major players in regulating virulence gene expression, which further form global gene regulatory networks. Sigma factors, including alternative sigma factors, are essential transcription initiation factors that direct RNA polymerase to bind specific promoter regions. On the

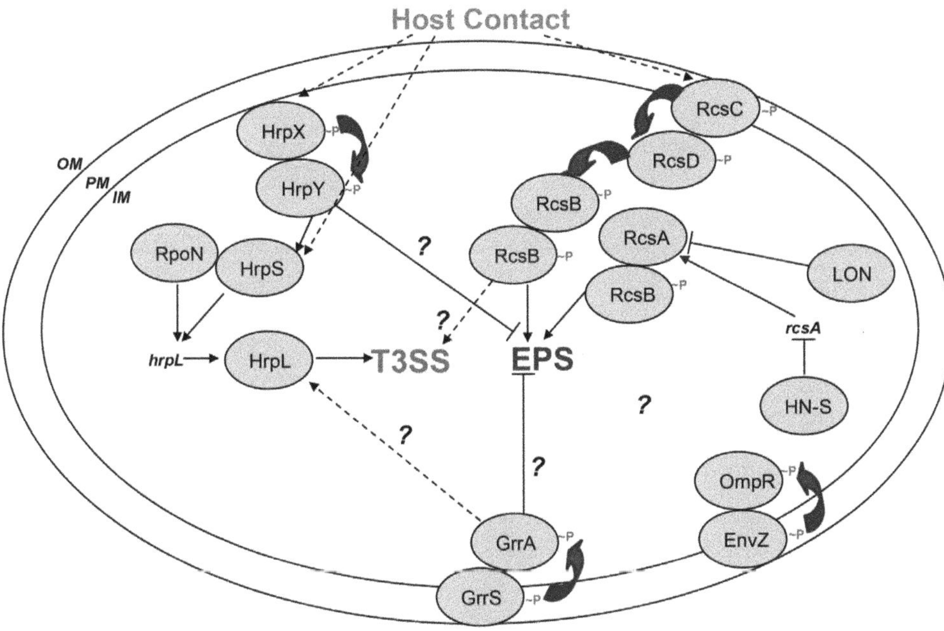

Figure 3.1 Simplified model of virulence factor regulation in *Erwinia amylovora*. Multiple signalling pathways coordinate the activity of pathogenicity factors in *E. amylovora* including the type III secretion system (T3SS) and exopolysaccharide (EPS) biosynthesis. A colour version of this figure is located in the plate section at the back of the book.

other hand, two-component systems represent major paradigms for signal transduction in prokaryotes and serve as basic stimulus–response coupling mechanisms to allow organisms to sense and respond to changes in many disparate environmental conditions. Furthermore, AraC-type regulators are key activators of T3SS gene expression, particularly in enterobacterial mammalian pathogens such as *E. coli* and *Salmonella,* and some plant pathogens such as *Ralstonia* and *Xanthomonas*. However, current evidence points to a minor role of AraC-type regulators and cell-density-dependent regulation of virulence factors in *E. amylovora.*

In *E. amylovora,* transcription of the *hrp*-T3SS genes is activated by the master regulator HrpL, a member of the ECF subfamily of sigma factors (Wei and Beer, 1995). Many studies point to the conclusion that most T3SS and effector genes (*eop1, eop3,* $avrRpt2_{Ea}$, *dspA/E,* and *hopC1*) are subject to direct HrpL regulation (McNally *et al.,* 2012; Nissinen *et al.,* 2007). A Hidden Markov model has identified about 30 *hrp* promoters in the genome of *E. amylovora* CFBP 1430, which contain the *hrp* box (GGAAC-N18/19-ACNNA) recognized by HrpL (McNally *et al.,* 2012; Bocsanczy *et al.,* 2012).

It has been reported that *hrpL* expression is activated by both HrpS (which belongs to the NtrC family of sigma 54 enhancer-binding proteins) and a two-component regulatory system HrpX (sensor) and HrpY (response regulator) (Wei *et al.,* 2000). Further domain structure analysis indicated that HrpX contains two PAS domains (initially found in PER, ARNT, and SIM proteins) within the N-terminal sensor region, suggesting that HrpX is a soluble and cytoplasmic protein that may sense intracellular signals and that other signalling pathways may also be involved in activating *hrpXY, hrpS,* or *hrpL* by sensing outside signals to regulate the T3SS. However, so far no such system has been reported and thus future research to identify extra regulators is warranted. Recent studies also found that *hrpXY* mutants remain virulent (Zhao *et al.*, 2009b), which is different from previous reports where Tn5-insertional *hrpXY* mutants were studied (Wei *et al.*, 2000). It is possible that Tn5 insertion could cause polar effects on the downstream genes such as *hrpS*. It has also been proposed that in *E. amylovora*, both HrpY and HrpS regulate *hrpL*, and that the effects of HrpY and HrpS are additive, which remains to be determined.

While early report suggests that HrpS, a member of the NtrC family of σ^{54} enhancer-binding proteins, only partially controls *hrpL* expression (Wei and Beer, 1995), our findings indicated that HrpS and alternative sigma factor RpoN (σ^{54}), but not stationary sigma factor RpoS (σ^{38}), are absolutely required for *hrpL* expression, which contains σ^{54} consensus sequences in its promoter region (Zhao *et al.*, 2009b, Li *et al.*, 2012). Further studies indicate that global regulators such as GrrSA two-component system may also be involved in regulating *hrp* gene expression (Li and Zhao, 2011).

Early studies have identified several key regulators of amylovoran biosynthesis, including RcsA (Bernhard *et al.*, 1990; Chatterjee *et al.*, 1990; Coleman *et al.*, 1990) and RcsB (Bereswell and Geider, 1997) as well as interactions between RcsA and RcsB (Kelm *et al.*, 1997; Wehland *et al.*, 1999, 2000). Both RcsA and RcsB belong to a unique enterobacterial-specific RcsBCD two-component system. The Rcs system is composed of three major proteins, RcsC, RcsD and RcsB; and involves several auxiliary proteins, RcsA and RcsF. However, the core of the Rcs system is the response regulator RcsB and the membrane-localized hybrid sensor kinase RcsC. The third member RcsD is also a membrane-bound sensor, and appears to be inactive as it lacks conserved residues in the active sites; however, it forms a complex with RcsC as well as RcsB based on structural studies (Schmoe *et al.*, 2011). The auxiliary protein RcsA is similar to RcsB but without the conserved aspartate residue in its N-terminal receiver domain. The outer membrane lipoprotein RcsF is involved in activation of RcsC.

The signal transduction mechanism of the Rcs system is an unusual multiple step phosphorelay, i.e. His-Asp-His-Asp. It is generally regarded that, upon sensing environmental stimuli, RcsC autophosphorylates and the phosphoryl group is then transferred to the receiver domain of RcsC and further transmitted to the phophoreceiver domain of RcsB via phosphotransmitter (Hpt) domain of RcsD. Phosphorylated RcsB could form RcsB–RcsB homodimers or interact with RcsA to form RcsAB heterodimers, which then bind to an 'RcsAB box' to regulate gene expression, including the promoter of *ams* operon involved in amylovoran biosynthesis. It has been recently demonstrated that the RcsCDB system is essential for virulence by positively controlling amylovoran biosynthesis in *E. amylovora* (Wang *et al.*, 2009, 2011a; Zhao *et al.*, 2009b). The *rcsB* and *rcsD* mutants are deficient in amylovoran production and cannot cause disease on host plants. Although the *rcsC* mutant produces high level of amylovoran *in vitro*, it remains non-pathogenic on host plants (Wang *et al.*, 2009, 2011a). The regulon of RcsBC has also been determined, which include many unique membrane protein-encoding genes (Wang *et al.*, 2012).

Besides the Rcs system, several two-component systems were previously identified to be induced during infection of host tissue in *E. amylovora* through an *in vivo* expression technology screen (IVET) (Zhao *et al.*, 2005). Genome wide screening of two-component system mutants has identified four groups of mutants which exhibited varying levels of amylovoran production *in vitro* (Wang *et al.*, 2011b; Zhao *et al.*, 2009b). These included GrrSA, EnvZ/

OmpR as well as the HrpXY system (Zhao *et al.*, 2009b). These findings indicate that two-component systems in *E. amylovora* play a major role in regulating amylovoran production, and evidence suggests that two component regulatory systems may form a regulatory network to govern the production of amylovoran.

Additional novel and global regulatory genes for amylovoran biosynthesis have also been identified through genetic screening. These included Lon protease, global regulator H-NS, RcsF, DjlA and AmyR (YbjN) (Eastgate *et al.*, 1995; Hildebrand *et al.*, 2006; Wang *et al.*, 2011b, 2012b). Both RcsF and DjlA are activators of RcsC, whereas H-NS binds to the promoter of *rcsA* and suppresses *rcsA* gene expression. In addition, the RcsA protein is subject to Lon-dependent degradation, which is a heat-shock protein. AmyR is an enterobacterial-specific orphan protein and has recently been characterized as a novel negative regulator of EPS production in both *E. coli* and *E. amylovora* (Wang *et al.*, 2011c, 2012b); however, the exact molecular mechanism of regulation is still unclear.

Concluding remarks and future prospects

The catalogue of *E. amylovora* research is an important component of our understanding of plant–microbe interactions. *E. amylovora* was identified by Thomas Burrill as the first bacterium to cause a plant disease and subsequently named after the father of plant bacteriology, Erwin F. Smith. Understanding HR in non-host plants continues to a be a major endeavour among plant pathologists, and HrpN, an *E. amylovora* harpin and translocator, was the first protein identified to function as a cell-free elicitor of HR. While many avenues are available for critical *E. amylovora* research, continued analyses of major virulence factors such as DspE and HrpN will play an important role in fundamentally understanding fire blight disease development and critical questions remain to be answered.

What is the biological significance of DspE interactions with DIPMs? What are the underlying mechanism responsible for DspE-induced plant cell death? Is DspE-mediated cell death a product of host defences exemplified by classic gene-for-gene interactions or does DspE-mediated cell death directly promote virulence typical of necrotrophic plant bacterial pathogens such as *Dickeya* spp. and *Pectobacterium* spp., two species closely related to *E. amylovora*. How does DspE suppress callose deposition in host interactions and what role does HrpN play in eliciting callose deposition? What host genetic determinants are responsible for defences triggered by HrpN, directly or indirectly?

In addition, recent revelations that AvrRpt2$_{Ea}$ exhibits quantitative gene-for-gene interactions with the apple species *Malus* X *robusta* 5, together with information that Eop1 and HopX1$_{Ea}$ also contribute quantitatively to host range and avirulence, indicate that *E. amylovora* is an importantly unique plant pathogen (Asselin *et al.*, 2011; Bocsanczy *et al.*, 2012; Vogt *et al.*, 2013). Disease development by *E. amylovora* exhibits some qualities of biotrophic plant bacterial pathogens as well as some qualities found in necrotrophic bacteria. While *E. amylovora* requires a functional T3SS like *Pseudomonas* spp. and *Xanthomonas* spp., it does not utilize comparatively large effector repertoires nor does *E. amylovora* produce the cocktail of cell wall-degrading enzymes found in *Dickeya* spp. and *Pectobacterium* spp.

Continuing to develop our understanding of how *E. amylovora* causes disease will likely be bolstered greatly by genomic technologies. A recent analysis of the *E. amylovora* pan-genome found that, while strains of *E. amylovora* display unique host specificity, the pan-genome of *E. amylovora* is remarkably conserved (Mann *et al.*,, 2013). A recent comparison of 12

E. amylovora genomes found that 89% of the *E. amylovora* pan-genome is represented by coding sequences conserved between strains (Mann *et al.*,, 2013). This degree of conservation is markedly higher than that exhibited by other bacterial plant pathogens and will be of great significance for plant pathologists interested in studying host specificity and virulence (Mann *et al.*, 2013).

In addition, the ultimate goal of fire blight research is to reduce the threat posed by *E. amylovora* to fruit production around the world. The recent identification of T3SS effectors that contribute to avirulence coupled with genomic data suggesting that host specificity in *E. amylovora* is likely due to a small number of genetic factors, means that continued research into *E. amylovora* pathogenesis is poised to produced translational results in the foreseeable future. In light of the spread of antibiotic resistant *E. amylovora* populations, the development of apple and pear cultivars with durable fire blight resistance will be a welcome solution to one of plant pathology's oldest problems.

References

Ádám, A.L., Pike, S., Hoyos, M.E., Stone, J.M., Walker, J.C., and Novacky, A. (1997). Rapid and transient activation of a myelin basic protein kinase in tobacco leaves treated with harpin from *Erwinia amylovora*. Plant Physiol. *115*, 853–861.

Aldridge, P., Metzger, M., and Geider, K. (1997). Genetics of sorbitol metabolism in *Erwinia amylovora* and its influence on bacterial virulence. Mol. Gen. Genet. *256*, 611–619.

Alfano, J.R., and Collmer, A. (2004). Type III secretion system effector proteins: double agents in bacterial disease and plants defense. Annu. Rev. Phytopathology *42*, 385–414.

Al-Karablieh, N., Weingart, H., and Ullrich, M.S. (2009). The outer membrane protein TolC is required for phytoalexin resistance and virulence of the fire blight pathogen *Erwinia amylovora*. Microb. Biotechnol. *2*, 465–475.

Asselin, J.E., Bonasera, J.M., Kim, J.F., Oh, C.-S., and Beer, S.V. (2011). Eop1 from a *Rubus* strain of *Erwinia amylovora* functions as a host-range limiting factor. Phytopathology *101*, 935–944.

Baker, C.J., Orlandi, E.W., and Mock, N.M. (1993). Harpin, an elicitor of the hypersensitive response into-bacco caused by *Erwinia amylovora*, elicits active oxygen production in suspension cells. Plant Physiol. *102*, 1341–1344.

Barbosa-Mendes, J.M., Filho, F., Filho, A.B., Harakava, R., Beer, S.V., and Mendes, B. (2009). Genetic transformation of *Citrus sinensis* cv. Hamlin with *hrpN* gene from *Erwinia amylovora* and evaluation of the transgenic lines for resistance to citrus canker. Sci. Hort. *122*, 109–115.

Barny, M.A. (1995). *Erwinia amylovora hrpN* mutants, blocked in harpin synthesis, express a reduced virulence on host plants and elicit variable hypersensitive reactions on tobacco. Eur. J. Plant Pathol. *101*, 333–340.

Barny, M.A., Guinebretière, M.H., Marçais, B., Coissac, E., Paulin, J.P., and Laurent, J. (1990). Cloning of a large gene cluster involved in *Erwinia amylovora* CFBP1430 virulence. Mol. Microbiol. *4*, 777–786.

Bellemann, P., and Geider, K. (1992). Localization of transposon insertions in pathogenicity mutants of*Erwinia amylovora* and their biochemical characterization. J. Gen. Microbiol. *138*, 931–940.

Bereswill, S., and Geider, K. (1997). Characterization of the *rcsB* gene from *Erwinia amylovora* and its influence on exopolysaccharide synthesis and virulence of the fire blight pathogen. J. Bacteriol. *179*, 1354–1361.

Bernhard, F., Poetter, K., Geider, K., and Coplin, D.L. (1990). The *rcsA* gene from *Erwinia amylovora*: identification, nucleotide sequence, and regulation of exopolysaccharide biosynthesis. Mol. Plant Microbe Interact. *3*, 29–437.

Berry, M.C., McGhee, G.C., Zhao, Y.F., and Sundin, G.W. (2009). Effect of a *waaL* mutation on lipopolysaccharide composition, oxidative stress survival, and virulence in *Erwinia amylovora*. FEMS Microbiol. Lett. *291*, 80–87.

Bocsanczy, A.M., Nissinen, R.M., Oh, C.-S., and Beer, S.V. (2008). HrpN of *Erwinia amylovora* functions in the translocation of DspA/E into plant cells. Mol. Plant Pathol. *9*, 425–434.

Bocsanczy, A.M., Schneider, D.J., DeClerck, G.A., Cartinhour, S., and Beer, S.V. (2012). HopX1 in *Erwinia amylovora* functions as an avirulence protein in apple and is regulated by HrpL. J. Bacteriol. *194*, 553–560.

Bogdanove, A.J., Wei, Z.M., and Beer, S.V. (1996). *Erwinia amylovora* secretes harpin via a type III pathway and contains a homolog of *yopN* of *Yersinia* spp. *178*, 1720–1730.

Bogdanove, A.J., Kim, J.F., Wei, Z., Kolchinsky, P., Charkowski, A.O., Conlin, A.K., Collmer, A., and Beer, S.V. (1998a). Homology and functional similarity of an *hrp*-linked pathogenicity locus, *dspEF*, of *Erwinia amylovora* and the avirulence locus *avrE* of *Pseudomonas syringae* pathovar tomato. Proc. Natl. Acad. Sci. U.S.A. *95*, 1325–1330.

Bogdanove, A.J., Bauer, D.W., and Beer, S.V. (1998b). *Erwinia amylovora* secretes DspE, a pathogenicity factor and functional AvrE homolog, through the Hrp (Type III secretion) pathway. J. Bacteriol. *180*, 2244–2247.

Bogs, J., and Geider, K. (2000). Molecular analysis of sucrose metabolism of *Erwinia amylovora* and influence on bacterial virulence. J. Bacteriol. *182*, 5351–5358.

Boller, T., and He, S.Y. (2009). Innate immunity in plants: an arms race between pattern recognition receptors in plants and effectors in microbial pathogens. Science *324*, 742–744.

Bonn, W.G., and van der Zwet, T. (2000). Distribution and economic importance of fire blight. In Fire Blight, the Disease and its Causative Agent, *Erwinia amylovora*, J.L. Vanneste, ed. (CAB International, Wallingford, UK), pp. 37–54.

Boureau, T., ElMaarouf-Bouteau, H., Garnier, A., Brisset, M.-N., Perino, C., Pucheu, I., and Barny, M.-A. (2006). DspA/E, a type III effector essential for *Erwinia amylovora* pathogenicity and growth *in planta*, induces cell death in host apple and nonhost tobacco plants. Mol. Plant Microbe Interact. *19*, 16–24.

Boureau, T., Siamer, S., Perino, C., Gaubert, S Patrit, O., Degrave, A., Fagard, M., Chevreau, E., and Barny, M.-A. (2011). The HrpN effector of *Erwinia amylovora*, which is involved in type III translocation, contributes directly or indirectly to callose elicitation on apple leaves. Mol. Plant Microbe Interact. *24*, 577–584.

Bugert, P., and Geider, K. (1995). Molecular analysis of the *ams* operon required for exopolysaccharide synthesis of *Erwinia amylovora*. Mol. Microbiol. *15*, 917–933.

Burse, A., Weingart, H., and Ullrich, M.S. (2004). The phytoalexin-inducible multidrug efflux pump AcrAB contributes to virulence in the fire blight pathogen, *Erwinia amylovora*. Mol. Plant Microbe Interact. *17*, 43–54.

Buttner, D., and Bonas, U. (2002). Port of entry – the type III secretion translocon. Trends Microbiol. *10*, 186–192.

Buttner, D., and He, S.Y. (2009). Type III protein secretion in plant-pathogenic bacteria. Plant Physiol. *150*, 1656–1664.

de Capdeville, G., Beer, S.V., Watkins, C.B., Wilson, C.L., Tedeschi, L.O., and Aist, J.R. (2003). Pre- and post-harvest harpin treatments of apples induce resistance to blue mold. Plant Dis. *87*, 39–44.

Cesbron, S., Paulin, J.-P., Tharaud, M., Barny, M.-A., and Brisset, M.-N. (2006). The alternative σ factor HrpL negatively modulates the flagellar system in the phytopathogenic bacterium *Erwinia amylovora* under *hrp*-inducing conditions. FEMS Microbiol. Lett. 257, 221–227.

Chatterjee, A., Chun, W., and Chatterjee, A.K. (1990). Isolation and characterization of an *rcsA*-like gene of *Erwinia amylovora* that activates extracellular polysaccharide production in *Erwinia* species, *Escherichia coli*, and *Salmonella typhimurium*. Mol. Plant Microbe Interact. *3*, 144–148.

Coleman, M., Pearce, R., Hitchin, F., Busfield, F., Mansfield, J.W., and Roberts, I.S. (1990). Molecular cloning, expression and nucleotide sequence of the *rcsA* gene of *Erwinia amylovora*, encoding a positive regulator of capsule expression: evidence for a family of related capsule activator proteins. J. Gen. Microbiol. *136*, 1799–1806.

Debroy, S., Thilmony, R., Kwack, Y.-B., Nomura, K., and He, S.Y. (2004). A family of conserved bacterial effectors inhibits salicylic acid-mediated basal immunity and promotes disease necrosis in plants. Proc. Natl. Acad. Sci. U.S.A. *101*, 9927–9932.

Degrave, A., Fagard, M., Perino, C., Brisset, M.N., Gaubert, S., Laroche, S., Patrit, O., and Barny, M.-A. (2008). *Erwinia amylovora* type three-secreted proteins trigger cell death and defense responses in *Arabidopsis thaliana*. Mol. Plant Microbe Interact. *21*, 1076–1086.

Degrave, A., Moreau, M., Launay, A., Barny, M.-A., Brisset, M.-N., Patrit, O., Taconnat, L., Vedel, R., and Fagard, M. (2013). The bacterial effector DspA/E is toxic in *Arabidopsis thaliana* and is required for multiplication and survival of fire blight pathogen. Mol. Plant Pathol. *14*, 506–517.

Dellagi, A., Brisset, M.-N., Paulin, J.-P., and Expert, D. (1998). Dual role of desferrioxamine in *Erwinia amylovora* pathogenicity. Mol. Plant Microbe Interact. *11*, 734–742.

Denning, W. (1794). On the decay of apple trees. In Transactions of the Society for the Promotion of Agriculture, Arts and Manufactures, R.R. Livingston, J. Lansing, S. van Rensselaer, S. de Witt, and J.B. Johnson, eds (Charles R. and George Webster, Albany, NY, USA), pp. 185–187.

Deslandes, L., and Rivas, S. (2012). Catch me if you can: bacterial effectors and plant targets. Trends Plant Sci. *17*, 644–655.

Dong H., Delaney, T.P., Bauer, D.W., and Beer, S.V. (1999). Harpin induces disease resistance in *Arabidopsis* through the systemic acquired resistance pathway mediated by salicylic acid and the *NIM1* gene. Plant J. *20*, 207–215.

Eastgate, J.A., Taylor, N., Coleman, M.J., Healy, B., Thompson, L., and Roberts, I.S. (1995). Cloning, expression and characterization of the lon gene of *Erwinia amylovora*: evidence for a heat shock response. J. Bacteriol. *177*, 932–937.

Eastgate, J.A., Thompson, L., Milner, J., Cooper, R.M., Pollitt, C.E., and Roberts, I.S. (1997). Identification of a nonpathogenic *Erwinia amylovora guaB* mutant. Plant Pathol. *46*, 594–599.

El-Maaroufa, H., Barny, M.A., Rona, J.P., and Bouteaub, F. (2001). Harpin, a hypersensitive response elicitor from *Erwinia amylovora*, regulates ion channel activities in *Arabidopsis thaliana* suspension cells. FEBS Lett. *497*, 82–84.

Falkenstein, H., Zeller, W., and Geider, K. (1989). The 29 kb plasmid, common in strains of *Erwinia amylovora*, modulates development of fire blight symptoms. Microbiology *135*, 2643–2650.

Gaudriault, S., Brisset, M.-N., and Barny, M.-A. (1998). HrpW of *Erwinia amylovora*, a new Hrp-secreted protein. FEBS Lett. *428*, 224–228.

Gaudriault, S., Paulin, J.P., and Barny, M.-A. (2002). The DspB/F protein of *Erwinia amylovora* is a type III secretion chaperone ensuring efficient intrabacterial production of the Hrp-secreted DspA/E pathogenicity factor. Mol. Plant. Pathol. 3, 313–320.

Geider, K. (2000). Exopolysaccharides of *Erwinia amylovora*: structure, biosynthesis, regulation, role in pathogenicity of amylovoran and levan. In Fire Blight: the Disease and its Causative Agent, *Erwinia amylovora*, J.L. Vanneste, ed. (CAB International, New York, NY, USA), pp. 117–140.

Geier, G., and Geider, K. (1993). Characterization and influence on virulence of the levansucrase gene from the fireblight pathogen *Erwinia amylovora*. Physiol. Mol. Plant Pathol. *42*, 387–404.

Gómez-Gómez, L., and Boller, T. (2000). FLS2: an LRR receptor-like kinase involved in the perception of the bacterial elicitor flagellin in *Arabidopsis*. Mol. Cell. *5*, 1003–1011.

Ham, J.H., Majerczak, D.R., Arroyo-Rodriguez, A.S., Mackey, D.M., and Coplin, D.L. (2006). WtsE, an AvrE-family effector protein from *Pantoea stewartii* subsp. *stewartii*, causes disease-associated cell death in corn and requires a chaperone protein for stability. Mol. Plant Microbe Interact. *19*, 1092–1102.

He, S.Y., and Jin, Q. (2003). The Hrp pilus: learning from flagella. Curr. Opin. Microbiol. *6*, 15–19.

Hildebrand, M., Aldridge, P., and Geider, K. (2006). Characterization of *hns* genes in *Erwinia amylovora*. Mol. Gen. Genom. *275*, 310–319.

Ingle, R.A., Carstens, M., and Denby, K.J. (2006). PAMP recognition and the plant-pathogen arms race. BioEssays *28*, 880–889.

Jin, Q., Hu, W., Brown, I., McGhee, G., Hart, P., Jones, A.L., and He, S.Y. (2001). Visualization of secreted Hrp and Avr proteins along the Hrp pilus during type III secretion in *Erwinia amylovora* and *Pseudomonas syringae*. Mol. Microbiol. *40*, 1129–1139.

Kachadourian, R., Dellagi, A., Laurent, J., Bricard, L., Kunesch, G., and Expert, D. (1996). Desferrioxamine-dependent iron transport in *Erwinia amylovora* CFBP1430: cloning of the gene encoding the ferrioxamine receptor FoxR. BioMetals 9, 143–150.

Kelm, O., Kiecker, C., Geider, K., and Bernhard, F. (1997). Interaction of the regulator proteins RcsA and RcsB with the promoter of the operon for amylovoran biosynthesis in *Erwinia amylovora*. Mol. Gen. Genet. *256*, 72–83.

Khan, M.A., Zhao, Y.F., and Korban, S.S. (2012). Molecular mechanisms of pathogenesis and resistance to the bacterial pathogen *Erwinia amylovora*, causal agent of fire blight disease in Rosaceae. Plant Mol. Biol. Rep. *30*, 247–260.

Kim, J.F., and Beer, S.V. (1998). HrpW of *Erwinia amylovora*, a new harpin that contains a domain homologous to pectate lyases of a distinct class. J. Bacteriol. *180*, 5203–5210.

Kim, J.F., Wei, Z.M., and Beer, S.V. (1997.) The *hrpA* and *hrpC* operons of *Erwinia amylovora* encode components of a type III pathway that secretes harpin. J. Bacteriol. *179*, 1690–1697.

Koczan, J.M., McGrath, M.J., Zhao, Y.F., and Sundin, G.W. (2009). Contribution of *Erwinia amylovora* exopolysaccharides amylovoran and levan to biofilm formation: implications in pathogenicity. Phytopathology 99, 1237–1244.

Koczan, J.M., Lenneman, B.R., McGrath, M.J., and Sundin, G.W. (2011). Cell surface attachment structures contribute to biofilm formation and xylem colonization by *Erwinia amylovora*. Appl. Environ. Microbiol. *77*, 7031–7039.

Kube, M., Migdoll, A.M., Muller, I., Kuhl, H., Beck, A., Reinhardt, R., and Geider, K. (2008). The genome of *Erwinia tasmaniensis* strain Et1/99, a non-pathogenic bacterium in the genus *Erwinia*. Environ. Microbiol. *10*, 2211–2222.

Kubori, T., Matsushima, Y., Nakamura, D., Uralil, J., Lara-Tejero, M., Sukhan, A., Gala, J.E., and Aizawa, S.-I. (1998). Supramolecular structure of the *Salmonella typhimurium* type III protein secretion system. Science *280*, 602–605.

Kvitko, B.H., Park, D.H., Velasquez, A.C., Wei, C.-F., Russell, A.B., Martin, G.B., Schneider, D.J., and Collmer, A. (2009). Deletions in the repertoire of *Pseudomonas syringae* pv. *tomato* DC3000 type III secretion effector genes reveal functional overlap among effectors. PLoS Pathog. *5*, e1000388.

Li, W.T., and Zhao, Y.F. (2011). Effect of EnvZ/OmpR and GrrS/GrrA systems on *Erwinia amylovora* virulence. Phytopathology *101*, S102.

Li, W.T., Ancona, V., and Zhao, Y.F. (2012). The role of sigma factors in regulating virulence gene expression in *Erwinia amylovora*. Phytopathology *102*, S470.

Lober, S., Jackel, D., Kaiser, N., and Hensel, M. (2006). Regulation of *Salmonella* pathogenicity island 2 genes by independent environmental signals. Int. J. Med. Microbiol. *296*, 435–447.

Longstroth, M. (2001). The fire blight epidemic in southwestern Michigan 2000. Compact Fruit Tree *34*, 16–19.

Lorang, J.M., and Keen, N.T. (1995). Characterization of *avrE* from *Pseudomonas syringae* pv. *tomato*: a *hrp*-linked avirulence locus consisting of at least two transcriptional units. Mol. Plant Microbe Interact. *8*, 49–57.

Lorenz, C., and Buttner, D. (2011). Secretion of early and late substrates of the type III secretion system from *Xanthomonas* is controlled by HpaC and the C-terminal domain of HrcU. Mol. Microbiol. *79*, 447–467.

McGhee, G.C., and Jones, A.L. (2000). Complete nucleotide sequence of ubiquitous plasmid pEA29 from *Erwinia amylovora* strain Ea88: gene organization and intraspecies variation. Appl. Environ. Microbiol. *66*, 4897–4907.

McGhee, G.C., Guasco, J., Bellomo, L.M., Blumer-Schuette, S.E., Shane, W.W., Irish-Brown, A., and Sundin, G.W. (2011). Genetic analysis of streptomycin-resistant (SmR) strains of *Erwinia amylovora* suggests that dissemination of two genotypes is responsible for the current distribution of SmR *E. amylovora* in Michigan. Phytopathology *101*, 182–191.

McManus, P.S., Stockwell, V.O., Sundin, G.W., and Jones, A.L. (2002). Antibiotic use in plant agriculture. Annu. Rev. Phytopathol. *40*, 443–465.

McNally, R.R., Toth, I., Cock, P., Pritchard, L., Hedley, P., Morris, J., Zhao, Y.F., and Sundin, G.W. (2012). Genetic characterization of the HrpL regulon of the fire blight pathogen *Erwinia amylovora* reveals novel virulence factors. Mol. Plant Pathol. *13*, 160–173.

Mann, R.A., Smits, T.H.M., Buhlmann, A., Blom, J., Goesmann, A., Frey, J.E., Plummer, K.M., Beer, S.V., Luck, J., Duffy, B., *et al.* (2013). Comparative genomics of 12 strains of *Erwinia amylovora* identifies a pan-genome with a large conserved core. PLoS Pathog. *8*, e55644.

Meng, X., Bonasera, J.M., Kim, J.F., Nissinen, R.M., and Beer, S.V. (2006). Apple proteins that interactwith DspA/E, a pathogenicity effector of *Erwinia amylovora*, the fire blight pathogen. Mol. Plant Microbe Interact. *19*, 53–61.

Metzger, M., Bellemann, P., Bugert, P., and Geider, K. (1994). Genetics of galactose metabolism of *Erwinia amylovora* and its influence on polysaccharide synthesis and virulence of the fire blight pathogen. J. Bacteriol. *176*, 450–459.

Nakka, S., Qi, M., and Zhao, Y.F. (2010). The *Erwinia amylovora* PhoPQ system is involved in resistance to antimicrobial peptide and suppresses gene expression of two novel type III secretion systems. Microbiol. Res. *165*, 665–673.

Nissinen, R.M., Ytterberg, A.J., Bogdanove, A.J., van Wijk, K.J., and Beer, S.V. (2007). Analyses of the secretomes of *Erwinia amylovora* and selected *hrp* mutants reveal novel type III secreted proteins and an effect of HrpJ on extracellular harpin levels. Mol. Plant Pathol. *8*, 55–67.

Norelli, J.L., Jones, A.L., and Aldwinckle, H.S. (2003). Fire blight management in the twenty-first century. Plant Dis. *87*, 756–765.

Nurnberger, T., and Kemmerling, B. (2006). Receptor protein kinases – pattern recognition receptors in plant immunity. Trends Plant Sci. *11*, 519–522.

Oh, C.-S., and Beer, S.V. (2005). Molecular genetics of *Erwinia amylovora* involved in the development of fire blight. FEMS Microbiol. Lett. 253, 185–192.

Oh, C.-S., Kim, J.F., and Beer, S.V. (2005). The Hrp pathogenicity island of *Erwinia amylovora* and identification of three novel genes required for systemic infection. Mol. Plant Pathol. *6*, 125–138.

Oh, C.-S., Martin, G.B., and Beer, S.V. (2007). DspA/E, a type III effector of *Erwinia amylovora*, is required for early rapid growth in *Nicotiana benthamiana* and causes NbSGT1-dependent cell death. Mol. Plant Pathol. *8*, 255–265.

Oh, C.-S., Carpenter, S.C.D., Hayes, M.L., and Beer, S.V. (2010). Secretion and translocation signals and DspB/F-binding domains in the type III effector DspA/E of *Erwinia amylovora*. Microbiology *156*, 1211–1220.

Ordax, M., Marco-Noales, E., Lopez, M.M., and Biosca, E.G. (2010). Exopolysaccharides favor the survival of *Erwinia amylovora* under copper stress through different strategies. Res. Microbiol. *161*, 549–555.

Pallen, M.J., Beatson, S.A., and Bailey, M. (2005). Bioinformatics, genomics and evolution of non-flagella type-III secretion systems: a Darwinian perspective. FEMS Microbiol. Rev. *29*, 201–229.

Penga, J.-L., Donga, H.-S., Donga, H.-P., Delaney, T.P., Bonasera, J.M., and Beer, S.V. (2003). Harpin-elicited hypersensitive cell death and pathogen resistance require the *NDR1* and *EDS1* genes. Physiol. Mol. Plant Pathol. *62*, 317–326.

Percival, G.C., Noviss, K., and Haynes, I. (2009). Field evaluation of systemic inducing resistance chemicals at different growth stages for the control of apple (*Venturia inaequalis*) and pear (*Venturia pirina*) scab. Crop Protect. *28*, 629–633.

Perino, C., Gaudriault, S., Vian, B., and Barny, M.A. (1999). Visualization of harpin secretion *in planta* during infection of apple seedlings by *Erwinia amylovora*. Cell. Microbiol. *1*, 131–141.

Pester, D., Milcevicova, R., Schaffer, J., Wilhelm, E., and Blumel, S. (2012). *Erwinia amylovora* expresses fast and simultaneously *hrp*/*dsp* virulence genes during flower infection on apple trees. PLoS ONE *7*, e32583.

Pozidis, C., Chalkiadaki, A., Gomez-Serrano, A., Stahlberg, H., Brown, I., Tampakaki, A.P., Lustig, A., Sianidis, G., Politou, A.S., Engel, A., *et al.* (2003). Type III protein translocase. J. Biol. Chem. *278*, 25816–25824.

Pusey, P.L., and Smith, T.J. (2008). Relation of apple flower age to infection of hypanthium by *Erwinia amylovora*. Plant Dis. *92*, 137–142.

Reboutier, D., Frankart, C., Briand, J., Biligui, B., Laroche, S., Rona, J.-P., Barny, M.-A., and Bouteau, F. (2007a). The $HrpN_{ea}$ harpin from *Erwinia amylovora* triggers differential responses on the nonhost *Arabidopsis thaliana* cells and on the host apple cells. Mol. Plant Microbe Interact. *20*, 94–100.

Reboutier, D., Frankart, C., Briand, J., Biligui, B., Rona, J.-P., Haapalainen, M., Barny, M.-A., and Bouteau, F. (2007b). Antagonistic action of harpin proteins: $HrpW_{ea}$ from *Erwinia amylovora* suppresses $HrpN_{ea}$-induced cell death in *Arabidopsis thaliana*. J. Cell Sci. *120*, 3271–3278.

Rezzonico, F., Braun-Kiewnick, A., Mann, R.A., Rodoni, B., Goesmann, A., Duffy, B., and Smits, T.H.M. (2012). Lipopolysaccharide biosynthesis genes discriminate between *Rubus*- and *Spiraeoideae*-infective genotypes of *Erwinia amylovora*. Mol. Plant Pathol. *13*, 975–984.

Sarowar, S., Zhao, Y., Soria-Guerra, R.E., Ali, S., Zheng, D., Wang, D., and Korban, S.S. (2011). Expression profiles of differentially regulated genes during the early stages of apple flower infection with *Erwinia amylovora*. J. Exp. Bot. *62*, 4851–4861.

Schmoe, K., Rogov, V.V., Yu, N., Lohr, F., Guntert, P., Bernhard, F., and Dotsch, V. (2011). Structural insights into *Rcs* phosphotransfer: the newly identified RcsD-ABL domain enhances interaction with the response regulator RcsB. Structure *19*, 577–587.

Sijam, K., Goodman, R.N., and Karr, A.L. (2008). The effect of salts on the viscosity and wilt-inducing capacity of the capsular polysaccharide of *Erwinia amylovora*. Physiol. Plant Pathol. *26*, 231–239.

Sinn, J.P., Oh, C.-S., Jensen, P.J., Carpenter, S.C.D., Beer, S.V., and McNellis, T.W. (2008). The C-terminal half of the HrpN virulence protein of the fire blight pathogen *Erwinia amylovora* is essential for its secretion and for its virulence and avirulence activities. Mol. Plant Microbe Interact. *21*, 1387–1397.

Sjulin, T.M., and Beer, S.V. (1978). Mechanism of wilt induction by amylovorin in cotoneaster shoots and its relation to wilting of shoots infected by *Erwinia amylovora*. Phytopathology *68*, 89–94.

Smits, T.H.M., Rezzonico, F., Kamber, T., Blom, J., Goesmann, A., Frey, J.E., and Duffy, B. (2010). complete genome sequence of the fire blight pathogen *Erwinia amylovora* CFBP 1430 and comparison to other *Erwinia* spp. Mol. Plant Microbe Interact. 23, 384–393.

Steinberger, E.M., and Beer, S.V. (1988). Creation and complementation of pathogenicity mutants of *Erwinia amylovora*. Mol. Plant Microbe Interact. *1*, 135–144.

Triplett, L.R., Melotto, M., and Sundin, G.W. (2009). Functional analysis of the N terminus of the *Erwinia amylovora* secreted effector DspA/E reveals features required for secretion, translocation, and binding to the chaperone DspB/F. Mol. Plant Microbe Interact. 22, 1282–1292.

Triplett, L.R., Wedemeyer, W.J., and Sundin, G.W. (2010). Homology-based modeling of the *Erwinia amylovora* type III secretion chaperone DspF used to identify amino acids required for virulence and interaction with the effector DspE. Res. Microbiol. *161*, 613–618.

Vencato, M., Tian, F., Alfano, J.R., Buell, C.R., Cartinhour, S., DeClerck, G.A., Guttman, D.S., Stavrinides, J., Joardar, V., Lindeberg, M., *et al.* (2006). Bioinformatics-enabled identification of the HrpL regulon and type III secretion system effector proteins of *Pseudomonas syringae* pv. *phaseolicola* 1448A. Mol. Plant Microbe Interact. *19*, 1193–1206.

Venecia, K., and Young, G.M. (2005). Environmental regulation and virulence attributes of the *Ysa* typeIII secretion system of *Yersinia enterocolitica* Biovar *1B*. Infect. Immun. *73*, 5961–5977.

Venisse, J.-S., Barny, M.-A., Paulin, J.-P., and Brisset, M.-N. (2003). Involvement of three pathogenicity factors of *Erwinia amylovora* in the oxidative stress associated with compatible interaction in pear. FEBS Lett. *537*, 198–202.

Vogt, I., Wohner, T., Richter, K., Flachowsky, H., Sundin, G.W., Wensing, A., Savory, E.A., Geider, K., Day, B., Hanke, M.-V., *et al.* (2013). Gene-for-gene relationship in the host–pathogen system *Malus* X *robusta* 5-*Erwinia amylovora*. New Phytol. *197*, 1262–1275.

Wang, D., Korban, S.S., and Zhao, Y.F. (2009). The *Rcs* phosphorelay system is essential for pathogenicity in *Erwinia amylovora*. Mol. Plant Pathol. *10*, 277–290.

Wang, D., Korban, S.S., Pusey, L., and Zhao, Y.F. (2011a). Characterization of the RcsC sensor kinase from *Erwinia amylovora* and other enterobacteria. Phytopathology *101*, 701–717.

Wang, D., Korban, S.S., Sundin, G.W., Clough, S., Toth, I., and Zhao, Y.F. (2011b). Regulatory genes and environmental regulation of amylovoran biosynthesis in *Erwinia amylovora*. Acta Hort. *896*, 195–202.

Wang, D.P., Calla, B., Vimolmangkang, S., Wu, X., Korban, S.S., Huber, S.C., Clough, S.J., and Zhao, Y.F. (2011c). The orphan gene *ybjN* conveys pleiotropic effects on multicellular behavior and survival of *Escherichia coli*. PLoS ONE *6*, e25293.

Wang, D.P., Qi, M.S., Calla, B., Korban, S.S., Clough, S.J., Cock, P., Sundin, G.W., Toth, I., and Zhao, Y.F. (2012a). Genome-wide identification of genes regulated by the Rcs phosphorelay system in *Erwinia amylovora*. Mol. Plant Microbe Interact. *25*, 6–17.

Wang D.P., Korban, S.S., Pusey, P.L., and Zhao, Y.F. (2012b). AmyR is a novel negative regulator of amylovoran production in *Erwinia amylovora*. PLoS ONE *7*, e45038.

Wang, L., and Beer, S.V. (2006). Application of signature-tagged mutagenesis to the study of virulence of *Erwinia amylovora*. FEMS Microbiol. Lett. *265*, 164–171.

Wehland, M., and Bernhard, F. (2000). The RcsAB box: characterization of a new operator essential for the regulation of exopolysaccharide biosynthesis in enteric bacteria. J. Biol. Chem. *275*, 7013–7020.

Wehland, M., Kiecker, C., Coplin, D.L., Kelm, O., Saenger, W., and Bernhard, F. (1999). Identification of an RcsA/RcsB recognition motif in the promoters of exopolysaccharide biosynthetic operons from *Erwinia amylovora* and *Pantoea stewartii* subspecies *stewartii*. J. Biol. Chem. *274*, 3300–3307.

Wei, Z.M., and Beer, S.V. (1993). HrpI of *Erwinia amylovora* functions in secretion of harpin and is a member of a new protein family. J. Bacteriol. *175*, 7958–7967.

Wei, Z.M., and Beer, S.V. (1995). HrpL activates *Erwinia amylovora hrp* gene transcritption and is a member of the ECF subfamily of σ factors. J. Bacterial. *177*, 6201–6210.

Wei, Z.M., Laby, R.J., Zumoff, C.H., Bauer, D.W., He, S.Y., Collmer, A., and Beer, S.V. (1992a). Harpin, elicitor of the hypersensitive response produced by the plant pathogen *Erwinia amylovora*. Science *257*, 85–88.

Wei, Z.M., Sneath, B.J., and Beer, S.V. (1992b). Expression of *Erwinia amylovora hrp* genes in response to environmental stimuli. J. Bacteriol. *174*, 1875–1882.

Wei, Z.M., Kim, J.F., and Beer, S.V. (2000). Regulation of hrp genes and type III protein secretion in *Erwinia amylovora* by HrpX/HrpY, a novel two component system, and HrpS. Mol. Plant Microbe Interact. *13*, 1251–1262.

Wilson, M., and Lindow, S.E. (1992). Interactions between the biological control agent *Pseudomonas fluorescens* A506 and *Erwinia amylovora* and pear blossoms. Phytopathology *83*, 117–132.

Winslow, C.-E.A., Broadhurst, J., Buchanan, R.E., Krumwiede, C., Rogders, L.A., and Smith, G.H. (1920). The families and genera of the bacteria. J. Bacteriol. *5*, 191–229.

Zhang, Y., Bak, D.D., Heid, H., and Geider, K. (1999). Molecular characterization of a protease secreted by *Erwinia amylovora*. J. Mol. Biol. *289*, 1239–1251.

Zhao, Y.F., and Qi, M.S. (2011). Comparative genomics of *Erwinia amylovora* and related *Erwinia* species – what do we learn? Genes *2*, 627–639.

Zhao, Y.F., Blumer, S.E., and Sundin, G.W. (2005). Identification of *Erwinia amylovora* genes induced during infection of immature pear tissue. J. Bacteriol. *187*, 8088–8103.

Zhao, Y., He, S.-Y., and Sundin, G.W. (2006). The *Erwinia amylovora avrRpt2*$_{EA}$ gene contributes to virulence on pear and AvrRpt2$_{EA}$ is recognized by *Arabidopsis* RPS2 when expressed in *Pseudomonas syringae*. Mol. Plant Microbe Interact. *19*, 644–654.

Zhao, Y.F., Sundin, G.W., and Wang, D.P. (2009a). Construction and analysis of pathogenicity island deletion mutants in *Erwinia amylovora*. Can. J. Microbiol. *55*, 457–464.

Zhao, Y.F., Wang, D., Nakka, S., Sundin, G.W., and Korban, S.S. (2009b). Systems-level analysis of two-component signal transduction systems in *Erwinia amylovora*: role in virulence, regulation of amylovoran biosynthesis and swarming motility. BMC Genomics *10*, 245.

Zhao, Y.F., Qi, M., and Wang, D. (2011). Evolution and function of flagella and non-flagella type III secretion systems in *Erwinia amylovora*. Acta Hort. *896*, 177–184.

van der Zwet, T., and Beer, S.V. (1999). Fire Blight – Its Nature, Prevention, and Control: A Practical Guide to Integrated Disease Management (Agricultural Research Service, Washington, DC, USA).

Plant-pathogenic *Acidovorax* Species

4

Tally Rosenberg, Noam Eckshtain-Levi and Saul Burdman

Abstract

The *Acidovorax* genus belongs to the Proteobacteria phylum, and comprises species that possess a variety of lifestyles, with its members inhabiting different environments, including soil, water and interactions with eukaryotic organisms. In the present chapter, we focus on *Acidovorax* species that are pathogenic on plants. *Acidovorax* species are able to cause disease to a variety of agriculturally and economically important crops. Moreover, some diseases caused by *Acidovorax* strains have been spreading worldwide in recent years and their importance is substantially increasing. One example is seedling blight and bacterial fruit blotch of cucurbit crops caused by *Acidovorax citrulli*, a disease that was barely known until the late 1980s' and today is one of the most serious threats to the cucurbit industry worldwide. Here we summarize taxonomical aspects of plant-pathogenic species of *Acidovorax* and describe major diseases caused by members of this genus, with emphasis on their epidemiology, control and basic aspects of pathogenesis and plant–pathogen interactions.

Introduction: the *Acidovorax* genus

The genus *Acidovorax* (Betaproteobacteria class, Burkholderiales order, Comamonadaceae family) comprises a variety of species adapted to different environments and lifestyles, including as free-living bacteria in soil and water, or in close, beneficial or deleterious interactions with eukaryotic organisms (Table 4.1).

Several *Acidovorax* strains have been isolated based on their efficient biodegradation activity. For example, *Acidovorax* sp. JS42 was isolated from a nitrobenzene-contaminated site, based on its ability to utilize 2-nitrotoluene, a toxic nitroarene compound (Haigler *et al.*, 1994; Rabinovitch-Deere and Parales, 2012).The interest on the biodegradation ability of this strain lead to its genome sequencing by the Joint Genome Institute in 2007 (http://genome.jgi-psf.org/aci_j/aci_j.info.html). An *Acidovorax avenae* strain, isolated from a polluted aquifer, was shown to perform aerobic degradation of chlorobenzenes (Monferran *et al.*, 2005). A similar ability was further attributed to an *Acidovorax facilis* strain (Nestler *et al.*, 2007). Another *Acidovorax* sp. strain was isolated from a polycyclic aromatic hidrocarbon (PAH)-contaminated soil. This strain was able to grow on a variety of PAHs as sole carbon sources, including phenanthrene, naphthalene, chrysene and others (Singleton *et al.*, 2009).

Acidovorax species like *A. temperans*, *A. facilis* and *A. delafieldii* have been reported as abundant members of microbial communities in activated sludge wastewater treatment plants. *A. temperans* CB2 was isolated from an activated sludge reactor treating municipal

Table 4.1 Plant-pathogenic and environmental species of *Acidovorax*

Species[1]	Features	References[2]
Plant pathogenic		
A. anthurii	Causal agent of bacterial leaf spot of anthurium	Gardan *et al.* (2000)
A. avenae	Pathogenic of various Poaceae species including maize, oat, wheat, barley, rye, sorghum, sugarcane, millet and others	Willems *et al.* (1992), Schaad *et al.* (2008)
A. cattleyae	Causal agent of leaf spot and bud rot on Cattleya, Dendrobium and Phalaenopsis orchids	Willems *et al.* (1992), Schaad *et al.* (2008)
A. citrulli	Causal agent of seedling blight and bacterial fruit blotch of cucurbits	Willems *et al.* (1992), Schaad *et al.* (2008)
A. konjaci	Causal agent of bacterial leaf blight of the carnivorous plant *Amorphophallus rivieri* vb. Konjac and blacterial black rot of radish	Willems *et al.* (1992), Myung *et al.* (2010)
A. oryzae	Pathogenic to rice (bacterial stripe disease); recently suggested as a new species distinguished from other *A. avenae* strains	Schaad *et al.* (2008)
A. valerianellae	Causal agent of bacterial spot disease of lamb's lettuce; recently reported to cause bacterial black spot disease in watermelon	Gardan *et al.* (2003), Han *et al.* (2012)
Environmental/non-pathogenic		
A. aerodenitrificans	Denitrifying species	The National Center for Biotechnology Information (NCBI)[3]
A. caeni	Denitrifying species, first reported from the activated sludge of a wastewater treatment plant	Heylen *et al.* (2008)
A. defluvii	First reported from a municipal wastewater treatment plant	Schulze *et al.* (1999)
A. delafieldii	Isolated from soil, water and various samples from clinical environments	Willems *et al.* (1990)
A. ebreus	First isolated from groundwater	Byrne-Bailey *et al.* (2010)
A. facilis	First isolated from lawn soil	Willems *et al.* (1990)
A. radicis	Isolated from wheat roots; potential plant growth-promoting rhizobacterium	Li *et al.* (2011a, 2012)
A. soli	First reported from a landfill soil	Choi *et al.* (2010)
A. temperans	First isolated from an activated sludge reactor treating municipal wastewater	Heijstra *et al.* (2009)
A. wohlfahrtii	Detected in the bryophyte *Tortula ruralis* and likely present in nephridia and egg capsules of some earthworms	Schramm *et al.* (2003), Opelt and Berg (2004), Davidson *et al.* (2010)

[1]This table does not include many *Acidovorax* sp. strains that have been characterized but were not classified into species.
[2]Recent references referring to the most current taxonomy are provided.
[3]This species was not validly published to date. The reference provided at the NCBI is Kim, K.K., Baek, S., and Lee, S. (Unpublished). *Acidovorax aerodenitrificans* sp. nov., novel aerobic denitrifying bacterium.

wastewater (Heijstra *et al.*, 2009). *A. delafieldii* strains were isolated from municipal activated sludge and were shown to possess broad degradation activity towards nitrophenol as well as nitrobenzene (Zhao and Ward, 1999). *Acidovorax* strains isolated from soil and water are able to degrade some commercial polyesters. For instance, *Acidovorax facilis* and *Acidovorax delafieldii* degrade poly-3-hydroxybutyrate (PHB) and poly-3-hydroxybutyrate-co-3-hydroxyvalerate (PHBV) in vivo (Mergaert and Swings, 1996).

Acidovorax species also inhabit the soil and the rhizosphere (Shivaji *et al.*, 2004; Tsai *et al.*, 2009). One strain, *Acidovorax radicis* N35, was recently isolated from surface-sterilized roots of wheat and was shown to possess plant-growth promotion potential (Li *et al.*, 2011a, 2012). Some phytopathogenic *Acidovorax* species, like *A. avenae* and *A. cattleyae*, were shown to possess antibacterial and antifungal activities (Hu and Young, 1998). While commercial use of pathogenic strains for biological control is not likely, Johnson *et al.* (2011) showed that a non-pathogenic mutant of *A. citrulli* retained seed colonization ability and significantly reduced seed-to-seedling transmission of watermelon, caused by the pathogenic strain.

Acidovorax species were also reported to form stable, host-specific symbiosis with various species of earthworms, where they were found to be among the dominant bacterial genera in earthworm nephridia. It was suggested that these bacteria play an important role in protein degradation during earthworm nitrogen excretion (Schramm *et al.*, 2003). *Acidovorax* strains were also detected in earthworm gut (Thakuria *et al.*, 2010) as well as in guts of insects (Delalibera *et al.*, 2005; Broderick and Lemaitre, 2012) and mammals (Benson *et al.*, 2010; Li *et al.*, 2011b). Strains of *Acidovorax* have also been isolated from various clinical sources (Willems *et al.*, 1990; Xu *et al.*, 2004; Sanapareddy *et al.*, 2012). Although their role as human or animal pathogens has not been established, at least two reports have shown *Acidovorax* strains as causal agent of human sepsis (Shetty *et al.*, 2005; Malkan *et al.*, 2009). In this chapter we focus on plant-pathogenic *Acidovorax* species including several aspects of plant diseases caused by these pathogens such as taxonomy, epidemiology, pathogen survival, disease control and basic aspects of pathogenicity and virulence.

Taxonomy of plant-pathogenic *Acidovorax* species

As seen in Table 4.1, the *Acidovorax* genus comprises 17 recognized species, of which seven are able to infect plants. The *Acidovorax* genus was proposed in 1990, and it initially comprised non-phythopathogenic species (Willems *et al.*, 1990). Most of the nowadays recognized *Acidovorax* plant-pathogenic species belonged for many years to the genus *Pseudomonas*. For instance, the cucurbit pathogenic bacterium (currently *A. citrulli*) was formerly classified as *Pseudomonas pseudoalcaligenes* subsp. *citrulli* (Schaad *et al.*, 1978), the konjac pathogen (today *A. konjaci*) was classified as *P. pseudoalcaligenes* subsp. *konjaci* (Goto, 1983), and strains infecting members of the Poaceae family were classified as *P. avenae* or *P. alboprecipitans* (Schaad *et al.*, 1975).

In 1992, several plant-pathogenic species were transferred for the first time to the *Acidovorax* genus (Willems *et al.*, 1992). This was based on the high level of phenotypic and rRNA cistron similarities to members of the already existing genus.Three subspecies were proposed within the *Acidovorax avenae* genus: the Poaceae pathogen *A. avenae* subsp. *avenae*, the *Cucurbitaceae* pathogen *A. avenae* subsp. *citrulli*, and the orchid pathogen, *A.*

avenae subsp. *cattleyae*. In the same study, the konjac pathogen was reclassified as *A. konjaci* (Willems *et al.*, 1992).

In 2008, DNA–DNA reassociation assays performed with various *A. avenae* strains from subspecies *avenae*, *citrulli* and *cattleyae* revealed the existence of four distinguished taxons. All four taxons could also be distinguished each from the other by sequence analyses of the 16S rDNA and the 16S-23S rDNA internally transcribed spacer, as well as by AFLP and phenotypic analyses. On the basis of these results, these subspecies were elevated to the species rank (namely *A. avenae*, *A. citrulli* and *A. cattleyae*), and a fourth species, *A. oryzae*, was proposed to distinguish between strains that infect rice from *A. avenae* infecting other Poaceae members (Schaad *et al.*, 2008).

More recently, new plant-pathogenic *Acidovorax* species have been described. This is the case of *A. anthurii* causing bacterial leaf spot of anthurium (Gardan *et al.*, 2000), and *A. valerianellae*, a pathogen of lamb's lettuce (Gardan *et al.*, 2003), that was recently proposed to cause bacterial black spot disease in watermelon (Han *et al.*, 2012).

Plant diseases caused by *Acidovorax* species

Plant diseases caused by *Acidovorax* species are described in Table 4.1. Below we summarize the most important diseases caused by these pathogens, with emphasis on their distribution, epidemiology, symptoms and control strategies.

Bacterial fruit blotch of cucurbits

Bacterial fruit blotch (BFB) is an economically important disease of cucurbits, caused by *A. citrulli*. Under high relative humidity and temperature, the bacterium causes seedling blight and fruit rot of a range of cucurbitaceous plants including watermelon, melon, squash, cucumber and pumpkin (Walcott *et al.*, 2004; Burdman and Walcott, 2012). This pathogen was first reported in 1965, when it was recovered from necrotic watermelon cotyledons at the US Department of Agriculture Plant Introduction Station in Georgia, USA. (Webb and Goth, 1965). At that time, the disease was restricted to seedlings and no symptoms were observed on fruits. The highly destructive potential of the disease was not realized until the late 1980s' when severe BFB outbreaks occurred in commercial watermelon fields in the Mariana Islands and in Florida (Wall and Santos, 1988; Somodi *et al.*, 1991). Later on, numerous BFB outbreaks occurred rapidly throughout south-eastern and mid-western USA between 1989 and 1994, leading to high yield losses (Latin and Rane, 1990; Black *et al.*, 1994; Evans and Mulrooney, 1991; Burdman and Walcott, 2012). After that, the rapid spread of BFB progressed in two directions: broadening of the host range to other cucurbits, and global expansion. To date, BFB outbreaks have been reported, mainly in watermelon and melon, in the Americas, Asia, Europe, the Middle and Far East, and Australia (Bahar and Burdman, 2010; Burdman and Walcott, 2012).

A. citrulli strains can be divided into at least two well-differentiated groups: group I includes strains that were mainly isolated from melon and other non-watermelon cucurbit plants, while group II includes the strains that cause typical BFB in watermelon. These groups can be distinguished by carbon substrate utilization, fatty acid methyl ester analysis, DNA fingerprinting and virulence on different hosts. Group I strains are moderately aggressive on a wide range of cucurbits, while group II strains are highly aggressive in watermelon but only mildly so on melon (Walcott *et al.*, 2000, 2004; Burdman *et al.*, 2005)

BFB affects the plant at all stages of its development; however, emerging seedlings and fruits are highly susceptible to the bacterium. Disease symptoms may appear on seedlings after 5–8 days after planting, depending on environmental conditions. For most cucurbits, initial seedling symptoms include water-soaking on the undersides of cotyledons (Burdman and Walcott, 2012). These lesions can extend along the veins of the cotyledons and later become necrotic, in severe cases they can cause seedling to collapse and die. However, sometimes, the symptoms may be very mild, or even invisible to the naked eye, and therefore escape detection. Leaf symptoms include extensive chlorosis, reddish-brown lesions along leaf veins and v-shaped lesions that extend from the margin to the base of the leaf; these may have little to no direct effect on the mature plant. The major economic losses result from the fruit rot phase of the disease. On watermelon and melon fruits, BFB symptoms appear as dark small, irregularly shaped, firm spots (Fig. 4.1A and B). As lesions on fruit age, the fruit pericarp might crack and bacterial ooze is released. Similar symptoms occur on pumpkin but not on infected cucumber (Langston *et al.*, 1999; Martin and O'Brien, 1999; Bahar and Burdman, 2010; Burdman and Walcott, 2012).

The primary inoculum for BFB epidemics comes from contaminated seed, where the pathogen remains on the seed coat and embryos (Rane and Latin, 1992; Dutta *et al.*, 2012). In many cases, before transfer to the field, seeds are sown in transplant facilities, which are usually characterized by mild temperatures, high humidity, and use of overhead irrigation. These conditions are ideal for spread, proliferation and establishment of *A. citrulli* in young seedlings (Latin and Hopkins, 1995). As plants grow in the field, the pathogen spreads to new leaves and neighbouring plants. Adult plants are relatively tolerant to BFB, but lesions

Figure 4.1 Symptoms of several diseases caused by *Acidovorax* species. Bacterial fruit blotch symptom in watermelon (A) and melon (B), caused by *A. citrulli*; bacterial brown stripe of maize, caused by *A. avenae* (C); leaf spot disease of *Cattleya*, caused by *A. cattleyae* (D); leaf symptoms caused by *Acidovorax* strains in the ornamental plants hydrangea (E), geranium (F) and lantana (G). (A) and (B), reproduced with permission from Burdman and Walcott (2012); (C), kindly provided by Takao Tsukiboshi (National Institute of Livestock and Grassland Science, Japan); (D), kindly provided by Terry Bottom (St Augustine Orchid Society, FL, USA); (E–G), kindly provided by Michael Munster (North Carolina State University, NC, USA). A colour version of this figure is located in the plate section at the back of the book.

on foliage provide a source of inoculum for infection of immature fruit. Before fruit ripen, the pathogen penetrates the stomata of the fruit producing the characteristic blotch symptoms. While it is clear that infested seed are important for BFB outbreaks, other possible sources of inoculum exist, including volunteer seedlings, cucurbit weeds and decaying infested debris (Langston *et al.*, 1999; Burdman and Walcott, 2012).

No resistant cultivars to BFB are available to date. In addition, chemical or physical seed treatments as well as field application of copper-based bactericides cannot completely eradicate the pathogen (Sowell and Schaad, 1979; Hopkins *et al.*, 1996; O'Brien and Martin, 1999; Burdman and Walcott, 2012). Therefore, the most important management strategy for BFB is pathogen exclusion by using pathogen-free seeds and seedlings. Currently, the standard seed test for BFB is the seedling grow out (SGO) assay in which about 30,000 seeds are planted and evaluated under favourable conditions for BFB development. Seedlings are visually inspected for BFB symptoms which will result in rejection of the entire lot if as little as one seedling is confirmed to be infected (Burdman and Walcott, 2012). The SGO assay suffers from several limitations; it is expensive, time- and space-consuming and requires the destructive use of a large numbers of seeds (Walcott, 2003; Bahar and Burdman, 2010). Therefore, various molecular approaches to detect the pathogen in infected seed have been developed primarily based upon the polymerase chain reaction (PCR) in order to enhance sensitivity while minimizing duration and costs (Walcott, 2003; Bahar *et al.*, 2008; Park *et al.*, 2008). However, the SGO assay is still the most common test used to assess *A. citrulli*-infested seeds.

Acidovorax diseases of cereal plants

The Poaceae family (also known as Gramineae or grass) is considered as the most important plant family for human economies. The first report for a bacterial disease in this family caused by a member of the genus *Acidovorax* was in the US in 1909, when Manns reported bacterial leaf blight (BLB) of oats (Willems *et al.*, 1992). Since then, the pathogen, *A. avenae* (formerly known as *A. avenae* subsp. *avenae*) has been reported to cause disease in other hosts of the Poaceae family and in many other parts of the world (Willems *et al.*, 1992; Schaad *et al.*, 2008). Besides BLB of oats, *A. avenae* has been reported to cause BLB, red stripe and brown stripe diseases in a wide variety of plants including sugarcane, millet, maize, sorghum, rye, foxtail bromegrass, wheatgrass, creeping bentgrass, and Fishtail palm (Rosen, 1922; Sumner and Schaad, 1977; Shakya, 1985; Saddler, 1994; Schaad *et al.*, 2008; Furuya *et al.*, 2009) (Table 4.1). *A. avenae* has also been reported to cause bacterial stalk rot of maize (Johnson *et al.*, 1949; Sumner and Schaad, 1977), although this disease has been mainly attributed to *Erwinia carotovora* pv. *zeae*. *A. avenae* was also reported to infect rice plants (Cottyn *et al.*, 1996; Li *et al.*, 2011c); however, as mentioned above, Schaad *et al.* (2008) suggested that the rice strains belong to a different species, with the name *A. oryzae* being proposed.

There are many similarities regarding environmental conditions for outbreaks, symptoms and control means among diseases caused by *A. avenae* (Saddler, 1994; Schaad *et al.*, 2008; Li *et al.*, 2011c). Generally, the initial symptoms are characterized by water-soaked, longitudinal stripes (Fig. 4.1C) or streaks that are formed in the leaves and may extend into the sheaths. Later on the lesion colour turns darker (Knauss *et al.*, 1978; Willems *et al.*, 1992; Saddler, 1994). The bacterium is seedborne and can survive on weeds near agricultural fields for long periods of time, which makes it difficult to eradicate (Gitaitis *et al.*, 1978). There is some evidence of resistant varieties for some plant species, and according to a few reports, use of antibiotics and bacteriocides have the potential to control these disease to some extent (Ullstrup, 1961; Saddler, 1994; Pataky *et al.*, 1997; Zia-ul-Hussnain *et al.*, 2011). In terms of pathogen detection in contaminated seeds, BIO-PCR is considered as the most precise method, although the use of selective media and serological tests are also common (Song *et al.*, 2000, 2004).

Bacterial brown spot of orchids

The orchid family (Orchidaceae) is a diverse and widespread family that comprises species that are amongst the most popular blooming potted plants worldwide (Hsu *et al.*, 2011). Bacterial brown spot (BBS) disease, caused by *A. cattleyae*, is considered as one of the major threats to the orchid industry. Although the majority of BBS reports have been associated with the *Phalaenopsis* genus, other genera from this family, including *Cattleya*, *Cypripedium*, *Dendrobium*, *Oncidium* and *Vanda*, can also be affected by the pathogen (Miller, 1990).

BBS was first reported about a century ago in Italy, and since then, the disease has spread to other countries, including US, Australia, Philippines, Portugal, the Netherlands and Taiwan (Saddler, 1994; Stovold *et al.*, 2001; Scortichini *et al.*, 2005). BBS symptoms generally begin as small soft water-soaked spots that later turn black in the leaf. Green-yellow haloes can often be seen around the spots (Fig. 4.1D), and eventually the spots expand rapidly, to cover and kill the whole leaf. BBS has a destructive potential also because the seedlings are extremely susceptible to the pathogen (Miller, 1990; Stovold *et al.*, 2001). The most effective strategy to combat BBS is through prevention, by decreasing seedling density in the greenhouse, minimizing manipulations during cultivation, avoiding overhead irrigation and keeping low temperature and humidity (Janse, 2006). However, when BBS occurs, common means of control include spraying with copper compounds and removal of infected areas with sterile tools (Janse, 2006).

Interestingly, in a study aiming to detect antifungal and antibacterial activity by different bacterial genera and species, *A. cattleyae* was the only *Acidovorax* species to show antibacterial activity against *Listeria innocua*. In addition, all tested *A. cattleyae* strains inhibited growth of the fungus *Rhodotorula mucilaginosa* (Hu and Young, 1998).

Other *Acidovorax* diseases

In this subsection we describe a few other diseases caused by *Acidovorax* species that, to date, are restricted to few a geographical locations.

The konjac plant (*Amorphophalus konjac*) is commonly used in Eastern Asia to produce a jelly product that has a texture similar to gelatin. Over a century ago there was a report about a disease that affects the foliage of konjac. The causal agent of this disease, named bacterial leaf blight (BLB), is *A. konjaci* (formerly *Pseudomonas pseudoalcaligenes* subsp. *konjaci*) (Goto, 1983). BLB symptoms are characterized by the formation of rectangular leaf spots limited by veins. Later on the whole leaf turns brown, and under humid conditions the disease progresses fast leading to wilting and/or drop-down of the leaves (Goto, 1983; Saddler, 1994). Partial control may be achieved using less susceptible varieties of konjac and by covering the ground with rice straw (Saddler, 1994). The disease was only reported in Japan and in North Carolina; however, the same species was shown to cause bacterial black rot of Korean radish (*Raphanus sativus*), as reported by Myung *et al.* (2010). This disease is characterized by brown-black rotting of the roots that lead to a reduction of the agricultural product value (Myung *et al.*, 2010).

A. valerianellae is the causal agent of bacterial black spot (BBS) disease of corn salad (*Valerianella locusta*). The disease was first reported in the early 1990s on corn salad fields in France, with symptoms being characterized by water soaked spots in the leaves that later gain a black appearance (Rat and Gardan, 1993). Later on BBS spread to other countries nearby such as Germany, Belgium and Austria (Grondeau *et al.*, 2001). Contaminated seeds are considered as the primary infection source of the disease, therefore most research around BBS has been focused on detection of the pathogen in corn salad seeds (Grondeau *et al.*, 2007; Grondeau and Samson, 2009; Thiele *et al.*, 2012). Interestingly, Han and colleagues recently reported BBS disease caused by this pathogen on watermelon plants in Korea (Han *et al.*, 2012).

The Lesser Antilles are a long island arc in the Caribbean Sea. In three of these islands – Martinique, Guadeloupe and Trinidad – a disease has been reported that affects ornamental plants from the *Anthurium* genus. The disease, bacterial leaf spot, is caused by *A. anthurii* (Gardan *et al.*, 2003). It was first described in 1989, and produces both localized and systemic symptoms (Prior and Rott, 1989). Local symptoms consist of necrotic lesions near the leaf margins and veins, that later enlarge, blacken and turn grey. From these lesions, the bacteria become systemic leading to general chlorosis and plant death (Gardan *et al.*, 2000). So far, the major management means to cope with the disease are utilization of pathogen-free planting material and sprinkle irrigation in order to minimize initial inoculum and pathogen dissemination, respectively.

Acidovorax strains have been also shown or suggested to cause leaf diseases in several other ornamental crops like geranium and petunia (M. Munster, personal communication); however, very little is known about the identity of the causal agents of these diseases as well as their distribution. Some examples of disease symptoms caused by *Acidovorax* spp. on ornamental plants are shown in Fig. 4.1E–G.

Pathogenicity and virulence factors of *Acidovorax* sp.

In contrast with the wide amount of knowledge accumulated on basic aspects of bacterial plant diseases caused by *Pseudomonas* spp., *Xanthomonas* spp., *Ralstonia solanacearum* and enterobacterial species, relatively little knowledge exist about virulence determinants of *Acidovorax* and the interaction between *Acidovorax* species and their plant hosts. As mentioned, most *Acidovorax* plant diseases started to gain importance in recent years and still, there are relatively few research groups investigating basic aspects of these diseases. Nevertheless, the recent sequencing and releases of the genomes of plant-pathogenic *A. citrulli* AAC00–1 (http://genome.jgi-psf.org/aciav/aciav.home.html) and *A. avenae* RS-1 (Xie *et al.*, 2011), as well as of several non-pathogenic *Acidovorax* species, provide a great contribution to the investigation of basic aspects of disease caused by these bacteria. Below we summarize the current knowledge about virulence- and pathogenicity-associated determinants of *Acidovorax* spp. Most information is about *A. citrulli*, which among plant-pathogenic *Acidovorax* has been the most studied.

Type III secretion

Many Gram-negative phytopathogenic bacteria utilize a type III secretion (T3S) system to secrete protein effectors directly into the host cell (Alfano and Collmer, 2004; Mansfield, 2009). Most T3S effectors characterized so far have been shown to contribute to virulence through modulation of host cellular processes and suppression of host defence responses (Mudgett, 2005; Grant *et al.*, 2006; Mansfield, 2009). Some T3S effectors were first identified as products of avirulence (*avr*) genes, which are recognized in resistant plants by the action of corresponding resistance (*R*) gene products. This recognition event results in effector-triggered immunity (ETI), which is generally associated with induction of a hypersensitive response (HR), a localized cell-death of the infected tissue that ultimately arrests growth of biotrophic pathogens (Jones and Dangl, 2006; Mansfield, 2009).

In many Gram-negative plant-pathogenic bacteria, the T3S system is required for pathogenicity on susceptible plants and HR induction in resistant plants. Therefore, the cluster of genes encoding the components of this secretion apparatus is named *hrp* (for hypersensitive response and pathogenicity). The genome sequence of the *A. citrulli* group II strain AAC00-1 as well as of *A. avenae* RS-1 (Xie *et al.*, 2011) revealed the existence of a Hrp-T3S system in these pathogens. Generation and characterization of *hrp* mutants of both group I and II strains of *A. citrulli* demonstrated that a functional Hrp-T3S system is required for both pathogenicity on cucurbit hosts and HR-inducing ability in leaves of non-host tobacco and tomato (Bahar and Burdman, 2010; Johnson *et al.*, 2011).

Annotation of the *A. citrulli* AAC00-1 genome revealed at least 11 putative T3S effectors showing homology to effectors from *Xanthomonas* species, *P. syringae* and *R. solanacearum*. Comparative analyses of effector genes from representative group I and II strains have revealed that strains from these groups significantly differ from each other in the sequence of most effectors. In addition, few effectors that are present in strain AAC00-1 and all assessed group II strains seem to be absent or non-functional in group I strains (Eckshtain-Levi *et al.*, in press). It is hypothesized that differences in the arsenal of T3S effectors are responsible, at least partially, for the observed differences in host preference among isolates from these groups.

Recently, Kondo *et al.* (2012) reported the genetic organization of the *hrp* gene cluster of *A. avenae* N1141. This strain is avirulent on rice plants, and the authors showed that one of the T3S effectors, HrpY, is responsible for ETI and HR elicitation by this strain in rice (Kondo *et al.*, 2012).

Type II secretion

Type II secretion (T2S) plays a role in pathogenicity of many Gram-negative bacteria by mediating translocation of several pathogenesis-related proteins from the bacterial cytoplasm to the extracellular environment (Douzi *et al.*, 2012). Secreted proteins include toxins and hydrolytic enzymes such as cellulases, pectatelyases, lipases and proteases, which promote virulence in various phytopathogenic bacterial species (Liu *et al.*, 2005; Hu *et al.*, 2007; Szczesny *et al.*, 2010).

T2S systems are built of 12 to 16 proteins, which form a secreton that spans the bacterial envelope. The secreton subunits are encoded by so-called general secretion pathway (*gsp*) genes that are often organized in large operons (Douzi *et al.*, 2012). *A. citrulli* AAC00-1 possesses two sets of *gsp* clusters. In this bacterium, deletion of both copies of *gspG* (*gspG1/G2*), encoding major pseudopilins, resulted in the loss of the ability to secrete endoglucanase (Johnson, 2010). This mutant also had a significantly reduced ability to colonize watermelon seedling cotyledons relative to the wild type, supporting that T2S enzymes may be virulence factors in *A. citrulli*.

The AAC00-1 *gspG1/G2* deletion mutant also displayed significant reductions in early watermelon seed colonization and seed-to-seedling BFB transmission (Johnson, 2010), suggesting that T2S proteins contribute to the bacterium's ability to colonize germinating watermelon seeds, prior to the infection of seedling tissue. Genomic analysis of AAC00-1 revealed the presence of endoglucanase, xylanase and pectatelyase. Nevertheless, assays with individual xylanase, pectatelyase and endoglucanse deletion mutants of AAC00-1, indicate that only endoglucanase contributes minimally to watermelon seed colonization (Johnson, 2010).

Type IV pili and polar flagella

Pathogenicity screens of a transposon mutant library generated in the background of the *A. citrulli* group I strain M6, revealed that a *pilM* mutant is significantly impaired in virulence on melon seed transmission assays relative to the wild type (Bahar *et al.*, 2009). *pilM* encodes a protein required for assembly of type IV pili (T4P). T4P are hair-like appendages found on the surface of a wide range of bacteria. They constitute an efficient device for a particular type of flagellum-independent surface motility, named twitching motility, and are involved in several other bacterial activities including adherence, colonization, biofilm formation, uptake of genetic material and virulence (Craig *et al.*, 2004; Nudleman and Kaiser, 2004). Despite the multiple functionality of T4P and their well-established role in pathogenicity of animal pathogenic bacteria, relatively little attention has been given to the role of T4P in plant-pathogenic bacteria (Burdman *et al.*, 2011). Further analysis of the *A. citrulli* M6 *pilM* mutant revealed that T4P is required for twitching motility and wild-type levels of biofilm formation in this pathogen. These findings were confirmed with the analysis of a marker exchange mutant of strain M6 impaired in *pilT*, a gene that is required for twitching motility (Bahar *et al.*, 2009, 2010). T4P were also shown to play an important role in surface attachment of the activated sludge strain *A. temperans* CB2 (Heijstra *et al.*, 2009). Recently, characterization of a *pilP* mutant of *A. avenae* RS-1 revealed that T4P is also important for biofilm formation and virulence of this pathogen (H. Liu *et al.*, 2012).

The contribution of T4P to colonization and virulence was shown to be especially important for vascular plant-pathogenic bacteria (i.e. those possessing the ability to colonize and spread via the plant xylem vessels), such as *R. solanacearum* (Liu *et al.*, 2001; Kang *et al.*, 2002) and *Xylella fastidiosa* (Meng *et al.*, 2005). It has been proposed that in these bacteria T4P might contribute to optimal establishment, colonization and spread via the host xylem vessels. In agreement with the above studies, it was demonstrated that *A. citrulli* possesses vascular colonization ability in melon seedlings, and that T4P is important for this process (Bahar *et al.*, 2009). Recently, Makizumi *et al.* (2011) showed that *A. citrulli* is able to colonize xylem vessels of squash leaves. Whether systemic colonization and spread play a role in fruit and seed infection by *A. citrulli* is still to be elucidated.

In the aforementioned pathogenicity screens with M6 random mutants, a mutant strain impaired in polar flagellum synthesis was also found to be impaired in virulence. Further characterization of a marker exchange mutant impaired in *fliC* (encoding flagellin) confirmed this finding (Bahar *et al.*, 2011). Polar flagellum was also shown to be important for wild-type levels of vascular colonization of melon seedlings by *A. citrulli*; however, in contrast to T4P, polar flagella were not shown to play an important role in biofilm formation, under tested conditions (Bahar *et al.*, 2010, 2011). Foliage inoculation experiments with mature melon plants demonstrated that polar flagellum contributes to virulence of *A. citrulli* at both pre- and post-penetration stages (Bahar *et al.*, 2011). In addition, we have observed that some *A. citrulli* strains have the tendency to lose polar flagella and swimming motility during storage. In agreement with the demonstrated role of polar flagellum in virulence of this bacterium, spontaneous mutants lacking polar flagellum synthesis ability are severely reduced in their virulence relative to their parental strains (Burdman, unpublished results).

In a recent pathogenicity screen of Tn5 random mutants in the background of the *A. citrulli* group I strain FC440, Liu and colleagues detected a mutant impaired in a flagellar hook-associated protein, with reduced virulence on cucumber cotyledons (J. Liu *et al.*, 2012). Interestingly, a phenotypic variant of the plant-growth promoting bacterium *Acidovorax radicis* N35, which was found to lose motility as well as the ability to synthesize polar flagella, was unable to promote growth of soil-grown barley plants, in contrast with the parental strain (J. Li *et al.*, 2012).

Flagellin of many bacterial species are known to act as pathogen-associated molecular patterns (PAMPs). They are recognized by plant pattern recognition receptors (PRRs), leading to induction of a moderate defence response called PAMP-triggered immunity (PTI) (Jones and Dangl, 2006; Boller and He, 2009). Flagellin was also shown to induce HR in several incompatible interactions (Taguchi *et al.*, 2003; Marutani *et al.*, 2005). Similarly, it has been shown that flagellin from *A. avenae* mediates HR, rapid generation of H_2O_2 and expression of defence genes in incompatible interactions with rice (Che *et al.*, 2000; Tanaka *et al.*, 2003, 2008). Moreover, transgenic rice plants expressing the *A. avenae* flagellin showed enhanced immune responses and enhanced resistance to the rice blast pathogen *Magnaporthe grisea* (Takakura *et al.*, 2008).

Genomic information as well as experimental evidence indicate that other plant-pathogenic species of *A. citrulli* produce both flagella and T4P; however, to the best of our knowledge, the role of these extracellular appendages in their pathogenicity has not been yet elucidated.

Quorum sensing

Bacteria are able to communicate with each other in a cell density-dependent form to coordinate the expression of several genes and behaviours (Bassler, 2002). This phenomenon, called quorum sensing (QS), is achieved through response to stimulatory concentrations of extracellular autoinducers. Typical autoinducers in Gram-negative bacteria are acylated homoserine lactones (AHLs) (Bassler, 2002; Ng and Bassler, 2009). The role of QS in virulence of plant-pathogenic bacteria has been extensively investigated in several species. Traits regulated by QS in plant-pathogenic bacteria include production of extracellular polysaccharides, degradative enzymes, siderophores and antibiotics, motility, biofilm formation,T3S and epiphytic fitness (Fray, 2002; von Bodman *et al.*, 2003). Development of approaches that aim at interfering with QS signalling may be helpful for control of phytobacterial diseases.

A. citrulli contains genes with high homology to *luxI* and *luxR*, encoding the AHL synthase and the AHL-dependent transcriptional protein, respectively. An *A. citrulli* mutant impaired in *aacI* (the *luxI* homologue) was shown to possess reduced virulence in pathogenicity tests on watermelon fruits relative to the wild type, thus supporting that QS is important for *A. citrulli* pathogenicity (Chen *et al.*, 2009). It was also reported that the autoinducer molecule of *A. citrulli* is N-3-oxo-octanoyl-L-homoserine lactone (3-oxo-C8-HSL) (Chen *et al.*, 2009), a finding that was recently confirmed by Fan *et al.* (2011), that characterized an *aacI* mutant in the background of a different strain. In this recent study, the mutant strain was shown to be unable of producing AHL molecules, had significantly reduced virulence in melon fruits and melon seedlings, and was significantly impaired in swimming motility. In contrast, under tested conditions, the mutant did not differ from the wild type in production of extracellular polysaccharides, biofilm formation and HR induction in tobacco (Fan *et al.*, 2011).

Concluding remarks

The *Acidovorax* genus comprises *several* species that are able to infect and cause disease in a wide range of plant families. Importantly, some diseases caused by *Acidovorax* strains have been spreading worldwide in recent years and their importance is substantially increasing. Despite the economic importance of some of the diseases caused by *Acidovorax* species, still very little is known about basic aspects of *Acidovorax*–plant interactions. A combination of basic and applied investigation is highly demmanded to gain insights into pathogenicity mechanisms and fitness traits of plant-pathogenic *Acidovorax* strains, and to develop more efficient tools to cope with the diseases they cause. A common fact for most *Acidovorax* diseases is the lack of knwoledge/availability of disease resistance sources. Development of resistant cultivars is potentially one of the most important strategies for management of diseases caused by these pathogens. In this regard, advances in our understanding of the role played by T3S effectors of *Acidovorax* strains may help identifying suitable resistance genes, which could facilitate the development of reliable resistance in commercial cultivars.

References

Alfano, J.R., and Collmer, A. (2004). Type III secretion system effector proteins: double agents in bacterial disease and plant defense. Annu. Rev. Phytopathol. *42*, 385–414.

Bahar, O., and Burdman, S. (2010). Bacterial fruit blotch: a threat to the cucurbit industry. Isr. J. Plant Sci. *58*, 19–31.

Bahar, O., Efrat, M., Hadar, E., Dutta, B., Walcott, R.R., and Burdman, S. (2008). New subspecies-specific polymerase chain reaction-based assay for the detection of *Acidovorax avenae* subsp *citrulli*. Plant Pathol. *57*, 754–763.

Bahar, O., Goffer, T., and Burdman, S. (2009). Type IV pili are required for virulence, twitching motility, and biofilm formation of *Acidovorax avenae* subsp *citrulli*. Mol. Plant Microbe Interact. *22*, 909–920.

Bahar, O., De La Fuente, L., and Burdman, S. (2010). Assessing adhesion, biofilm formation and motility of *Acidovorax citrulli* using microfluidic flow chambers. FEMS Microbiol. Lett. *312*, 33–39.

Bahar, O., Levi, N., and Burdman, S. (2011). The cucurbit pathogenic bacterium *Acidovorax citrulli* requires a polar flagellum for full virulence before and after host-tissue penetration. Mol. Plant Microbe Interact. *24*, 1040–1050.

Bassler, B.L. (2002). Small talk. Cell-to-cell communication in bacteria. Cell *109*, 421–424.

Benson, A.K., Kelly, S.A., Legge, R., Ma, F., Low, S.J., Kim, J., Zhang, M., Oh, P.L., Nehrenberg, D., Hua, K., *et al.* (2010). Individuality in gut microbiota composition is a complex polygenic trait shaped by multiple environmental and host genetic factors. Proc. Natl. Acad. Sci. U.S.A. *107*, 18933–18938.

Black, M.C., Isakeit, T., and Barnes, L.W. (1994). First report of bacterial fruit blotch of watermelon in Texas. Plant Dis. *78*, 831.

von Bodman, S.B., Bauer, W.D., and Coplin, D.L. (2003). Quorum sensing in plant-pathogenic bacteria. Annu. Rev. Phytopathol. *41*, 455–482.

Boller, T., and He, S.Y. (2009). Innate immunity in plants: an arms race between pattern recognition receptors in plants and effectors in microbial pathogens. Science *324*, 742–744.

Broderick, N.A., and Lemaitre, B. (2012). Gut-associated microbes of *Drosophila melanogaster*. Gut Microbes 3, 1–16.

Burdman, S., and Walcott, R. (2012). *Acidovorax citrulli*: generating basic and applied knowledge to tackle a global threat to the cucurbit industry. Mol. Plant Pathol. *13*, 805–815.

Burdman, S., Kots, N., Kritzman, G., and Kopelowitz, J. (2005). Molecular, physiological, and host-range characterization of *Acidovorax avenae* subsp. *citrulli* isolates from watermelon and melon in Israel. Plant Dis. *89*, 1339–1347.

Burdman, S., Bahar, O., Parker, J.K., and De La Fuente, L. (2011). Involvement of type IV pili in pathogenicity of plant-pathogenic bacteria. Genes 2, 706–735.

Byrne-Bailey, K.G., Weber, K.A., Chair, A.H., Bose, S., Knox, T., Spanbauer, T.L., Chertkov, O., and Coates, J.D. (2010). Completed genome sequence of the anaerobic iron-oxidizing bacterium *Acidovorax ebreus* strain TPSY. J. Bacteriol. *192*, 1475–1476.

Che, F.S., Nakajima, Y., Tanaka, N., Iwano, M., Yoshida, T., Takayama, S., Kadota, I., and Isogai, A. (2000). Flagellin from an incompatible strain of *Pseudomonas avenae* induces a resistance response in cultured rice cells. J. Biol. Chem. *275*, 32347–32356.

Chen, T., Qian, G.L., Yang, X.L., Ma, Y.J., Hu, B.S., and Liu, F.Q. (2009). Detection of a quorum sensing signal molecule of *Acidovorax avenae* subsp. *citrulli* and its regulation of pathogenicity. Chin. J. Agric. Biotechnol. *6*, 49–53.

Choi, J.H., Kim, M.S., Roh, S.W., and Bae, J.W. (2010). *Acidovorax soli* sp. nov., isolated from landfill soil. Int. J. Syst. Evol. Microbiol. *60*, 2715–2718.

Cottyn, B., Cerez, M.T., Outryve, M.F.V., Barroga, J., Swings, J., and Mew, T.W. (1996). Bacterial diseases of rice. I. Pathogenic bacteria associated with sheath rot complex and grain discoloration of rice in the Philippines. Plant Dis. *80*, 429–436.

Craig, L., Pique, M.E., and Tainer, J.A. (2004). Type IV pilus structure and bacterial pathogenicity. Nature *2*, 363–378.

Davidson, S.K., Powell, R.J., and Stahl, D.A. (2010). Transmission of a bacterial consortium in *Eisenia fetida* egg capsules. Environ. Microbiol. *12*, 2277–2288.

Delalibera, I., Handelsman, J., and Raffa, K.F. (2005). Contrasts in cellulolytic activities of gut microorganisms between the wood borer, *Saperda vestita* (Coleoptera: Cerambycidae), and the bark beetles, *Ips pini* and *Dendroctonus frontalis* (Coleoptera: Curculionidae). Environ. Entomol. *34*, 541–547.

Douzi, B., Filloux, A., and Voulhoux, R. (2012). On the path to uncover the bacterial type II secretion system. Philos. Trans. R. Soc. Lond. B Biol. Sci. *367*, 1059–1072.

Dutta, B., Avci, U., Hahn, M.G., and Walcott, R.R. (2012). Location of *Acidovorax citrulli* in infested watermelon seeds is influenced by the pathway of bacterial invasion. Phytopathology *102*, 461–468.

Eckstain-Levi, N., Munitz, T., Živanović, M., Traore, S.M., Spröer, C., Zhao, B., Welbaum, G., Walcott, R., Sikorski, J., and Burdman, S. (2014). Comparative analysis of type III secreted effector genes reflects divergence of *Acidovorax citrulli* strains into three distinct lineages. Phytopathology *in press* (http://dx.doi.org/10.1094/PHYTO-12-13-0350-R).

Evans, T.A., and Mulrooney, R.P. (1991). First report of watermelon fruit blotch in Delaware. Plant Dis. *75*, 1074.

Fan, J., Qian, G., Chen, T., Zhao, Y., Liu, F., Walcott, R.R., and Hu, B. (2011). The acyl-homoserine lactone (AHL)-type quorum sensing system affects growth rate, swimming motility and virulence in *Acidovorax avenae* subsp. *citrulli*. World J. Microbiol. Biotechnol. *27*, 1155–1166.

Fray, R.G. (2002). Altering plant–microbe interaction through artificially manipulating bacterial quorum sensing. Ann. Bot. *89*, 245–253.

Furuya, N., Ito, T., and Tsuchiya, K. (2009). Occurrence of bacterial brown stripe of creeping bentgrass on golf course green in Kyushu. J. Fac. Agr. Kyushu Univ. *54*, 13–17.

Gardan, L., Dauga, C., Prior, P., Gillis, M., and Saddler, G.S. (2000). *Acidovorax anthurii* sp. nov., a new phytopathogenic bacterium which causes bacterial leaf-spot of anthurium. Int. J. Syst. Evol. Microbiol. *50*, 235–246.

Gardan, L., Stead, D.E., Dauga, C., and Gillis, M. (2003). *Acidovorax valerianellae* sp. nov., a novel pathogen of lamb's lettuce *Valerianella locusta* (L.) Laterr. Int. J. Syst. Evol. Microbiol. 53, 795–800.

Gitaitis, R.D., Stall, R.E., and Strandberg, J.O. (1978). Dissemination and survival of *Pseudomonas alboprecipitans* ascertained by disease distribution. Phytopathology *68*, 227–231.

Goto, M. (1983). *Pseudomonas pseudoalcaligenes* subsp. *konjaci* subsp. nov., the causal agent of bacterial leaf blight of konjac (*Amorphophalus konjac* Koch). Int. J. Syst. Bacteriol. *33*, 539–545.

Grant, S.R., Fisher, E.J., Chang, J.H., Mole, B.M., and Dangl, J.L. (2006). Subterfuge and manipulation: type III effector proteins of phytopathogenic bacteria. Annu. Rev. Microbiol. *60*, 425–449.

Grondeau, C., and Samson, R. (2009). Detection of *Acidovorax valerianellae* in corn-salad seeds, seed transmission of the pathogen and disease development in the field. Plant Pathol. *58*, 846–852.

Grondeau, C., Samson, R., and Gardan, L. (2001). Description and identification of the new phytopathogenic bacterium causing bacterial black spot of corn-salad (*Valerianella locusta*), as *Acidovorax valerianellae*. (2000). In Proceedings of the 10th International Conference on Plant-pathogenic bacteria, D.H. DeBoer, ed. (Kluwer Academic Publisher, Dordrecht, The Netherlands), pp. 165–166.

Grondeau, C., Manceau, C., and Samson, R. (2007). A semiselective medium for the isolation of *Acidovorax valerianellae* from soil and plant debris. Plant Pathol. *56*, 302–310.

Haigler, B.E., Wallace, W.H., and Spain, J.C. (1994). Biodegradation of 2-nitrotoluene by *Pseudomonas* sp. strain JS42. Appl. Environ. Microbiol. *60*, 3466–3469.

Han, Y.K., Han, K.S., Lee, S.C., Kim, S., and Lee, J. (2012). First report of bacterial black spot disease in watermelon caused by *Acidovorax valerianellae* in Korea. Plant Dis. *96*, 759–760.

Heijstra, B.D., Pichler, F.B., Liang, Q., Blaza, R.G., and Turner, S.J. (2009). Extracellular DNA and type IV pili mediate surface attachment by *Acidovorax temperans*. Anton. Leeuw. Int. J. Gen. Mol. Microbiol. *95*, 343–349.

Heylen, K., Lebbe, L., and De Vos, P. (2008). *Acidovorax caeni* sp. nov., a denitrifying species with genetically diverse isolates from activated sludge. Int. J. Syst. Evol. Microbiol. *58*, 73–77.

Hopkins, D.L., Cucuzza, J.D., and Watterson, J.C. (1996). Wet seed treatments for the control of bacterial fruit blotch of watermelon. Plant Dis. *80*, 529–532.

Hsu, C.C., Chung, Y.L., Chen, T.C., Lee, Y.L., Kuo, Y.T., Tsai, W.C., Hsiao, Y.Y., Chen, Y.W., Wu, W.L., and Chen, H.H. (2011). An overview of the *Phalaenopsis* orchid genome through BAC end sequence analysis. BMC Plant Biol. *11*, 3.

Hu, F.P., and Young, J.M. (1998a). Biocidal activity in plant pathogenic *Acidovorax, Burkholderia, Herbaspirillum, Ralstonia* and *Xanthomonas* spp. J. Appl. Microbiol. *84*, 263–271.

Hu, J., Qian, W., and He, C. (2007). The *Xanthomonas oryzae* pv. *oryzae eglXoB* endoglucanase gene is required for virulence to rice. FEMS Microbiol. Lett. *269*, 273–279.

Janse, J.D. (2006). Phytobacteriology: Principles and Practice (CABI Publishing, Wallingford, UK).

Johnson, A.G., Robert, A.L., and Cash, L. (1949). Bacterial leaf blight and stalk rot of corn. J. Agr. Res. *78*, 719–732.

Johnson, K.L. (2010). Elucidation of the host–pathogen interactions that influence seed-to-seedling transmission of *Acidovorax citrulli*. (PhD thesis: Department of Plant Pathology, University of Georgia, Athens, GA, USA).

Johnson, K.L., Minsavage, G.V., Le, T., Jones, J.B., and Walcott, R.R. (2011). Efficacy of a nonpathogenic *Acidovorax citrulli* strain as a biocontrol seed treatment for bacterial fruit blotch of cucurbits. Plant Dis. *95*, 697–704.

Jones, J.D.G., and Dangl, J.L. (2006). The plant immune system. Nature *444*, 323–329.

Kang, Y., Liu, H., Genin, S., Schell, M.A., and Denny, T.P. (2002). *Ralstonia solanacearum* requires type 4 pili to adhere to multiple surfaces and for natural transformation and virulence. Mol. Microbiol. *46*, 427–437.

Knauss, J.F., Miller, J.W., and Virgona, R.J. (1978). Bacterial blight of Fishtail palm, a new disease. Proc. Fla. State Hort. Soc. *91*, 245–247.

Kondo, M., Yoshida, Y., Miyata, C., Fujiwara, S., Nakajima, Y., Hirai, H., Takayama, S., Isogai, A., and Che, F.S. (2012). Genetic organization of the hrp gene cluster in *Acidovorax avenae* strain N1141 and a novel effector protein that elicits immune responses in rice (*Oryza sativa* L.). Biosci. Biotechnol. Biochem. *76*, 129–138.

Langston Jr. D.B., Walcott, R.R., Gitaitis, R.D., and Sanders Jr. F.H. (1999). First report of a fruit rot of pumpkin caused by *Acidovorax avenae* subsp. *citrulli* in Georgia. Plant Dis. *83*, 199.

Latin, R.X., and Hopkins, D.L. (1995). Bacterial fruit blotch of watermelon: the hypothetical exam question becomes reality. Plant Dis. *79*, 761–765.

Latin, R.X., and Rane, K.K. (1990). Bacterial fruit blotch of watermelon in Indiana. Plant Dis. *74*, 331.

Li, B., Liu, B.P., Yu, R.R., Tao, Z.Y., Wang, Y.L., Xie, G.L., Li, H.Y., and Sun, G.C. (2011c). Bacterial brown stripe of rice in soil-less culture system caused by *Acidovorax avenae* subsp *avenae* in China. J. Gen. Plant Pathol. *77*, 64–67.

Li, D., Rothballer, M., Schmid, M., Esperschuetz, J., and Hartmann, A. (2011a.) *Acidovorax radicis* sp. nov., a wheat-root-colonizing bacterium. Int. J. Syst. Evol. Microbiol. *61*, 2589–2594.

Li, D., Rothballer, M., Engel, M., Hoser, J., Schmidt, T., Kuttler, C., Schmid, M., Schloter, M., and Hartmann, A. (2012). Phenotypic variation in *Acidovorax radicis* N35 influences plant growth promotion. FEMS Microbiol. Ecol. *79*, 751–762.

Li, Q., Zhang, Q., Wang, C., Tang, C., Zhang, Y., Li, N., and Li, J. (2011b). Fish oil enhances recovery of intestinal microbiota and epithelial integrity in chronic rejection of intestinal transplant. PLoS ONE *6*, e20460.

Liu, H., Kang, Y., Genin, S., Schell, M.A., and Denny, T.P. (2001). Twitching motility of *Ralstonia solanacearum* requires a type IV pilus system. Microbiology *147*, 3215–3229.

Liu, H.L., Zhang, S.P., Schell, M.A., and Denny, T.P. (2005). Pyramiding, unmarked deletions in *Ralstonia solanacearum* shows that secreted proteins in addition to plant cell-wall-degrading enzymes contribute to virulence. Mol. Plant Microbe Interact. *18*, 1296–1305.

Liu, H., Tian, W.X., Ibrahim, M., Li, B., Zhang, G.Q., Zhu, B., and Xie, G.L. (2012). Characterization of *pilP*, a gene required for twitching motility, pathogenicity, and biofilm formation of *Acidovorax avenae* subsp. *avenae* RS-1. Eur. J. Plant Pathol. *134*, 551–560.

Liu, J., Luo, S.Z., Zhang, Q., Wang, Q.H., Chen, J.F., Guo, A.G., and Shan, W.X. (2012). Tn5 transposon mutagenesis in *Acidovorax citrulli* for identification of genes required for pathogenicity on cucumber. Plant Pathol. *61*, 364–374.

Makizumi, Y., Igarashi, M., Gotoh, K., Murao, K., Yamamoto, M., Udonsri, N., Ochiai, H., Thummabenjapone, P., and Kaku, H. (2011). Genetic diversity and pathogenicity of cucurbit-associated *Acidovorax*. J. Gen. Plant Pathol. *77*, 24–32.

Malkan, A.D., Strollo, W., Scholand, S.J., and Dudrick, S.J. (2009). Implanted-port-catheter-related sepsis caused by *Acidovorax avenae* and methicillin-sensitive *Staphylococcus aureus*. J. Clin. Microbiol. *47*, 3358–3361.

Mansfield, J.W. (2009). From bacterial avirulence genes to effector functions via the *hrp* delivery system: an overview of 25 years of progress in our understanding of plant innate immunity. Mol. Plant Pathol. *10*, 721–734.

Martin, H.L., and O'Brien, R.G. (1999). First report of *Acidovorax avenae* subsp. *citrulli* as a pathogen of cucumber. Plant Dis. *83*, 965.

Marutani, M., Taguchi, F., Shimizu, R., Inagaki, Y., Toyoda, K., Shiraishi, T., and Ichinose, Y. (2005). Flagellin from *Pseudomonas syringae* pv. *tabaci* induced *hrp*-independent HR in tomato. J. Gen. Plant Pathol. *71*, 289–295.

Mergaert, J., and Swings, J. (1996). Biodiversity of microorganisms that degrade bacterial and synthetic polyesters. J. Ind. Microbiol. Biotechnol. *17*, 463–469.

Miller, J.W. (1990). Bacterial Brown Spot of Orchid Caused by *Pseudomonas cattleyae* (Department of Agricultural and Consumer Services, Gainesville, FL, USA).

Monferran, M.V., Echenique, J.R., and Wunderlin, D.A. (2005). Degradation of chlorobenzenes by a strain of *Acidovorax avenae* isolated from a polluted aquifer. Chemosphere *61*, 98–106.

Mudgett, M.B. (2005). New Insights to the function of phytopathogenic bacterial type III effectors in plants. Annu. Rev. Plant Biol. *56*, 509–531.

Myung, I.S., Lee, Y.K., Jeong, I.H., Moon, S.Y., Lee, S.W., and Shim, H.S. (2010). A new disease, bacterial black rot of Korean radish, caused by *Acidovorax konjaci*. New Dis. Rep. *22*, 26.

Nestler, H., Kiesel, B., Kaschabek, S.R., Mau, M., Schlomann, M., and Balcke, G.U. (2007). Biodegradation of chlorobenzene under hypoxic and mixed hypoxic-denitrifying conditions. Biodegradation *18*, 755–767.

Ng, W.L., and Bassler, B.L. (2009). Bacterial quorum-sensing network architectures. Annu. Rev. Genet. *43*, 197–222.

Nudleman, E., and Kaiser, D. (2004). Pulling together with type IV pili. J. Mol. Microbiol. Biotechnol. *7*, 52–62.

O'Brien, R.G., and Martin, H.L. (1999). Bacterial blotch of melons caused by strains of *Acidovorax avenae* subsp *citrulli*. Aust. J. Exp. Agr. *39*, 479–485.

Opelt, K., and Berg, G. (2004). Diversity and antagonistic potential of bacteria associated with bryophytes from nutrient-poor habitats of the Baltic Sea coast. Appl. Environ. Microbiol. *70*, 6569–6579.

Park, Y.H., Lee, Y.J., Choi, Y.W., Son, B.G., and Kang, J.S. (2008). Evaluations of PCR primers used in the detection of *Acidovorax avenae* subsp. *citrulli* causing bacterial fruit blotch (BFB) in cucurbits. Hort. Environ. Biotechnol. *49*, 325–331.

Pataky, J.K., duToit, L.J., and Kerns, M.R. (1997). Bacterial leaf blight on shrunken-2 sweet corn. Plant Dis. *81*, 1293–1298.

Prior, P., and Rott, P. (1989). Bacterial leaf spot of anthurium (*Anthurium* spp.) caused by a *Pseudomonas* sp. in the french West Indies. J. Phytopathol. *124*, 215–224.

Rabinovitch-Deere, C.A., and Parales, R.E. (2012). Three types of taxis used in the response of *Acidovorax* sp. strain JS42 to 2-nitrotoluene. Appl. Environ. Microbiol. *78*, 2306–2315.

Rane, K.K., and Latin, R.X. (1992). Bacterial fruit blotch of watermelon: association of the pathogen with seed. Plant Dis. *76*, 509–512.

Rat, B., and Gardan, L. (1993). Mâche – la tache bactérienne due à un *Pseudomonas* sp. Fruits Légumes *111*, 42–43.
Rosen, H.R. (1922). A bacterial disease of foxtail (*Chaetochloa lutescens*). Ann. Mo. Bot. Gard. *9*, 333–388.
Saddler, G.S. (1994). IMI descriptions of fungi and bacteria, set 122, nos. 1211–1220. Mycopathologia *128*, 39–63.
Sanapareddy, N., Legge, R.M., Jovov, B., McCoy, A., Burcal, L., Araujo-Perez, F., Randall, T.A., Galanko, J., Benson, A., Sandler, R.S., *et al.* (2012). Increased rectal microbial richness is associated with the presence of colorectal adenomas in humans. ISME J. *6*, 1858–1868.
Schaad, N.W., Kado, C.I., and Sumner, D.R. (1975). Synonymy of *Pseudomonas avenae* Manns 1905 and *Pseudomonas alboprecipitans* Rosen 1922. Int. J. Syst. Bacteriol. *25*, 133–137.
Schaad, N.W., Sowell, G., Goth, R.W., Colwell, R.R., and Webb, R.E. (1978). *Pseudomonas pseudoalcaligenes* subsp. *citrulli* subsp. nov. Int. J. Syst. Bacteriol. *28*, 117–125.
Schaad, N.W., Postnikova, E., Sechler, A., Claflin, L.E., Vidaver, A.K., Jones, J.B., Agarkova, I., Ignatov, A., Dickstein, E., and Ramundo, B.A. (2008). Reclassification of subspecies of *Acidovorax avenae* as *A. avenae* (Manns 1905) emend., *A. cattleyae* (Pavarino, 1911) comb. nov., *A. citrulli* (Schaad *et al.*, 1978) comb. nov., and proposal of *A. oryzae* sp. nov. Syst. Appl. Microbiol. *31*, 434–446.
Schramm, A., Davidson, S.K., Dodsworth, J.A., Drake, H.L., Stahl, D.A., and Dubilier, N. (2003). *Acidovorax*-like symbionts in the nephridia of earthworms. Environ. Microbiol. *5*, 804–809.
Schulze, R., Spring, S., Amann, R., Huber, I., Ludwig, W., Schleifer, K.H., and Kampfer, P. (1999). Genotypic diversity of *Acidovorax* strains isolated from activated sludge and description of *Acidovorax defluvii* sp. nov. Syst. Appl. Microbiol. *22*, 205–214.
Scortichini, M., D'Ascenzo, D., and Rossi, M.P. (2005). New record of *Acidovorax avenae* subsp. *cattleyae* on orchid in Italy. J. Plant Pathol. *87*, 244–244.
Shakya, D.D., Vinther, F., and Mathur, S.B. (1985). World wide distribution of a bacterial stripe pathogen of rice identified as *Pseudomonas avenae*. J. Phytopathol. *114*, 256–259.
Shetty, A., Barnes, R.A., Healy, B., and Groves, P. (2005). A case of sepsis caused by *Acidovorax*. J. Infection *51*, 171–172.
Shivaji, S., Reddy, G.S.N., Aduri, R.P., Kutty, R., and Ravenschlag, K. (2004). Bacterial diversity of a soil sample from Schirmacher Oasis, Antarctica. Cell. Mol. Biol. *50*, 525–536.
Singleton, D.R., Ramirez, L.G., and Aitken, M.D. (2009). Characterization of a polycyclic aromatic hydrocarbon degradation gene cluster in a phenanthrene-degrading *Acidovorax* strain. Appl. Environ. Microbiol. *75*, 2613–2620.
Somodi, G.C., Jones, J.B., Hopkins, D.L., Stall, R.E., Kucharek, T.A., Hodge, N.C., and Watterson, J.C. (1991). Occurrence of a bacterial watermelon fruit blotch in Florida. Plant Dis. *75*, 1053–1056.
Song, W.Y., Kim, H.M., and Hwang, C.Y. (2000). A new selective medium for detecting *Acidovorax avenae* subsp. *avenae* in rice seeds. Plant Pathol. J. *16*, 236–241.
Song, W.Y., Kim, H.M., Hwang, C.Y., and Schaad, N.W. (2004). Detection of *Acidovorax avenae* ssp. *avenae* in rice seeds using BIO-PCR. J. Phytopathol. *152*, 667–676.
Sowell, G., and Schaad, N.W. (1979). *Pseudomonas pseudoalcaligenes* subsp. *citrulli* on watermelon – seed transmission and resistance of plant introductions. Plant Dis. Rep. *63*, 437–441.
Stovold, G., Bradley, J., and Fahy, P. (2001). *Acidovorax avenae* subsp. *cattleyae* (*Pseudomonas cattleyae*) causing leafspot and death of *Phalaenopsis* orchids in New South Wales. Australas. Plant Pathol. *30*, 73–74.
Sumner, D.R., and Schaad, N.W. (1977). Epidemiology and control of bacterial leaf blight of corn. Phytopathology *67*, 1113–1118.
Szczesny, R., Jordan, M., Schramm, C., Schulz, S., Cogez, V., Bonas, U., and Buttner, D. (2010). Functional characterization of the Xcs and Xps type II secretion systems from the plant pathogenic bacterium *Xanthomonas campestris* pv. *vesicatoria*. New Phytol. *187*, 983–1002.
Taguchi, F., Shimizu, R., Inagaki, Y., Toyoda, K., Shiraishi, T., and Ichinose, Y. (2003). Post-translational modification of flagellin determines the specificity of HR induction. Plant Cell Physiol. *44*, 342–349.
Takai, R., Isogai, A., Takayama, S., and Che, F.S. (2008). Analysis of flagellin perception mediated by flg22 receptor OsFLS2 in rice. Mol. Plant Microbe Interact. *21*, 1635–1642.
Takakura, Y., Che, F.S., Ishida, Y., Tsutsumi, F., Kurotani, K., Usami, S., Isogai, A., and Imaseki, H. (2008). Expression of a bacterial flagellin gene triggers plant immune responses and confers disease resistance in transgenic rice plants. Mol. Plant Pathol. *9*, 525–529.
Tanaka, N., Che, F.S., Watanabe, N., Fujiwara, S., Takayama, S., and Isogai, A. (2003). Flagellin from an incompatible strain of *Acidovorax avenae* mediates H_2O_2 generation accompanying hypersensitive cell death and expression of PAL, Cht-1, and PBZ1, but not of Lox in rice. Mol. Plant Microbe Interact. *16*, 422–428.

Thakuria, D., Schmidt, O., Finan, D., Egan, D., and Doohan, F.M. (2010). Gut wall bacteria of earthworms: a natural selection process. ISME J. *4*, 357–366.

Thiele, K., Smalla, K., Kropf, S., and Rabenstein, F. (2012). Detection of *Acidovorax valerianellae*, the causing agent of bacterial leaf spots in corn salad *Valerianella locusta* (L.) Laterr., in corn salad seeds. Lett. Appl. Microbiol. *54*, 112–118.

Tsai, S.H., Selvam, A., Chang, Y.P., and Yang, S.S. (2009). Soil bacterial community composition across different topographic sites characterized by 16S rRNA gene clones in the Fushan Forest of Taiwan. Bot. Stud. *50*, 57–68.

Ullstrup, A.J. (1961). Corn Diseases in the United States and Their Control (US Department of Agriculture, U.S. Government Printing Office, Washington, DC, USA).

Walcott, R.R. (2003). Detection of seedborne pathogens. HortTechnology *13*, 40–47.

Walcott, R.R., Langston Jr. D.B., Sanders Jr. F.H., and Gitaitis, R.D. (2000). Investigating intraspecific variation of *Acidovorax avenae* subsp. *citrulli* using DNA fingerprinting and whole cell fatty acid analysis. Phytopathology *90*, 191–196.

Walcott, R.R., Fessehaie, A., and Castro, A.C. (2004). Differences in pathogenicity between two genetically distinct groups of *Acidovorax avenae* subsp. *citrulli* on cucurbit hosts. J. Phytopathol. *152*, 277–285.

Wall, G.C., and Santos, V.M. (1988). A new bacterial disease on watermelon in the Mariana Islands. Phytopathology *78*, 1605.

Webb, R.E., and Goth, R.W. (1965). A seedborne bacterium isolated from watermelon. Plant Dis. Rep. *49*, 818–821.

Willems, A., Falsen, E., Pot, B., Jantzen, E., Hoste, B., Vandamme, P., Gillis, M., Kersters, K., and Deley, J. (1990). *Acidovorax*, a new genus for *Pseudomonas facilis*, *Pseudomonas delafieldii*, E. Falsen (EF) group 13, EF group 16, and several clinical isolates, with the species *Acidovorax facilis* comb. nov., *Acidovorax delafieldii* comb. nov., and *Acidovorax temperans* sp. nov. Int. J. Syst. Bacteriol. *40*, 384–398.

Willems, A., Goor, M., Thielemans, S., Gillis, M., Kersters, K., and De Ley, J. (1992). Transfer of several phytopathogenic *Pseudomonas* species to *Acidovorax* as *Acidovorax avenae* subsp. *avenae* subsp. nov., comb. nov., *Acidovorax avenae* subsp. *citrulli*, *Acidovorax avenae* subsp. *cattleyae*, and *Acidovorax konjaci*. Int. J. Syst. Bacteriol. *42*, 107–119.

Xie, G.L., Zhang, G.Q., Liu, H., Lou, M.M., Tian, W.X., Li, B., Zhou, X.P., Zhu, B., and Jin, G.L. (2011). Genome sequence of the rice-pathogenic bacterium *Acidovorax avenae* subsp. *avenae* RS-1. J. Bacteriol. *193*, 5013–5014.

Xu, J., Moore, J.E., Millar, B.C., Alexander, H.D., McClurg, R., Morris, T.C., and Rooney, P.J. (2004). Improved laboratory diagnosis of bacterial and fungal infections in patients with haematological malignancies using PCR and ribosomal RNA sequence analysis. Leuk. Lymphoma *45*, 1637–1641.

Zhao, J.S., and Ward, O.P. (1999). Microbial degradation of nitrobenzene and mono-nitrophenol by bacteria enriched from municipal activated sludge. Can. J. Microbiol. *45*, 427–432.

Zia-ul-Hussnain, S., Haque, M.I., Mughal, S.M., Shah, K.N., Irfan, A., Afghan, S., Shahazad, A., Batool, A., Khanum, P., Hussain, K., *et al.* (2011). Isolation and biochemical characterizations of the bacteria (*Acidovorax avenae* subsp. *avenae*) associated with red stripe disease of sugarcane. Afr. J. Biotechnol. *10*, 7191–7197.

The Interactions Between Gram-positive Pathogens and Plant Hosts

5

Elizabeth A. Savory, Allison L. Creason, Olivier M. Vandeputte, Edward W. Davis II and Jeff H. Chang

Abstract

Gram-positive bacteria are scientifically interesting and economically important pathogens of many plants. Despite the small number of reported disease causing species, these bacteria display a wide range of lifestyles on plants, from being xylem-limited and practically obligatory, to opportunistic environmental pathogens. In line with the diversity of lifestyles, their evolutionary paths, genome characteristics, and virulence mechanisms are also highly variable. In this chapter, the attributes of plant-pathogenic *Clavibacter*, *Streptomyces*, and *Rhodococcus* are discussed in the context of models derived from Gram-negative phytopathogens to highlight key differences. The Gram-positive phytopathogens present a number of opportunities to contribute new insights into the mechanisms of plant–microbial interactions and the evolution of a host-adapted lifestyle in bacteria.

Introduction

Gram-positive pathogens are among the top causal agents of infectious diseases in humans. The Gram-positive pathogenic members of the pyogenic cocci group, *Staphylococcus aureus*, *S. pyogenes*, and *S. pneumoniae*, for example, are responsible for a large fraction of bacterial infections that vary from mild to life-threatening diseases (Willems *et al.*, 2011). Endospore-forming bacteria, such as species of *Bacillus* and *Clostridium*, and the related, but non-spore forming *Listeria monocytogenesis*, are well known for their toxins and fatal diseases (McKenney *et al.*, 2013; Peck *et al.*, 2011; Williams *et al.*, 2012). *Mycobacterium tuberculosis* is a devastating pathogen, causing more than a million deaths a year (http://www.who.int/). *Corynebacterium diphtheriae* once caused a major number of deaths among children until the introduction of effective vaccines in the 1920s, which still remains a part of a standard childhood vaccination regimen (Adler *et al.*, 2013).

Gram-positive plant pathogens do not share the same notoriety despite causing economically important diseases that severely compromise food and plant production (Table 5.1). In addition, Gram-positive pathogens appear to have taken a different path towards host adaptation as compared to Gram-negative phytopathogens. As such, from the perspective of a basic biologist, Gram-positive pathogens have the potential to offer new insights into the mechanisms and evolution of bacterial virulence, as well as to the plant immune system, its vulnerabilities, and its evolved counter-adaptations to pathogens. However, despite the

Table 5.1 Characteristics of Gram-positive phytopathogens

	Disease or symptoms[1]	Host(s)[1]	NCBI RefSeq number[1]	Genome size (Mb)[1]	# Coding sequence[1]	GC%[1]	Primary virulence determinants[1]
Clavibacter							
C. michiganensis subsp. *michiganensis*	Bacterial wilt and canker	Tomato	NC_009480 (Chromosome)	3.4	3078	72.5	Serine protease
			NC_009478.1 (pCM1)	0.27	28	67.6	Endoglucanase
			NC_009479.1 (pCM2)	0.70	68	66.5	
C. michiganensis subsp. *sepidonicus*	Ring rot	Potato	NC_010407.1 (Chromosome)	3.4	3116	72.4	
			NC_010399.1 (pCS1)	0.50	67	67.5	
			NC_010408.1 (pCSL1)	0.95	117	68.8	
C. michiganensis subsp. *insidiosus*	Wilting and stunting	Alfalfa					
C. michiganensis subsp. *nebraskensis*	Wilt and Blight	Corn	NC_020891.1	3.06	2817	73	
C. michiganensis subsp. *tesselarius*	Leaf freckles and spots	Wheat					
Leifsonia							
L. xyli subsp. xyli	Ratoon stunting	Sugarcane	NC_006087.1	2.58	2030	67.7	
Streptomyces							
S. scabies (SYN, S. scabiei)	Scab	Potato, carrot, radish, beet, and peanut	NC_013929.1	10.15	8746	71.5	All *Streptomyces* spp. that cause scab symptoms synthesize thaxtomin A
S. acidiscabies	Scab		NZ_AHBF00000000	11.0	10,070	70.4	
S. turgidiscabies	Scab		NZ_AEJB00000000	10.83	10,069	69.9	

	Disease or symptoms[1]	Host(s)[1]	NCBI RefSeq number[1]	Genome size (Mb)[1]	# Coding sequence[1]	GC%[1]	Primary virulence determinants[1]
S. europaeiscabiei	Scab						
S. luridiscabiei	Scab						
S. niveiscabiei	Scab						
S. puniciscabiei	Scab						
S. reticuliscabiei	Scab						
S. stelliscabiei	Scab						
S. ipomoeae	Soil rot or pox	Sweet potato					Thaxtomin C
Rhodococcus							
R. fascians	Leafy galls, fasciation	Wide range	20 isolates; Creason *et al*. (2014)	5.1–5.9	5000–5800	~64.4	Cytokinins
			NC_021080 (pFiD188)	0.2	184	61.8	
Curtobacterium							
C. flaccumfaciens pv. *flaccumfaciens*	Bacterial leaf spot	Common bean	Not available				
C. flaccumfaciens pv. *betae*	Bacterial leaf spot						
C. flaccumfaciens pv. *oortii*	Bacterial leaf spot						
C. flaccumfaciens pv. *poinsettiae*	Bacterial leaf spot						
C. flaccumfaciens pv. *basellae*	Bacterial leaf spot						
C. flaccumfaciens pv. *ilicis*	Bacterial leaf spot						
C. flaccumfaciens pv. *beticola*	Bacterial leaf spot						

Table 5.1 Continued

	Disease or symptoms[1]	Host(s)[1]	NCBI RefSeq number[1]	Genome size (Mb)[1]	# Coding sequence[1]	GC%[1]	Primary virulence determinants[1]
Rathayibacter							
R. rathayi	Gumming disease	Orchard grass	Not available				
R. iranicus	Gumming disease	Wheat					
R. tritici	Gumming disease	Wheat					
R. toxicus	Gumming disease	Ryegrass					

[1]See associated text for references; some genome statistics were from http://www.ncbi.nlm.nih.gov/genome/

efforts of several research groups, there is still a significant void in our understanding of these important pathogens. This gap is not a reflection of lack of importance or interest, but more likely a consequence of the difficulties in working with challenging Gram-positive bacteria-host pathosystems and their often unique host-adapted lifestyles. Advances in technologies, the development of genome resources for both Gram-positive phytopathogens and their hosts, and use of surrogate model hosts all contribute to offset some of the challenges and make the characterization of Gram-positive pathogens a slightly less daunting task.

In the following, we will introduce the Gram-positive phytopathogens and discuss the current understanding of their virulence mechanisms and host interactions with particular attention given to pathogenic *Rhodococcus* species. In addition, we will describe the available genome resources for each species as well as their contributions towards understanding the evolution of virulence for Gram-positive plant pathogens. The discussion will be developed around a framework based on the interactions between model Gram-negative pathogens and plants, with the explicit goal of highlighting fundamental differences between these two groups of bacteria and their interactions with plants. We direct the reader to some excellent reviews that cover the three main pathogenic species of Gram-positive bacteria in greater details (Bignell *et al.*, 2010a; Eichenlaub and Gartemann, 2011; Stes *et al.*, 2011).

The Gram-positive phytopathogens

The cell wall-containing Gram-positive pathogens of plants are all members of the Actinobacteria phylum (Fig. 5.1), one of the largest taxonomical units recognized within the domain Bacteria (Gao and Gupta, 2012). Actinobacteria display great variation in both motility and morphology, with shapes and structures ranging from coccoid to highly differentiated branched mycelium. In addition, many members of the Actinobacteria form spores. Actinobacteria also inhabit a diversity of ecosystems and, most notably, are dominant in the rhizospheres and phyllospheres of many plant species, including an enrichment in the endophytic compartments of *Arabidopsis thaliana* roots (Bulgarelli *et al.*, 2012, 2013; Lundberg *et al.*, 2012). These observations suggest that plants establish distinctive bacterial communities in the soil, including and possibly preferentially recruiting members of the Actinobacteria. In fact, several genera of Actinobacteria have been isolated as endophytic symbionts from a variety of plants and are suggested to confer benefits to plants by either promoting growth of the host or suppressing diseases caused by phytopathogens (Seipke *et al.*, 2012).

An exceptionally small number of species within Actinobacteria have tipped the balance of the host–microbe interaction and independently evolved the ability to pathogenize plants to derive greater benefit from their hosts. *Clavibacter* and *Leifsonia* spp. are found within the suborder Micrococcineae in the order Actinomycetales. These pathogens colonize the xylem and are host-specific at the species or subspecies level. *C. michiganensis* is the only species within its genus and is made up of non-spore forming bacteria (Eichenlaub and Gartemann, 2011). Subspecies of this pathogen cause systemic vascular diseases in monocots or dicots, with the most studied being *michiganensis* (*Cmm*) and *sepedonicus* (*Cms*) that cause bacterial wilt and canker of tomato, and ring rot of potato, respectively. *C. michiganensis* is an economically important pathogen since no resistant cultivars have been identified (Eichenlaub and Gartemann, 2011). This is underscored by its recognition as the only Gram-positive bacterium to receive honorable mention in a survey of plant pathologists for their 'Top 10'

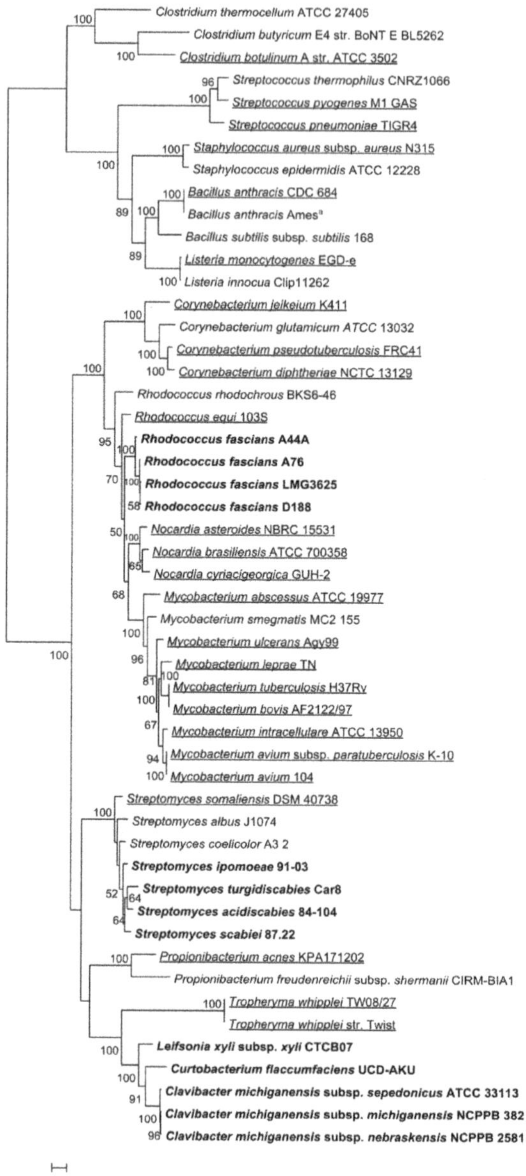

Figure 5.1 Phylogeny of Gram-positive plant pathogens and selected species. Using TBLASTN, amino acid sequences for *rpoB*, *secY*, and *ychF* were identified from the genome sequences of Gram-positive plant pathogens, select mammalian pathogens, type strains, and related non-pathogenic members of each genera. The sequences were aligned using MAFFT-L-INS-i (Katoh and Standley, 2013) and concatenated. The most appropriate model of substitution for phylogenetic analysis of each protein was selected using the BIC selection method implemented in ProtTest 3 (Darriba *et al.*, 2011). A maximum-likelihood phylogenetic tree was generated using RAxML 7.4.6 (Stamatakis, 2006). Plant-pathogenic strains are bolded and mammalian pathogenic strains are underlined. The interior node values indicate the percentage of bootstraps in agreement out of 1000; bootstrap values below 50 are not shown. Isolates in the bottom group are members of Actinobacteria. The scale bar represents the mean number of amino acid substitutions per site.

most scientifically/economically important bacterial pathogens (Mansfield *et al.*, 2012). *Leifsonia xyli* subsp. *xyli* has only been reported on monocots and is the causative agent of ratoon stunting, the most economically important disease of cultivated sugarcane (Davis *et al.*, 1980). This bacterium additionally has the honour of being the first Gram-positive phytopathogen to have its genome sequence determined (Monteiro-Vitorello *et al.*, 2004).

Streptomyces is the largest genus within the Actinobacteria class. These are the most recognized Actinobacteria because of their important contributions to carbon cycling, causing the 'earthy' odour of soil, and their prolific production of secondary metabolites that have seen widespread use in medicine and industry (Challis and Hopwood, 2003; Chater *et al.*, 2010; Takasuka *et al.*, 2013). *Streptomyces*, including its pathogenic members, are spore forming, filamentous bacteria that develop a complex network of mycelia (Flärdh and Buttner, 2009). Over 900 species have been described with a very minor fraction characterized as being virulent towards plants (Loria *et al.*, 2006). These pathogenic members are not particularly host specific and can cause scab disease on monocots and dicots, typically infecting laterally expanding underground plant tissues (Lerat *et al.*, 2009). While there are approximately 10 phytopathogenic species, the best characterized and economically important is *S. scabies* (synonym *S. scabei*), the causative agent of scab on root and tuber crops, such as potato, carrot, beets, and peanuts (Loria *et al.*, 2006). The next most studied and geographically distributed are *S. turgidiscabies* and *S. acidiscabies* (Lerat *et al.*, 2009).

Rhodococcus is a diverse genus of non-spore-forming, non-motile bacteria within the suborder Corynebacterineae. Its members are mostly environmental bacteria that display tremendous capacity to degrade organic compounds and are targeted as candidates for bioremediation of anthropomorphically disturbed environments (Larkin *et al.*, 2005). There are only two reported *Rhodococcus* species that are pathogens, *R. equi* a pathogen of animals and opportunistic pathogen of humans, and the plant pathogen *R. fascians* (Bargen and Haas, 2009; Stes *et al.*, 2011). Over the past two decades, characterization of a single isolate has contributed to major advances in understanding the virulence of *R. fascians*. Isolate D188 infects laterally expanding foliar tissues to cause growth abnormalities on a reportedly large and expanding number of hosts, including both dicots and monocots, as well as woody and herbaceous plants (Putnam and Miller, 2007). Its impact on agriculture is not as renowned as that of *Clavibacter* and *Streptomyces*, but this may be a factor of poor diagnosis as well as its apparent exclusiveness to the nursery industry, despite its ability to infect crop plants.

Finally, related to *Clavibacter* are *Curtobacterium flaccumfaciens* (Fig. 5.1) and *Rathayibacter toxicus* (Zgurskaya *et al.*, 1993). There are six pathovars of *C. flaccumfaciens*, with the most detrimental and characterized being *C. flaccumfaciens* pv. *flaccumfaciens*, a vascular pathogen that causes bacterial wilt of common bean (Agarkova *et al.*, 2012). *C. flaccumfaciens* pv. *flaccumfaciens* is a quarantined pathogen in many countries and the causal agent of the most important bacterial disease of dry beans in the United States (Agarkova *et al.*, 2012). This pathogen is seed transmitted and infects systemically within the xylem, causing wilting and necrotic lesions with bright yellow borders (Agarkova *et al.*, 2012; Harveson and Schwartz, 2007). Little is known about its virulence mechanisms and its genome has not been determined. *R. toxicus* is a nematode-vectored pathogen that causes gumming disease of annual grasses (Dorofeeva *et al.*, 2002). This pathogen is of interest because of the production of corynetoxins that can cause neurological disorders and death to livestock if consumed (Dorofeeva *et al.*, 2002). Estimates based on pulsed-field gel electrophoresis suggest *R. toxicus* has a linear chromosome of approximately 2.3 Mb (Agarkova *et al.*, 2006). Very little else is known regarding this pathogen.

In summary, there are few described Gram-positive plant pathogens but they display great diversity in several regards. Some of these pathogens cause economically important diseases and some can infect a tremendously diverse number of host species. Additionally, Gram-positive pathogens can infect a variety of plant tissues through various colonization schemes and have evolved a wide array of virulence strategies with different outcomes. Hereafter, we will focus on the three main genera of Gram-positive pathogens, *Clavibacter*, *Streptomyces*, and *Rhodococcus*, with some attention given to *Leifsonia*.

Infection of plant tissue by Gram-positive phytopathogens

C. michiganensis is a biotrophic pathogen that infects through wounds, hydathodes, or by contaminated seeds (Eichenlaub and Gartemann, 2011). *Cmm* preferentially inhabits xylem, attaching to spiral secondary wall thickenings of protoxylem and spreads systemically throughout the whole plant to reach apical regions in approximately two weeks (Chalupowicz *et al.*, 2012). Given its xylem-limited lifestyle, the pathogen is predicted to derive nutrients, likely in the form of carboxylic acids and sugars, from within the xylem vessels of its host (Gartemann *et al.*, 2008). Despite its classification as a biotroph, *Cmm* will eventually degrade the cell walls of xylem vessels and surrounding parenchymatic cells, resulting in cankerous lesions that typify its disease (Eichenlaub and Gartemann, 2011). *L. xyli* subspecies are also xylem-limited pathogens but little is known about their infection mechanism or mode of entry. *L. xyli subsp. xyli* has only been identified on cultivated sugarcane and is potentially an opportunist or recently host-adapted pathogen that takes advantage of wounding caused by machines or instruments used during harvesting to gain entry (Hoy *et al.*, 1999). It is suggested that the stunting disease symptoms are self-inflicted by a substance produced by the host in response to infection that blocks vessel flow and limits growth. This is not unlike Pierce's Disease of grapevine, where the host produces tyloses as well as pectin-rich gels and crystals in response to *Xylella fastidiosa* that are implicated in occluding the xylem (Sun *et al.*, 2013). Though, the contribution of host-produced substances in Pierce's disease is unresolved and still controversial.

Streptomyces are similar to fungi in regards to having complex multicellular structures and developmentally regulated life cycles (Flärdh and Buttner, 2009). During the vegetative lifecycle the cells grow in a polarized fashion *via* tip extension, forming branched hyphae that eventually develop into a mycelium network. This growth pattern allows the bacterium to extend into substrates to access nutrients. In response to nutrient stress, *Streptomyces* will develop specialized aerial hyphae that form spores to be dispersed into the environment. The best studied of the phytopathogenic *Streptomyces* species, *S. scabies*, infects only rapidly growing and expanding portions of the potato tuber and other root and tuber crops. Like other members of *Streptomyces*, its pathogenic members are also successful saprophytes and can persist in the soil. When a spore comes in contact with a host, it will germinate and infect. The primary route of invasion is through lenticels but *S. scabies* will also take advantage of wounds to penetrate into potatoes or other root and tuber crops (Snowdon, 2010). At this stage, the pathogen maintains a cycle of host cell death as a means to derive nutrients from the host. Corky lesions will form and coalesce into large scabs. Though common scab of potato does not typically affect yields, the presence of lesions certainly affects appearance and market value of crops (Lerat *et al.*, 2009). *S. scabies* and other *Streptomyces* spp. that cause scab symptoms on expanding underground tissues are also able to infect fibrous roots; however, these infections lead to root browning, stunting, and seedling death (Loria *et al.*, 2006).

R. fascians causes growth abnormalities such as fasciation, witches' brooms, or leafy galls (Putnam and Miller, 2007). *R. fascians* is a successful epiphyte of plant surfaces, primarily on the lamina of leaves, with a preference for epidermal cell wall junctions (Cornelis *et al.*, 2001). The mechanism by which *R. fascians* infects host tissues is still unresolved. *R. fascians* does not appear to colonize preferentially near stomata or trichomes. Instead, plant cells in the vicinity of *R. fascians* colonies give the appearance of collapsing and the bacteria is suggested to 'ingress' into host tissues (Vereecke *et al.*, 2002). At the onset of symptom development, few *R. fascians* bacteria are observed inside the plant, but numbers increase over time with groups of bacteria penetrating and growing extracellularly up to four cell layers deep by two weeks after infection (Cornelis *et al.*, 2001). There is no evidence to suggest that *R. fascians* invades vascular tissues though it has been suggested that some *R. fascians* cells may be surrounded by a plant cell wall, implicating the potential for intracellular pathogen growth (Cornelis *et al.*, 2001). The latter is unlikely since intracellular growth has never been observed for any other foliar plant-associated bacteria. Nonetheless, it cannot be completely dismissed because species of *Frankia*, which are members of Actinobacteria that can fix nitrogen, are internalized by host root cells during infection (Pawlowski and Demchenko, 2012). In mature leafy galls, *R. fascians* cells were found throughout proliferating tissues, with bacteria surrounding the meristematic zones rather than being localized inside these regions (Cornelis *et al.*, 2002).

PAMP-triggered immunity

The plant immune system is an important barrier that host-adapted microbes must overcome. Beyond the constitutive barriers that contribute to physically impede invasion by potential pathogens, plants have a variety of inducible immune responses. PAMP- or the synonymous MAMP-triggered immunity (PTI or MTI) is one of the first layers encountered by microbes (Jones and Dangl, 2006). Detection of potential pathogens by PTI is similar to mammalian innate immunity, with perception of PAMPs or MAMPs, pathogen/pattern- or microbe-associated molecular patterns, respectively, by plasma membrane-localized pattern recognition receptors (Schwessinger and Ronald, 2012). We choose to use the PAMP acronym in reference to pattern-associated molecular patterns to reinforce the notion that these patterns are conserved across a broad spectrum of bacteria as opposed to exclusively to pathogenic bacteria. Furthermore, implicit to the definition of PAMPs is their essentiality to microorganisms and the constraint against rapid evolution. This constraint is opposed by strong diversifying selection to evade detection by pattern recognition receptors (McCann *et al.*, 2012). Examples of conserved patterns of bacteria include flagella, elongation factor, and most notably, components of the bacterial cell wall (Boller and He, 2009). Perception of PAMPs by pattern recognition receptors leads to the induction of a number of signalling events that consists of a rapid burst in calcium and reactive oxygen species, activation of a mitogen activated protein kinase cascade, and synthesis of immunity-associated hormones (Nicaise *et al.*, 2009). The hormones represent signalling nodes that coordinate changes in gene expression to mount an effective immune response (Robert-Seilaniantz *et al.*, 2011a).

Eubacteria can be divided into two major groups, referred to as Gram-positive or Gram-negative, based on their abilities to retain crystal violet dye. This ability is determined by structural differences between the two groups of bacteria (Esko *et al.*, 2009). Both contain lipoproteins as well as peptidoglycan in their cell walls, an array of polysaccharides that plays an essential role in conferring rigidity and structure to cells (Ray *et al.*, 2013). In Gram-negative bacteria, peptidoglycan is found in the periplasmic space and bounded by two membranes. The outer membrane is asymmetrical, containing lipopolysaccharide (LPS) in the outer leaflet and lipoproteins, which are covalently attached to peptidoglycan in the inner leaflet. Gram-positive bacteria lack the outer membrane and have instead a thick peptidoglycan layer that is cross-linked to lipoteichoic acid. It is this thick peptidoglycan layer that is stained by the crystal violet dye; the much thinner layer found in Gram-negative bacteria does not retain the dye. Mycobacteria are generally classified as Gram-positive bacteria, but they contain an unusual outer membrane, called mycomembrane, that has features more similar to cell walls of Gram-negative bacteria (Niederweis *et al.*, 2010). The peptidoglycan layer found in Mycobacteria is covalently linked to long fatty acids called mycolic acid *via* an arabinogalactan polymer. This unusual arrangement includes other lipids, forming an asymmetrical outer membrane. *Rhodococcus* spp., presumably including the phytopathogenic *R. fascians*, also have mycolic acids in their cell envelope (Hsu *et al.*, 2011; Yam *et al.*, 2011).

The essentiality of the cell wall and its location to the outer surface of cells makes its constituents excellent targets of innate immune systems. The LPS found in the outer leaflet of Gram-negative bacteria contains Lipid A (endotoxin), which can cause fever and septic shock in mammals. LPS is also a PAMP recognized by mammalian Toll-like receptor 4 and triggers an immune response (Esko *et al.*, 2009). LPS can also trigger the plant immune response, inducing expression of nitric oxide synthase and other defence genes (Zeidler *et al.*, 2004). Lipoproteins, which are ubiquitously found in the cell walls of all Eubacteria, are recognized primarily by mammalian Toll-like receptor 2 (Ray *et al.*, 2013). Both peptidoglycan and Gram-positive lipoteichoic acids are also recognized by Toll-like receptor 2, but the elicitation of innate immunity in mammals by lipoteichoic acid is still somewhat controversial (Nizet and Esko, 2009; Ray *et al.*, 2013). Similarly, lipoglycans of Gram-positive bacteria and Mycobacteria also elicit immunity in a Toll-like receptor 2-dependent manner (Underhill *et al.*, 1999).

Elicitation of PTI by Gram-positive phytopathogens has not been well studied, but its potential has been demonstrated. *S. aureus* is an animal pathogen and rhizosphere-associated Gram-positive bacterium that is reportedly able to infect plants (Prithiviraj *et al.*, 2005). Both heat-killed *S. aureus* and peptidoglycan purified from this bacterium are capable of inducing a multitude of changes, including synthesis of signalling molecules and expression of defence genes in *Arabidopsis* that are indicative of PTI (Gust *et al.*, 2007). In addition, purified peptidoglycan from *Streptomyces* was also shown to elicit early plant immune responses (Gust *et al.*, 2007). Microarray analysis of susceptible tomato plants infected with *Cmm* revealed the differential induction of slightly more than 100 genes relative to mock-inoculated plants (Balaji *et al.*, 2008). The most represented category of differentially expressed genes was defence-related. Independent silencing of six of these host genes induced by *Cmm* resulted in increased susceptibility, consistent with these genes having a role in immunity (Balaji *et al.*, 2011). These observations were also consistent with proteomic-based methods that revealed the accumulation of defence-associated proteins (Savidor *et al.*, 2012). Plant cells in the vicinity of *R. fascians* colonies showed increased autofluorescence, which was interpreted as evidence for the biosynthesis of phenolic compounds and induction of a plant immune response in response to infection (Cornelis *et al.*, 2001; Vereecke *et al.*, 1997). Finally, even Gram-positive mutualistic *Frankia* induce the expression of defence-associated genes during infection of compatible hosts (Hocher *et al.*, 2011). This observation underscores the potential for microbes, regardless of the outcome of their symbiosis, to elicit PTI.

Countering PTI

Gram-negative phytopathogenic bacteria have up to six different secretion systems that can be used to deploy proteins to counter immune signalling or manipulate host cells during infection. A subset, the type III, IV, and VI secretion systems, not only bypass both membranes of the bacterium, but can also engage the membrane of a host, giving the pathogen the opportunity to directly inject proteins, called effectors, into host cells. The type III secretion system (T3SS) is the most intensively studied secretion system of bacterial pathogens. It also happens to be the most complex secretion system, typically encoded by a large cluster of genes that contribute to its regulation, assembly, and function (Cornelis, 2006). Depending on the species and isolate, plant-pathogenic Gram-negative bacteria directly inject as few as five to approximately 70 type III effector proteins (T3Es) into host cells (Baltrus *et al.*, 2011; Chang *et al.*, 2005; Guttman *et al.*, 2002; McNally *et al.*, 2012; Mukaihara *et al.*, 2010; Petnicki-Ocwieja *et al.*, 2002; Roden *et al.*, 2004). To date, hundreds of T3Es have been identified from a variety of Gram-negative species that include mutualists, pathogens, and commensals, but only a limited number of T3Es have been functionally characterized. Nevertheless, current models suggest that these important virulence proteins function to target host proteins associated with PTI to counter early immune signalling (Feng and Zhou, 2012). As a matter of fact, despite differences in pathogenic lifestyles, effectors of Gram-negative bacterial, oomycete, and fungal pathogens, all seemingly converge on a common set of host targets that function in host immunity (Dou and Zhou, 2012; Mukhtar *et al.*, 2011).

Pseudomonas aeruginosa PA14 is an opportunistic plant pathogen that bears mentioning. *P. aeruginosa* PA14, like other Gram-negative pathogens, encodes a T3SS. But unlike those that have been discussed, *P. aeruginosa* PA14 is a multihost pathogen that can cause disease in immune-compromised mammals, caterpillar, *Caenorhabditis elegans, Drosophila melanogaster* and *Arabidopsis* (Plotnikova *et al.*, 2000). Interestingly, the T3SS is not universally required on all of its hosts and is dispensable for infection of both *C. elegans* and Arabidopsis (Miyata *et al.*, 2003). Additionally, there is no strong evidence that PAMPs of *P. aeruginosa* PA14 are sufficient to elicit innate immune responses in *C. elegans* (Irazoqui *et al.*, 2010).

With the clear differences in cell wall architecture between the two groups of bacteria, it is not surprising that Gram-positive bacteria lack most of the type I–VI secretion systems capable of bypassing two bacterial membranes. The one exception is the type IV secretion system but it appears to have a different set of components and architecture. Moreover, this secretion system has a prominent role in bacterial conjugation. As it stands, there are no known examples of proteins that are translocated directly from a Gram-positive bacterium directly into a host cell. In phytopathogenic *Rhodococcus,* there is not even evidence for secreted virulence proteins. But its ability to actively suppress PTI has been suggested based on results from transcriptome profiling of infected plants (Depuydt *et al.*, 2009).

Conclusions were based on the observation that infected plants did not accumulate H_2O_2, an early response of plant immunity (Mittler *et al.*, 2004). Also, several commonly induced genes involved in the production, scavenging of, or protection against reactive oxygen species (ROS) and markers of immunity were all down-regulated (Depuydt *et al.*, 2009). Though suggestions that *R. fascians* actively suppress PTI are appealing, conclusions should be viewed with caution as changes in immune responses typically occur much earlier than the 7–24 days post inoculation that were examined by Depuydt and colleagues (Katagiri and Tsuda, 2010; Navarro *et al.*, 2004; Tao *et al.*, 2003; Torres and Dangl, 2005). If pathogenic *Rhodococcus* did actively suppress PTI, given the nature of the study, it would be premature to speculate on the potential role of effector proteins in mediating this response.

Effector-triggered immunity

Plants have a second layer of immunity known as effector-triggered immunity (ETI) that differs from PTI in two important regards. Perception is mediated by resistance (R) proteins that typically recognize effector proteins translocated into host cells (Jones and Dangl, 2006). Unlike pattern recognition receptors, most R proteins do not function as receptors that directly interact with the effector. Rather, most characterized R proteins monitor host cell homeostasis, surveying for an effector-dependent modification, such as a phosphorylation or protein cleavage event, to a 'guarded' target protein (Axtell and Staskawicz, 2003; Chung *et al.*, 2011; Liu *et al.*, 2011; Mackey *et al.*, 2003; Shao *et al.*, 2003). The second significant difference between ETI and PTI lies in the robustness of the response, with ETI occurring faster and with greater amplitude and often manifesting as a localized programmed cell death called the hypersensitive response (Mur *et al.*, 2008).

ETI has been implicated but not confirmed against Gram-positive phytopathogens. Cell-free culture supernatants of *Cmm* and *Cms* cause a cell death response when infiltrated directly into leaves of *Nicotiana tabacum* (Alarcón *et al.*, 1998; Nissinen *et al.*, 1997). The protein-like characteristics of the elicitor and the implied secretion from the bacteria led the authors to conclude the plant response is indicative of a hypersensitive response and therefore ETI. However, secreted, but non-effector proteins can also lead to cell death (Alfano *et al.*, 1996; Bauer *et al.*, 1995). Moreover, the pathogen protein was infiltrated into plant tissue but likely elicited a host response from outside the host cell, which is inconsistent with the intracellular location of most R proteins.

The interaction between plants and their microbial pathogens can be modelled as a co-evolutionary arms race (Stavrinides *et al.*, 2008). The interacting partners are under constant conflict, predicted to be shaped by natural selection to rapidly evolve traits to maximize their own benefit while minimizing harm caused by the other. For many host–pathogen interactions, *R* and effector genes are at the frontline of this conflict. Hosts gain new resistance specificities through *R* gene duplication and diversification, which is consistent with the observation that *R* genes are often clustered in plant genomes (Mcdowell and Simon, 2006). The pathogen also presents evidence for rapid evolution. Effector collections are hypervariable with evidence for rapid gene loss and gain that is reflective of avoidance of ETI balanced with maintenance of sufficiency for dampening PTI, respectively (Baltrus *et al.*, 2011). The only known plant resistance to *R. fascians*, which is described in greater detail in following sections, is suggested to be a prenylated isoflavanone produced by the host to quench pathogen signalling mediated by one of its known virulence factors (Rajaonson *et al.*, 2011). This is one of the few lines of evidence consistent with an evolutionary arms race between host and Gram-positive plant pathogen.

Secretion of proteases and plant cell wall-degrading enzymes

In addition to the secretion systems that allow for direct interactions with host cells, Gram-negative phytopathogenic bacteria also have a type I secretion system that can export proteins across both the inner and outer membranes in one step (Delepelaire, 2004). The type II and V secretion systems, in contrast, secrete proteins in two steps, relying on the universal Sec and two-arginine (Tat) systems to first deliver proteins into periplasmic space, then secreting the proteins past the outer membrane (Cianciotto, 2005; Henderson *et al.*, 2004). A critical difference between the Sec and Tat pathways is the state of the secreted proteins, with the Tat pathway devoted to folded proteins (Natale *et al.*, 2008). The necessity of the Tat secretion system in the virulence of a Gram-positive phytopathogen has been demonstrated. A *tat* deficient mutant of *S. scabies* failed to elicit symptoms similar to those caused by the wild type pathogen on roots of Arabidopsis seedlings, though not surprisingly, the mutant also exhibited several pleiotropic effects (Joshi *et al.*, 2010). However, independent mutants affected in seven genes encoding candidate Tat-secreted proteins were reduced in virulence, but not in *in vitro* growth, of the pathogen. Based on homologies, the secreted proteins were inferred to be involved in degrading the plant cell wall, interacting with host cells, countering antimicrobials, and in bacterial transport. In all, these data suggest that *S. scabies* relies on the universal Tat secretion system to secrete virulence proteins that modify the host cell.

Gram-positive bacteria appear to have a novel apparatus referred to as ESX or a type VII secretion system, which is required for full virulence of several mammalian pathogens such as *M. tuberculosis*, *M. bovis*, *S. aureus*, and *B. anthracis* (Burts *et al.*, 2005; Garufi *et al.*, 2008; Pym *et al.*, 2003; Stanley *et al.*, 2003). As a matter of fact, the necessity of the ESX system was first unknowingly demonstrated as early as the 1920s when an attenuated strain of *M. bovis* was identified following serial passages and used in human vaccines against tuberculosis (Abdallah *et al.*, 2007). Characterizations of this mutant many decades later revealed the deletion of an ~10 kb region that encompassed the ESX-encoding region. However, several non-pathogenic bacteria have homologues of this apparatus, suggesting it is not exclusive for host associations or virulence (Abdallah *et al.*, 2007). *S. scabies* is the only Gram-positive phytopathogen in which the ESX system has been characterized (Fyans *et al.*, 2013). Independent mutant strains affected in four different genes associated with the putative ESX system were not affected in virulence but mutants affected in *esxA* and *esxB*, which encode small proteins with predicted similarities to those secreted *via* the ESX system, exhibited growth abnormalities. The latter is consistent with suggestions that the ESX system is a general secretion system of Gram-positive bacteria.

The plant cell wall is a matrix of various polysaccharides and a dynamic barrier that is actively remodelled and reinforced to oppose pathogen infection (Underwood, 2012). Many Gram-negative phytopathogenic bacteria, regardless of whether they are biotrophic or necrotrophic, use secretion systems to secrete a variety of cell wall-degrading enzymes. The soft rot pathogens of *Pectobacterium*, previously known as *Erwinia*, secrete proteases and a battery of degradative enzymes in T1SS- and T2SS-dependent manners, respectively (Toth and Birch, 2005). Interestingly, despite its arsenal of degradative enzymes,

Pectobacterium requires a functional T3SS for optimal virulence (Holeva *et al.*, 2004). Even the xylem-inhabiting *Ralstonia* and hemibiotrophic pathogens use plant cell wall-degrading enzymes. In each of these cases, the expression of the degradative enzyme encoding genes are coordinated by or with expression of the T3SS genes that are essential for their virulence (Jha *et al.*, 2007; Liu *et al.*, 2008; Wang *et al.*, 2008; Szczesny *et al.*, 2010; Valls *et al.*, 2006). The necessity of cell wall-degrading enzymes for virulence is challenging to demonstrate because their genes are often redundant and any individual gene is dispensable.

Given the absence of effector secretion systems in Gram-positive bacteria, proteases and cell wall-degrading enzymes are obvious virulence candidates. Mining of the *Cmm* and *Cms* genome sequences has revealed the potential for a suite of plant cell wall hydrolysing enzymes, including a polygalacturonase, a pectate lyase, xylanases, cellulases, and other endoglucanases (Bentley *et al.*, 2008; Gartemann *et al.*, 2008). Few, however, have been functionally validated. *Cmm* strain NCPPB382 includes two small circular plasmids, pCM1 and pCM2, which have been implicated in virulence. Strains lacking individual plasmids or both were delayed or completely compromised, respectively, in causing wilting symptoms on tomato plants (Meletzus *et al.*, 1993). Although, all three strains grew to similar levels *in planta* as wild type *Cmm*, suggesting that pCM1 and pCM2 are not necessary for infection when the bacteria are delivered using artificial laboratory methods. More recently, analysis using confocal microscopy to visualize GFP-expressing *Cmm* strains *in planta* suggested that the plasmids were necessary for systemic movement and/or efficient colonization distal to the primary site of inoculation (Chalupowicz *et al.*, 2012). Mapping of the wilt-determinant locus on pCM1 revealed it to be the *celA* gene, which encodes a modular protein belonging to the family of endo-β-1,4-glucanases, having an N-terminal secretion signal sequence, a catalytic domain, a cellulose binding domain, and an expansin-like domain (Jahr *et al.*, 2000). The wilt-determinant found on the pCM2 plasmid is *pat-1*, which encodes a protein with sequence homology to serine proteases (Dreier *et al.*, 1997). The potential for *pat-1* to encode a *bona fide* protease was supported by the demonstration that a variant with a substitution mutant in the putative catalytic domain failed to restore wilting symptoms (Burger *et al.*, 2005).

Genes encoding putative proteases that target antimicrobial plant compounds, as well as concanamycins, expansins, and cutinases have all been identified from genome sequences of *Streptomyces* spp. (Bignell *et al.*, 2010b). As is the case with putative virulence genes in *Clavibacter*, few have been experimentally validated. The *nec1* gene from *S. scabies* was discovered based on its sufficiency to confer the ability to colonize and cause necrosis on potato tuber discs upon the non-pathogenic soil-inhabiting *S. lividans* (Bukhalid and Loria, 1997). Subsequent studies demonstrated that Nec1 of *S. turgidiscabies* was secreted and that secreted fractions were sufficient to cause necrosis on potato tuber discs (Joshi *et al.*, 2007b). A *nec1* mutant of *S. turgidiscabies* caused milder symptoms on non-host plants but whether *nec1* is necessary for virulence of either *Streptomyces* species on potato hosts was untested. Additionally, challenging the necessity of *nec1* in virulence is that its gene was not detected in all pathogenic isolates identified in a survey for scab-causing *Streptomyces* isolates (Wanner, 2006).

Cmm, but not *Cms*, and *Streptomyces* pathogens have a *tomA* gene, whose translated products are homologous to a class of glycosyl hydrolase family 10 saponin-detoxifying enzymes called tomatinases (Bentley *et al.*, 2008; Bignell *et al.*, 2010b; Kaup *et al.*, 2005). Saponins, such as the α-tomatine found in tomato, are antimicrobial glycosylated steroids or steroidal alkaloids that are constituents of plant immunity (Bouarab *et al.*, 2002). They are considered phytoanticipins, antimicrobial compounds that are preformed prior to pathogen attack, whose mode of action is to disrupt membrane integrity by forming complexes with sterols, particularly those with 3β-hydroxy groups (González-Lamothe *et al.*, 2009). Several fungal pathogens encode saponin-detoxifying enzymes that are members of the glycosyl hydrolase family that can remove one or more sugar residues from saponins (Bouarab *et al.*, 2002). *Fusarium oxysporum* f. sp. *lycopersici* and *Cladosporium fulvum* both encode glycosyl hydrolase family 10 members that are functional detoxifying enzymes necessary for full virulence on host plants (Pareja-Jaime *et al.*, 2008). In both *Cmm* and *S. scabies*, their respective TomA proteins were capable of hydrolysing α-tomatine *in vitro* but mutants were not affected in their growth on tomatoes (Kaup *et al.*, 2005; Seipke and Loria, 2008). However, mutants did exhibit α-tomatine-dependent differences *in vitro*. Growth of the *Cmm* mutant on agar plates was inhibited and growth of aerial mycelia of the *S. scabies* mutant was affected. Thus, the potential for *tomA* to be a *bona fide* virulence gene is still unresolved. Further, given that bacteria do not have sterols in their membranes, TomA likely confers a different or additional function to these bacteria.

In *R. fascians*, degradative enzymes necessary for virulence have not been identified. Other than the linear plasmid encoding biosynthetic proteins, which are described in greater details below, the only other locus implicated in virulence is a chromosomally encoded gene with predicted housekeeping functions. The *vic* locus was identified in a random mutagenesis screen, in which its mutant was compromised in its ability to stunt growth of tobacco seedlings (Vereecke *et al.*, 2002). Analysis of the translated coding sequence revealed extensive homology to proteins involved in the glyoxylate shunt of the Krebs cycle, a bypass of the TCA cycle that allows organisms to use compounds with two carbons when glucose is not available. Notably, induction of this pathway is an early response of *M. tuberculosis* during internalization by macrophages (Lorenz and Fink, 2002). It is currently unknown whether the *vic* locus of *R. fascians* has a 'classic' virulence function in perturbing the host or is simply expressed to accommodate changes in nutrient availability. Nonetheless, the identification of this gene as necessary for virulence is consistent with the hypothesized evolution of pathogenic *Rhodococcus*, as described in following sections (Letek *et al.*, 2010).

Virulence molecules of Gram-positive pathogens

Pathogens belonging to *Streptomyces* and *Rhodococcus* are related to environmental strains with demonstrable metabolic pliability and the ability to produce prolific quantities of secondary metabolites (Flärdh and Buttner, 2009; Larkin *et al.*, 2005). With this in mind, it should not be surprising that some of the more outstanding examples of virulence molecules for these Gram-positive pathogens are not proteins, but are in fact metabolites.

All scab-causing *Streptomyces* synthesize thaxtomin, a phytotoxin which thus far appears to be unique to pathogenic members of this genus (Loria *et al.*, 2008). Thaxtomin is necessary for pathogenicity, as deletion mutants affected in the thaxtomin-encoding locus fail to cause necrosis and exhibit reduced mycelial development on potato tubers (Healy *et al.*, 2000). The precise mode of action of thaxtomin remains unknown but its plant target appears to be cellulose synthase. The distribution and movement of cellulose synthase and the incorporation of ^{14}C-isotope in cell wall fractions are all altered following treatment with thaxtomin (Bischoff *et al.*, 2009; Scheible *et al.*, 2003) In addition, applications of thaxtomin resulted in ectopic lignification and deposition of callose, which are additional signs of reduced plant cell wall integrity. In all, the plant responses paralleled those observed in response to known cellular synthesis inhibitors, suggesting that thaxtomin directly affects cellulase synthases (Bischoff *et al.*, 2009). Thaxtomin also causes a rapid programmed cell death in plant cells (Duval *et al.*, 2005). But based on the comparative analyses of molecular markers and morphological as well as transcriptome-wide changes in response to a known cellulose synthase inhibitor, the thaxtomin-induced cell death response was not considered either a 'defence-like' response or a direct effect of the phytotoxin (Duval and Beaudoin, 2009; Duval *et al.*, 2005).

A significant amount of effort has been devoted to determining the structure and novel biosynthetic pathway of thaxtomins. Eleven structures have been described, with thaxtomin A being the major metabolite (King and Calhoun, 2009). *S. ipomoeae*, in contrast, produces a less modified form known as thaxtomin C that is necessary for its virulence on sweet potato (Guan *et al.*, 2012). These phytotoxins share a common backbone of a cyclic dipeptide consisting of a phenylalanine and a nitro-tryptophan but vary in their methyl and hydroxyl groups (Healy *et al.*, 2000). TxtD is a nitric oxide synthase that generates nitric oxide from L-arginine, a necessary step for adding a moiety to thaxtomins that is essential for its phytotoxic activity (Kers *et al.*, 2004). The TxtD-produced nitric oxide, along with oxygen, are used by TxtE, a cytochrome P450-type monoxygenase to catalyse the nitration of L-tryptophan (Barry *et al.*, 2012). This nitrated tryptophan moiety is subsequently used by TxtB, which together with TxtA, synthesize the thaxtomin backbone (Johnson *et al.*, 2009). These two proteins are both non-ribosomal peptide synthases (NRPSs), large multienzyme complexes that synthesize peptides much like an assembly line, independent of a mRNA (Gross and Loper, 2009). The variable groups are likely added after cyclization, as exemplified by the hydroxylation step catalysed by TxtC (Healy *et al.*, 2002).

As is the case for many virulence gene clusters, a transcriptional regulator is located in close proximity to thaxtomin biosynthetic genes that regulates their expression. The *txtR* gene encodes a member of the AraC/XylS family of transcriptional regulators (Joshi *et al.*, 2007a). This family of transcriptional regulatory proteins, in general, sense environmental signals *via* interactions with diffusible or transported molecules to either induce and/or repress DNA-binding (Gallegos *et al.*, 1997). Indeed, *txtR* was shown to be necessary for expression of *txtA-D*, production of nitric oxide, and virulence of *S. scabies*. Moreover, a protein fusion of maltose-binding protein and TxtR could bind to a cellobiose affinity column, suggesting that degradation products of the host cell wall were directly perceived by TxtR to induce thaxtomin synthesis (Joshi *et al.*, 2007a). Suberin, another plant polymer, has also been implicated in inducing taxtomin gene expression and synthesis (Lerat *et al.*, 2010).

S. scabies 87–22 has a 31 kb long cluster of genes with homology and an organization that suggest the potential for biosynthesis of a coronafacic acid-like molecule (Bignell *et al.*, 2010b). This molecule is an intermediate in the biosynthesis of coronatine (Rangaswamy *et al.*, 1998). Coronatine is a mimic of jasmonic acid (JA) that is synthesized by many hemibiotrophic *Pseudomonas syringae* pathovars. Coronatine binds the host Skp1-cullin-F box protein E3 ubiquitin ligase complex to promote the degradation of jasmonate signalling repressor proteins (Katsir *et al.*, 2008; Pauwels and Goossens, 2011). JA is a stress-associated plant hormone central to plant immunity that, in combination with ethylene, controls immunity to necrotrophic pathogens (Glazebrook, 2005). Furthermore, JA is classically viewed as an antagonist of salicyclic acid (SA), which regulates immunity to biotrophic and hemibiotrophic pathogens. Thus, by synthesizing coronatine, *P. syringae* leverages the relationship between the two hormone signalling pathways to its advantage. The de-repression of the JA pathway by coronatine depresses SA signalling and compromises immunity against biotrophic pathogens (Glazebrook, 2005). Expression patterns of the coronafacic acid-like cluster of genes in *S. scabies* are consistent with a role in pathogenesis and independent knockout mutants showed subtle difference in disease phenotypes. In particular, the Δ*cfa6* mutant did not cause the tissue necrosis or swelling of root tips that was observed after inoculations with wild type bacteria. Despite this, the mutants were not affected in overall virulence and were still capable of causing severe stunting and eventual death of shoots and roots, similar to wild-type *S. scabies* (Bignell *et al.*, 2010b).

Growth promoting hormones such as auxins and cytokinins are central players in plant development and are seemingly good targets for pathogens to manipulate to cause growth abnormalities and potentiate nutrient acquisition from the host. The Gram-negative pathogens *Agrobacterium tumefaciens*, *P. savastanoi*, and *Pantoea agglomerans* are good examples of pathogens that co-opt plant hormone signalling (Barash and Manulis-Sasson, 2009). However, *A. tumefaciens* is a unique pathogen that relies on its ability to transfer a T-DNA to genetically modify its host to upset hormone balance (Akiyoshi *et al.*, 1984). In *P. savastanoi* and *P. agglomerans*, the cytokinin biosynthesis genes are present in pathogenicity islands (see below), suggestive of an important function in virulence. However, mutation of the cytokinin biosynthesis genes failed to render either bacteria non-pathogenic, instead resulting in pathogens that caused reduced knot or gall disease symptoms without affecting their *in planta* growth (Barash and Manulis-Sasson, 2007; Bardaji *et al.*, 2011; Manulis *et al.*, 1998). This is in contrast to mutations in T3SS-encoding loci that render the bacteria non-pathogenic (Frederick *et al.*, 2001; Sisto *et al.*, 2004).

Auxins and cytokinins also promote susceptibility and immunity, respectively (Choi *et al.*, 2010; Navarro *et al.*, 2006). Network modelling supported by experimental data demonstrated that the growth-promoting hormones converge on the SA pathway, with auxins repressing SA levels and signalling and cytokinins functioning synergistically with SA to promote plant immunity and accumulation of antimicrobials (Naseem *et al.*, 2012; Robert-Seilaniantz *et al.*, 2011b). Moreover, bacterial virulence proteins have the potential to manipulate auxin signalling pathways to promote susceptibility (Chen *et al.*, 2007). Therefore, perturbation of hormone signalling pathways by pathogens with the purpose of causing growth abnormalities must be done in a delicate manner to avoid inciting a more robust immune response.

The synthesis of cytokinins by *R. fascians* has been intensively scrutinized as a virulence mechanism as early as the mid-1960s and revisited several times each decade thereafter (Eason *et al.*, 1996; Helgeson and Leonard, 1966; Klämbt *et al.*, 1966; Matsubara *et al.*, 1968; Murai *et al.*, 1980; Rathbone and Hall, 1972; Scarbrough *et al.*, 1973; Thimann and Sachs, 1966). However, to date, the source of the cytokinins and their precise roles in virulence are still unresolved. One of the early clues that cytokinins may be involved came from the demonstration that colonies of *R. fascians* (at the time, referred to as *Corynebacterium fascians*) or even killed *R. fascians* applied to senescent oat leaves, resulted in an increased retention of chlorophyll (Thimann and Sachs, 1966). These are the so-called 'green islands' that correlate with localized increases in cytokinin levels (Walters *et al.*, 2008). The isolation and identification of the first cytokinin from spent liquid growth cultures of *R. fascians* was described soon thereafter, with the most active compound identified as N^6-isopentenyladenosine, or 2iP (Helgeson and Leonard, 1966; Klämbt *et al.*, 1966).

The relevance of the findings was immediately challenged by the suggestion that 2iP was derived from tRNA and not necessarily synthesized *de novo* for virulence (Matsubara *et al.*, 1968). There are 75 tRNA modifications that have been described, and tRNAs that recognize codons beginning with uracil are almost always modified with 2iP or its derivative (Agris *et al.*, 2007; Phizicky and Hopper, 2010). Detectable cytokinin activity, mostly associated with 2iP, can be found in acid and enzymatic hydrolysates of tRNAs extracted from *in vitro* grown *R. fascians* (Matsubara *et al.*, 1968). Moreover, radioactive 2iP from tRNAs was readily recovered from *R. fascians* cultures fed with ^{14}C-labelled adenine (Rathbone and Hall, 1972). However, 2iP only accounted for 20% of the cytokinin activity recovered, indicating the potential for other cytokinins that may have *bona fide* roles in bacterial virulence. Indeed, two other active fractions were identified with one including 2-methylthio-*cis*-zeatin (Armstrong *et al.*, 1976; Klämbt *et al.*, 1966). Additionally, Armstrong and colleagues also speculated that *R. fascians* synthesized io^{6}Ado and i^{6}Ado as well as methylthio derivatives of isopentenyl adenine and isopentenyl adenosine. This observation, along with the unusually potent activity, led the authors to suggest that this Gram-positive pathogen synthesizes a spectrum of cytokinins that in total are responsible for the disease symptoms induced on infected plants (Armstrong *et al.*, 1976). But yet again, challenging these conclusions was the demonstration that cis-zeatin riboside could also be isolated from tRNAs of *R. fascians* (Einset and Skoog, 1977). Moreover, the substantially more biologically active trans-zeatin can also be isolated from tRNAs of prokaryotes (Koenig *et al.*, 2002).

The identification of the *fas* locus, a six gene operon encoded on the linear plasmid, pFiD188, as necessary for causation of disease symptoms by D188 appeared to confirm an essential role of *R. fascians*-synthesized cytokinins in virulence (Crespi *et al.*, 1992, 1994). Naturally occurring cytokinins are adenines with either an isoprenoid or aromatic side chain at the N^6 position (Frébort *et al.*, 2011). The synthesis of the former type starts with the transfer of an isoprenoid moiety from either dimethylallyl pyrophosphate or (E)-4-hydroxy-3-methyl-but-2-enyl diphosphate to adenine in its nucleotide form or bound to RNA. This key step is mediated by one of two classes of isopentenyl transferases (IPTs), the adenylate or tRNA IPTs, the former of which are the major source of cytokinins. In *R. fascians*, the IPT is encoded by *fasD* (Crespi *et al.*, 1992). The presence of an IPT gene in *R. fascians* is not unique, as homologues are found in the genome of other microbes (Kakimoto, 2003). Interestingly, however, FasD does not cluster with any of the three groups of IPTs, microbial IPTs, or plant adenylate or tRNA IPTs (Frébort *et al.*, 2011; Kakimoto, 2003).

The *fas* operon encodes additional genes with functions in different aspects of cytokinin biosynthesis. This collection of genes is thus far unique to *R. fascians* and *S. turgidiscabies* (Crespi *et al.*, 1994; Kers *et al.*, 2005; Pertry *et al.*, 2010). Isopentenyladenine can be hydroxylated by a cytochrome P450 monooxygenase, which is potentially encoded by *fasA*, to produce zeatin-type cytokinins. Active cytokinins are free base forms and in rice, for example, are activated by *LONELY GUY* (*LOG*), which encodes a protein with phosphoribohydrolase activity that releases a 5′ monophosphate (Kurakawa *et al.*, 2007). In *R. fascians*, this activity may be conferred by the protein product of *fasF*, a homologue of *LOG*. Finally, *fasE* encodes a putative cytokinin dehydrogenase that irreversibly degrades cytokinins by cleaving the N^6 side chain and freeing the adenine. In *S. turgidiscabies*, a cluster of *fas* genes is located adjacent to the thaxtomin biosynthesis gene cluster and share between

30 and 70% identity to the *fas* genes of *R. fascians* (Kers *et al.*, 2005). It is interesting that the thaxtomin-deficient mutant of *S. turgidiscabies* could promote growth abnormalities on plants, suggesting the *fas* homologues are functional in this necrotizing pathogen (Joshi and Loria, 2007).

However, the use of more advanced methods has not provided unequivocal results to resolve conflicts regarding the source and effect of cytokinins in *R. fascians* virulence. Most notably, a plasmid-cured strain that fails to cause disease symptoms produces the same spectrum of cytokinins as the wild type strain, albeit at slightly lower concentrations (Pertry *et al.*, 2010, 2009). Interestingly, single deletion mutants suggest that *fasF,* which is inferred to activate cytokinins based on its homology to *LOG,* has the least effect on the development of symptoms by *R. fascians,* as compared to wild type or mutants of *fasA, D and E,* the latter of which is inferred to inactivate cytokinins (Pertry *et al.*, 2010). Also challenging the necessity of *fas* and cytokinins in the pathogenesis of *R. fascians* is the failure to identify complete and functional *fas* operons in draft genome sequences of demonstrably virulent isolates (Creason *et al.*, 2014). In all, the obvious connection between the production of cytokinins by *R. fascians* and provocation of cytokinin-like changes in host morphology is too enticing to ignore, as is the case with other Gram-negative phytopathogens, phytopathogenic fungi, and Gram-negative and Gram-positive plant growth promoting bacteria that are suggested to synthesize cytokinins (Frugier *et al.*, 2008; Koenig *et al.*, 2002; Ortíz-Castro *et al.*, 2008; Pallai *et al.*, 2012; Walters *et al.*, 2008). However, it is still unresolved whether these microbial-synthesized cytokinins are *bona fide* virulence factors that manipulate plant-signalling pathways to affect host cell growth.

As previously described, if *R. fascians* does in fact synthesize cytokinins to perturb host cell growth, it must also address the potential that it could also incite SA-mediated immunity, which given its biotrophic-like lifestyle, would be presumed to be operational against *R. fascians* (Naseem and Dandekar, 2012). One possible solution is to also synthesize and secrete auxins to antagonize SA signalling. Indeed, *R. fascians* can produce and secrete substantial amounts of auxins (Vandeputte *et al.*, 2005). Moreover, auxin levels were optimally produced in a pFiD188-dependent manner and in *in vitro* conditions conducive to virulence gene activation suggesting auxins are produced during infection.

Adjacent to the *fas* locus is the *att* biosynthetic locus (Crespi *et al.*, 1992; Francis *et al.*, 2012; Maes *et al.*, 2001). Data suggest that the product of the *att* locus may be a small molecule that regulates the transition of *R. fascians* from epiphytic to endophytic growth. Mutants of *att* and a strain lacking the linear plasmid are successful epiphytes, but are compromised in virulence when inoculated onto plants in the absence of wounding (Maes *et al.*, 2001; Nikolaeva *et al.*, 2009). Additionally, extracts derived from plants infected with *att* mutants fail to induce *att* or *fas* gene expression, unlike the extracts from plants infected with wild type *R. fascians* (Maes *et al.*, 2001; Pertry *et al.*, 2009). The translated sequences of the *att* locus have homology to β-lactam biosynthesis proteins, which together with the previously described observation, suggest the product of the *att* locus has a function in regulating gene expression. This is not unusual as many reports suggest that antibiotics can act as signalling molecules in natural environments (Bader *et al.*, 2003; Brodhagen *et al.*, 2004; Coulthurst *et al.*, 2005; Da Silva *et al.*, 2002; de Bruijn *et al.*, 2007; Duerkop *et al.*, 2009; Goh *et al.*, 2002; Kidarsa *et al.*, 2011; Linares *et al.*, 2006; Maurhofer *et al.*, 2004; McGowan *et al.*, 2005; McNabe *et al.*, 2011; O'Brien and Wright, 2011; Romero *et al.*, 2011; Ryan and Dow, 2008). Additionally, whereas the *fas* locus is tightly regulated in *R. fascians,* its expression was constitutive in *S. turgidiscabies* under the treatment conditions tested (Joshi and Loria, 2007). This is of important note since neither the *att* genes nor the *fasR* gene are present in *S. turgidiscabies*. FasR, like TxtR, is a member of the AraC/XylS transcriptional regulator family, which are often induced by binding small molecules, suggesting a possible link between the product of the *att* locus and regulation of FasR (Gallegos *et al.*, 1997).

The only known form of resistance to *R. fascians* is *via* interference with *att* signalling (Rajaonson *et al.*, 2011). The potential for an evolutionary arms race between host and pathogen that centres on this locus strongly suggest an important function for *att* in pathogen virulence. *Dalbergia* is a genus of plant species naturally resistant to *R. fascians*. Its members produce perbergin, a prenylated isoflavanone with the novel chemical structure (Z)-3-(2,3-dihydroxy-4-methoxyphenyl)-8-(3,7-dimethylocta-2,6-dienyl)-5,7-dihydroxy-2,3-dihydrochromen-4-one. It was demonstrated that perbergin specifically inhibited the expression of *att* and *fas* reporter gene fusions in *in vitro* grown cells but its effects decreased if applied following the induction of *att* autoregulation. Thus, a plausible explanation is that perbergin competes with the regulator of the *att* locus and its structure may provide clues to the structure of the molecule that regulates virulence gene expression in *R. fascians*. The immunity reinforces the importance of Gram-positive pathogen virulence as a selective force on the evolution of plants and underscores the potential for many yet to be discovered immunity mechanisms.

A third locus on pFiD188 was also identified based on a *R. fascians* mutant that conferred a hypervirulent phenotype capable of inducing the development of larger and faster forming galls (Crespi *et al.*, 1992). The *hyp* locus is located ~10 kb away from the *att* locus and is predicted to encode a putative esterase/lipase (Francis *et al.*, 2010). It has, however, not been further examined since its original discovery.

Evolution of a host-adapted lifestyle

Finished and draft genome sequences are essential for contemporary methods of research. Computational mining coupled to functional screens and traditional bacterial genetics have contributed to advance our understanding of bacterial virulence mechanisms and elicitation of plant immunity (Studholme *et al.*, 2011). Moreover, comparative genomics and, in particular, comparisons deeply within pathogenic species of Gram-negative bacteria have helped resolve taxonomical relationships and provided new insights into the evolution and migration of pathogen populations (Baltrus *et al.*, 2011; Bart *et al.*, 2012; Cai *et al.*, 2011; Genin and Boucher, 2004).

Genome sequences have been published, released, or are in the draft stage for members of *Clavibacter*, *Streptomyces*, *Rhodococcus*, and *Leifsonia* (Table 5.1; NC_013929.1, NZ_AEJB00000000; Creason *et al.*, 2014; Bentley *et al.*, 2008; Gartemann *et al.*, 2008; Monteiro-Vitorello *et al.*, 2004). The amount of homology that can be detected between the different species of Gram-positive phytopathogens is provided as a circular ideogram (Fig. 5.2). One obvious conclusion is that more closely related species have higher amounts of homology. Genes homologous between *Cmm* and *Cms* represent more than 25% of all homologues found in all six pair wise comparisons to other species. Similarly, within the *Streptomyces* genus, more than 30% of the identified homologous genes between all possible pair wise comparisons were to another member of the genus. However, the three species have only approximately 4000 orthologous genes, that represent between 40% and 45% of

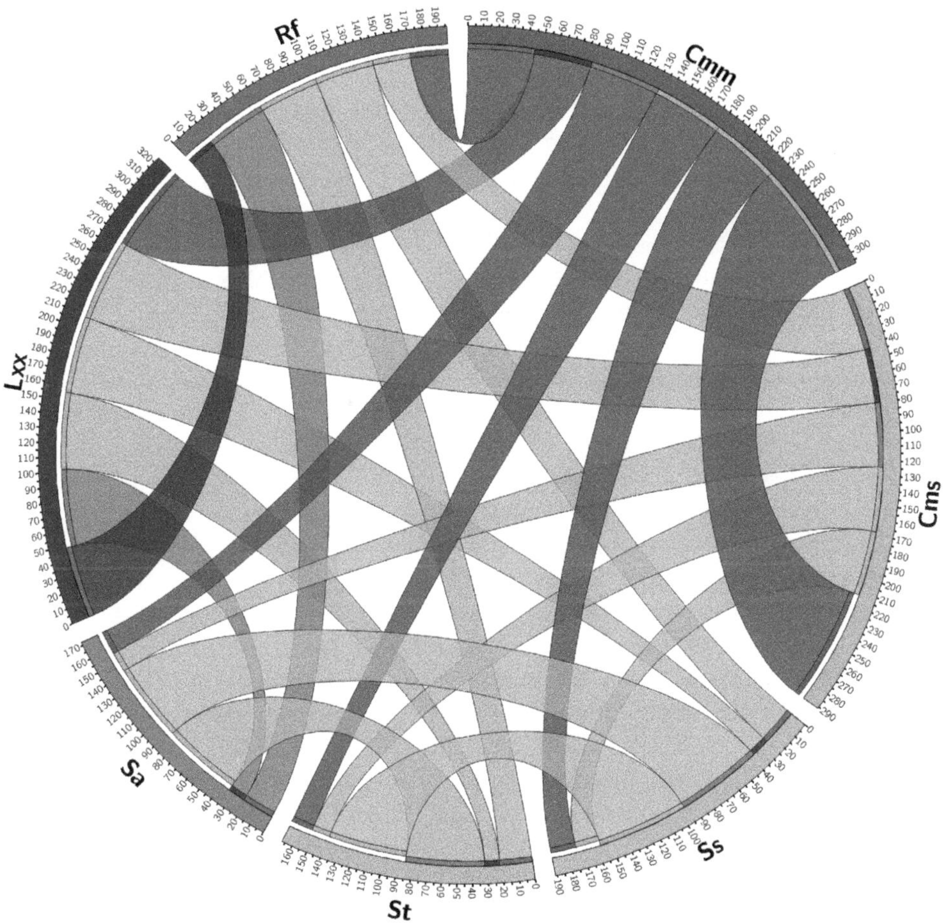

Figure 5.2 Visualization of genome-wide homology between Gram-positive plant pathogens. The outer track is expressed as a ratio of the total number of pairs of homologues divided by number of genes in a genome, e.g. there were three times as many pairs of homologues based on six pair wise comparisons, as genes in the genome of *Cmm*. The interior ribbons connect the genome comparisons. Ribbons vary in width to depict the extent of homology relative to genome size, e.g. pairs of homologues represented ~45% and ~25% of the *Cmm* and *Rf* genomes, respectively. Homologues were identified using reciprocal best BLAST hit analysis. Genomes were assigned arbitrary colours: red, *Clavibacter michiganesis* subsp. *michiganensis* (*Cmm*); yellow, *Clavibacter michiganensis* subsp. *sepedonicus* (*Cms*); lime green, *Streptomyces scabies* (*Ss*); mint green, *Streptomyces turgidiscabies* (*St*); light blue, *Streptomyces acidiscabies* (*Sa*); dark blue, *Leifsonia xyli* subsp. *xyli* (*Lxx*); purple, *Rhodococcus fascians* D188 (*Rf*). The figure was generated using Circos (Krzywinski *et al*., 2009). A colour version of this figure is located in the plate section at the back of the book.

their respective coding sequences. Interestingly, despite that *L. xyli* subsp. *xyli* is related to and shares a common xylem-limited lifestyle with *Clavibacter*, the per cent of homologous genes it shares is not biased to either *Cmm* or *Cms* but is rather similar, ranging between 50% and 60%, to all compared genome sequences. *R. fascians*, in contrast, shares more homologous genes in common to members of *Streptomyces* than either *Clavibacter* or *L. xyli* subsp.

xyli. It is also of interesting note that based on the variation in genome sizes and reported analyses, each of the species seems to occupy different positions along the continuum of symbiosis (Table 5.1). In general, specialized niche-adapted pathogens have small genomes, environmental bacteria have large genomes, and facultative bacteria have intermediate sized genomes (Ochman and Davalos, 2006). These different tendencies also correlate with differences in other characteristics such as GC content, evidence for horizontal gene transfer and mobile genetic elements.

Cmm, *Cms*, and *L. xyli* subsp. *xyli* have small genomes. The former two are high GC-containing bacteria with chromosomes of ~3.3 Mb and two plasmids that contribute minimally to genome size (Bentley *et al.*, 2008; Gartemann *et al.*, 2008). The genes hypothesized to be involved in host recognition, colonization, and engaging the host immune system are predicted to be on the chromosomes whereas those involved in symptom development seem to be located on plasmids (Gartemann *et al.*, 2008). Comparisons of their genomes suggested that *Cmm* and *Cms* are in different stages of pathogen evolution (Bentley *et al.*, 2008; Gartemann *et al.*, 2008). In *Cmm*, there are very few cases of pseudogenes and insertion sequence (IS) elements that when viewed in light of the metabolic potential of *Cmm*, led the authors to speculate that this bacterium is a recent pathogen still in the process of adapting to its host-associated lifestyle (Bentley *et al.*, 2008). *Cms*, in contrast, had greater than a hundred pseudogenes that likely compromised its metabolic potential as well as IS elements that often bordered the many short regions syntenic to *Cmm*, indicative of contributions to large-scale genome rearrangements (Bentley *et al.*, 2008; Gartemann *et al.*, 2008). These observations are consistent with its more restricted adaptation to near exclusivity to vasculature tissue of plants and support a recent bottlenecking event that led to the speciation of *Cms*. In *L. xyli* subsp. *xyli* CTCB07, despite the high ~68% GC content, the genome is only ~2.6 Mb in size; bacteria with small genomes tend to have a low GC percentage (Bentley and Parkhill, 2004; Monteiro-Vitorello *et al.*, 2004). More striking is its unusually low density of 80% coding sequence, as 13% of the genes were pseudogenes. The genome decay and loss of function in genes typically associated with free-living bacteria is like that of *Cms* and similarly suggests a bottlenecking event and a more specialized niche adaptation towards plant vasculature. The degradation of the genome and the remaining genes likely conserved because of their essentiality may explain the observation that more than half of its genome is homologous to those of other Gram-positive pathogens (Fig. 5.2).

S. acidiscabies 84-104, *S. scabies* 87-22, and *S. turgidiscabies* Car8 are high GC-containing bacteria with exceptionally large genomes that exceed 10 Mb. The descriptions and comparisons of their genomes have yet to be reported. Likewise, no genome sequence has been reported for *R. fascians*, but high-quality draft genome sequences have been generated for >20 isolates of *R. fascians* that included the well-studied D188 isolate, culture collection isolates, and many pathogenic and non-pathogenic isolates collected from infected plant tissues sampled across the United States and Europe (Creason *et al.*, 2014). All isolates have a high GC content but despite their relatedness to environmental *Rhodococcus* that have large genomes, the pathogenic *Rhodococcus* have intermediate sized genomes of ~5.8 Mb typical of facultative host-associated microbes. To our surprise, the preliminary analysis of the genomes revealed the likelihood that these sequenced isolates in fact represent two, if not more, distinct species of plant-pathogenic *Rhodococcus* and may challenge current thoughts that *R. fascians* has a very broad host range. This conclusion is supported by the phylogenetic relationship among four isolates that represent the genetic diversity of the sequenced plant-pathogenic *Rhodococcus* (Fig. 5.1).

The distributed genome hypothesis posits that pathogens have access to a large 'environmental' gene pool from which they can acquire new traits (Ehrlich *et al.*, 2008). Access to these genes is driven *via* transduction, conjugation, and transformation, three mechanisms of horizontal (aka lateral) gene transfer (HGT) that, following integration and selection, can lead to genome innovation (Thomas and Nielsen, 2005). Virulence can be further refined through recombination events that generate novel combinations of genes as well as gene loss and genome degradation, processes that are necessary for actualization of a host-adapted lifestyle (Koonin, 2009; Maurelli *et al.*, 1998; Parkhill *et al.*, 2003). Evidence for genome degradation has already been described and interpreted as an evolutionary process that led to the pathogenic lifestyle of *Cms* and *L. xyli* subsp. *xyli* (Bentley *et al.*, 2008; Monteiro-Vitorello *et al.*, 2004).

Genes acquired horizontally from disparate species often retain distinguishing hallmarks that differentiate them from the rest of the genome and make them relatively easy targets for identification. The most obvious include an association with 'DNA vehicles' and their remnant sequences, mobile genetic elements such as plasmids, IS elements, phage, and integrative conjugative elements (Koonin, 2009; Maurelli *et al.*, 1998; Parkhill *et al.*, 2003). Additionally, k-mers, codon usage, and GC% can also be retained as source signatures and exhibit significant variances from genome averages (Karlin *et al.*, 1998).

Indeed, the few characterized and predicted virulence genes of Gram-positive phytopathogens are primarily located on 'virulence' plasmids and within long stretches of horizontally acquired sequences, so-called genomic or more specifically, pathogenicity islands (PAIs). *Cmm* strain NCPPB382 includes two small circular plasmids, pCM1 and pCM2, that when both evicted, results in a strain that fails to cause any wilting symptoms on tomato plants (Meletzus *et al.*, 1993). As previously described pCM1 and pCM2 carry *celA* and *pat-1*, respectively (Dreier *et al.*, 1997; Jahr *et al.*, 2000). In the chromosome of *Cmm*, there are as many as 20 regions with a GC content lower than the average of 72.6% (Gartemann *et al.*, 2008). Of particular interest is the *chp/tom* PAI that when deleted, compromises the ability of the mutant to grow *in planta*, cause wilting symptoms, and move throughout the plant (Chalupowicz *et al.*, 2012). Because the *Cmm* strain cured of pCM1 and pCM2 failed to elicit any wilting symptoms, no chromosomally encoded genes are sufficient for causing symptoms (Meletzus *et al.*, 1993). Thus, the most parsimonious explanations are that genes of the *chp/tom* PAI are necessary for *in planta* growth or alternatively, necessary for regulating plasmid-borne virulence genes. The *chp* subregion is 79 kb long with an unusually low coding density of only 44 predicted genes and atypical codon usage. Seven of the genes are homologous to *pat-1* of pCM2 and designated as Chp (chromosomal homology to pat-1) serine proteases (Burger *et al.*, 2005; Gartemann *et al.*, 2008). A mutant affected in *chpC* was reported to be reduced in local and systemic host colonization, implicating its necessity for infection (Chalupowicz *et al.*, 2012; Gartemann *et al.*, 2008). Moreover, three homologues of *chp* are inferred to be pseudogenes based on the presence of premature termination codons. This is consistent with the rapid evolution of virulence genes and the potential for an evolutionary arms race. Two other *pat-1* homologues were also identified on pCM2. The *tom* subregion is 50 kb long and has 35 predicted genes mostly involved in metabolism. But this region also includes *tomA*, predicted to encode a tomatinase that contributes to the detoxification of a host-produced alkaloid (Kaup *et al.*, 2005).

Cms does not have a single large island equivalent to the *chp/tom* PAI since the genome has undergone a number of rearrangements. Consistent with this is the observation that homologues of the *chp/tom* PAI are found distributed across the genome of *Cms* (Bentley *et al.*, 2008; Nissinen *et al.*, 2009). Furthermore, *Cms* contains a homologue of the *tomA* gene but *in silico* characterizations suggest it has been pseudogenized. *Cms* also carries a circular and a linear plasmid with two and one *pat-1* homologues, respectively. Eight other chromosomal homologues were also identified (Bentley *et al.*, 2008). As is the case in *Cmm*, many of the *Cms pat-1* homologues are predicted to lack N-terminal encoding signal sequences or are pseudogenized based on the presence of premature termination codons. Briefly, the genome of *Leifsonia xyli* pv *xyli* strain CTCB07 has four regions with signatures indicative of HGT with three encoding homologues of *Cmm* virulence genes (Monteiro-Vitorello *et al.*, 2004). Additionally, all four regions contain numerous prophage-, transposase-, and IS element-related genes.

Streptomyces presents a convincing example of genome innovation in the evolution of plant pathogenesis in Gram-positive bacteria. The *txtD* and *nec1* genes delineate a chromosomal region of 325 kb that, based on the lower GC% and the inclusion of *tomA* and *fas* virulence gene homologues, was suggested to be part of a larger ~660 kb PAI in *S. turgidiscabies* (Kers *et al.*, 2005). The *fas* genes are flanked by two transposable element genes of the IS110 family (Bignell *et al.*, 2010a; Joshi and Loria, 2007; Loria *et al.*, 2006). Moreover, all or part of the PAI could be mobilized and was sufficient to confer virulence to the otherwise non-pathogenic *S. diastatochromogenes* species but not *S. coelicolor*. Subsequent sequence analyses revealed a putative tyrosine site-specific recombinase and an excisionase encoding gene that are often present in integrative conjugative elements (Huguet-Tapia *et al.*, 2011). In Gram-negative bacteria, integrative conjugative elements (ICEs) also typically encode for conjugative machinery most commonly related to T4SS (Wozniak and Waldor, 2010). ICEs of other Actinomycetes are different and tend to encode a FtsK-SpoIIE domain encoding gene for conjugation and appear to integrate in a tRNA gene (Poele *et al.*, 2008). The PAI of *S. turgidiscabies* does not meet either of these criteria, which at this point is by no means definitive evidence against the PAI being an ICE given that ICEs are understudied in Gram-positive bacteria (Kers *et al.*, 2005).

The ability of *R. fascians* D188 to cause growth abnormalities on its hosts is dependent upon an ~200 kb conjugative linear plasmid (Crespi *et al.*, 1992; Desomer *et al.*, 1988). Analysis of the pFiD188 plasmid sequence has led to suggestions that it is a composite molecule that resulted from recombination with horizontally acquired fragments (Francis *et al.*, 2012). The plasmid has four regions, designated R1, R2, R3 and R6 that are syntenic with plasmids of environmental *Rhodococcus* isolates. Three unique regions, coding for 65% of the 184 predicted genes, were identified in pFiD188 (U1, located between regions R3 and R2; U2 between regions R2 and R6; and U3 at the end of pFiD188 after region R6). The U regions are proposed to encode pathogenicity functions specific to *R. fascians* whereas the conserved R regions encode loci required for plasmid maintenance (Francis *et al.*, 2012).

Several regions of pFiD188 were predicted to be acquired via HGT events (Francis *et al.*, 2012). Consistent with conclusions regarding function, only one potentially horizontally acquired region, HGT1, comprised mainly of genes encoding conserved hypothetical proteins, is in a conserved region (i.e. R1). The four other putatively acquired regions are located in U regions with HGT2 and HGT5 flanked at one of their ends by putative transposase or integrase/recombinase genes (pFi_037 and pFi_141, respectively) which may have been involved in the acquisition of these regions (Francis *et al.*, 2012). HGT3 includes the well-characterized virulence loci, *fasR* and *fas*, as well as two methyl transferase genes (Francis *et al.*, 2012). Consistent with the mobility of this region, *fasD* was also found on either a smaller linear plasmid or circular plasmid within a collection of 18 virulent *R. fascians* isolates (Stange *et al.*, 1996).

The distributed genome hypothesis may not apply to all plant pathogens. In *P. aeruginosa* PA14, a survey of a non-redundant transposon library, affected in approximately 80% of non-essential genes, identified nearly 170 genes necessary for virulence. Most were found to be core to *P. aeruginosa* with no bias towards genomic islands (Feinbaum *et al.*, 2012). It was also subsequently demonstrated that *P. aeruginosa* PA14 co-opted a housekeeping molecule, trehalose, for virulence on *Arabidopsis* (Djonović *et al.*, 2013). This non-reducing disaccharide is a common molecule for conferring tolerance to osmotic stress (Strøm and Kaasen, 1993). A similar evolutionary mechanisms of co-option has also been suggested for the equine and opportunistic human pathogen *R. equi* (Letek *et al.*, 2010). In contrast to genome decay as suggested for subspecies of *C. michiganensis*, or substantial HGT events indicative of the distributed genome hypothesis, *R. equi* is suggested to have evolved pathogenicity *via* limited but key acquisition events and co-option of Actinobacterial genes (Letek *et al.*, 2010). One of the innovation events was acquisition of the ~80 kb virulence plasmid (Takai *et al.*, 2000). Evidence of co-option comes from the observed crosstalk between plasmid and chromosomal metabolic genes, which when genetically disrupted, compromised virulence (Letek *et al.*, 2010). As such, members of *Rhodococcus* may already be potentiated towards rapid niche adaptation and a few, but important horizontally acquired genes may be all that is necessary for a host-associated lifestyle. Indeed, similar conclusions can be drawn from preliminary analyses of chromosome sequences from the isolates of plant-pathogenic *Rhodococcus* (Creason *et al.*, 2014). The limited association of putative virulence genes with HGT combined with observations that genetically divergent species of *Rhodococcus* are pathogenic to plants and a protein involved in glyoxylate shunt contributes to virulence, are consistent with such a mechanism of pathogen evolution. Similarly, in *C. michiganensis*, plasmids are suggested to be essential for symptom development, but genes required for other aspects of pathogenesis are located on the chromosome (Gartemann *et al.*, 2008).

Summary

The characterization of Gram-positive plant-pathogenic bacteria is an important step necessary for the understanding of several economically important diseases of crop plants. Equally significant are the contributions that Gram-positive pathogens will have towards a more complete understanding of plant–microbe interactions. Without argument, the discovery and characterization of pathogen effectors that function within host cells have had critical contributions to our understanding of the molecular interactions during host infection and the dissection of host immunity. The potential that studying Gram-positive bacteria will yield new insights is born out of the possibility that these pathogens do not appear to employ translocated effectors and that current models describing plant–bacteria interactions do not seem to fit this group of pathogens. This, along with the observation that all cell-walled Gram-positive pathogens are members of the Actinobacteria, point to the possibility that these pathogens may be more engaged in a chemical warfare (Bednarek, 2012). Similar observations have been reported for *P. aeruginosa* PA14 and *C. elegans*

(Cezairliyan *et al.*, 2013; Troemel *et al.*, 2006). As such, studying Gram-positive bacteria has the potential to contribute to a deeper understanding of small molecule signalling and secondary metabolites in immunity.

Genome sequences have been generated for representative members of Gram-positive phytopathogens and comparisons between species have yielded new insights into the evolution of these bacteria. In *Rhodococcus*, and perhaps *Streptomyces*, large pan genomes accessible by horizontal gene transfer may not be instrumental for their pathogenic lifestyle, but rather for their environmental lifestyles. Instead, current evidence based on a limited number of genome sequences suggest that genetic pliability is critical and driven by few, but key gene acquisitions (Letek *et al.*, 2010).

Genomes can be highly variable in size and content, even within a single bacterial species (Doolittle and Zhaxybayeva, 2009). Thus, comparisons of genome sequences from distantly related organisms only allow for general conclusions. This is exemplified by the observations that 75% of the genes common to pathogenic *Streptomyces* are also present in non-pathogenic members of this genus (Huguet-Tapia and Loria, 2012). Moreover, conclusions on species uniqueness also need to be viewed with a cautious eye since the genome sequence of a single isolate may not sufficiently represent the species, or its pan-genome (Medini *et al.*, 2008). Advances in next generation sequencing technology need to be leveraged for deep draft-genome sequencing within species of Gram-positive pathogens for stronger insights into their virulence and evolution.

Acknowledgements

Work in the Chang lab is supported by DOE, NSF, NIH, USDA, and the Agricultural Research Foundation. Elizabeth A. Savory is supported by USDA NIFA post-doctoral fellowship #2013-03453. This material is based upon work supported by the National Science Foundation Graduate Research Fellowship under Grant No. DGE-1314109 to Edward W. Davis II. Any opinion, findings, and conclusions or recommendations expressed in this material are those of the authors(s) and do not necessarily reflect the views of the National Science Foundation. Edward W. Davis II is also supported by a Provost's Distinguished Graduate Fellowship from Oregon State University. Olivier M. Vandeputte is a Post-doctoral Researcher of the FRS-FNRS (*Fonds de la Recherche Scientifique*, Belgium) and supported by grants from FRS-FNRS and the *Fonds David et Alice Van Buuren* (Belgium).

References

Abdallah, A.M., Gey van Pittius, N.C., Champion, P.A.D., Cox, J., Luirink, J., Vandenbroucke-Grauls, C.M.J.E., Appelmelk, B.J., and Bitter, W. (2007). Type VII secretion – mycobacteria show the way. Nat. Rev. Microbiol. *5*, 883–891.

Adler, N.R., Mahony, A., and Friedman, N.D. (2013). Diphtheria: forgotten, but not gone. Intern. Med. J. *43*, 206–210.

Agarkova, I.V., Vidaver, A.K., Postnikova, E.N., Riley, I.T., and Schaad, N.W. (2006). Genetic characterization and diversity of *Rathayibacter toxicus*. Phytopathology *96*, 1270–1277.

Agarkova, I.V., Lambrecht, P.A., Vidaver, A.K., and Harveson, R.M. (2012). Genetic diversity among *Curtobacterium flaccumfaciens* pv. *flaccumfaciens* populations in the American high plains. Can. J. Microbiol. *58*, 788–801.

Agris, P.F., Vendeix, F.A.P., and Graham, W.D. (2007). tRNA's Wobble decoding of the genome: 40 years of modification. J. Mol. Biol. *366*, 1–13.

Akiyoshi, D.E., Klee, H., Amasino, R.M., Nester, E.W., and Gordon, M.P. (1984). T-DNA of *Agrobacterium tumefaciens* encodes an enzyme of cytokinin biosynthesis. Proc. Natl. Acad. Sci. U.S.A. *81*, 5994–5998.

Alarcón, C., Castro, J., Muñoz, F., Arce-Johnson, P., and Delgado, J. (1998). Protein(s) from the Gram-positive bacterium *Clavibacter michiganensis* subsp. *michiganensis* induces a hypersensitive response in plants. Phytopathology *88*, 306–310.

Alfano, J.R., Bauer, D.W., Milos, T.M., and Collmer, A. (1996). Analysis of the role of the *Pseudomonas syringae* pv. *syringae* HrpZ harpin in elicitation of the hypersensitive response in tobacco using functionally non-polar *hrpZ* deletion mutations, truncated HrpZ fragments, and *hrmA* mutations. Mol. Microbiol. *19*, 715–728.

Armstrong, D.J., Scarbrough, E., and Skoog, F. (1976). Cytokinins in *Corynebacterium fascians* cultures: isolation and identification of 6-(4-hydroxy-3-methyl-cis-2-butenylamino)-2-methylthiopurine. Plant Physiol. *58*, 749–752.

Axtell, M., and Staskawicz, B. (2003). Initiation of RPS2-specified disease resistance in *Arabidopsis* is coupled to the AvrRpt2-directed elimination of RIN4. Cell *112*, 369–377.

Bader, M.W., Navarre, W.W., Shiau, W., Nikaido, H., Frye, J.G., McClelland, M., Fang, F.C., and Miller, S.I. (2003). Regulation of *Salmonella typhimurium* virulence gene expression by cationic antimicrobial peptides. Mol. Microbiol. *50*, 219–230.

Balaji, V., Mayrose, M., Sherf, O., Jacob-Hirsch, J., Eichenlaub, R., Iraki, N., Manulis-Sasson, S., Rechavi, G., Barash, I., and Sessa, G. (2008). Tomato transcriptional changes in response to *Clavibacter michiganensis* subsp. *michiganensis* reveal a role for ethylene in disease development. Plant Physiol. *146*, 1797–1809.

Balaji, V., Sessa, G., and Smart, C.D. (2011). Silencing of host basal defense response-related gene expression increases susceptibility of *Nicotiana benthamiana* to *Clavibacter michiganensis* subsp. *michiganensis*. Phytopathology *101*, 349–357.

Baltrus, D.A., Nishimura, M.T., Romanchuk, A., Chang, J.H., Mukhtar, M.S., Cherkis, K., Roach, J., Grant, S.R., Jones, C.D., and Dangl, J.L. (2011). Dynamic evolution of pathogenicity revealed by sequencing and comparative genomics of 19 *Pseudomonas syringae* isolates. PLoS Pathog. *7*, e1002132.

Barash, I., and Manulis-Sasson, S. (2007). Virulence mechanisms and host specificity of gall-forming *Pantoea agglomerans*. Trends Microbiol. *15*, 538–545.

Barash, I., and Manulis-Sasson, S. (2009). Recent evolution of bacterial pathogens: the gall-forming *Pantoea agglomerans* case. Annu. Rev. Phytopathol. *47*, 133–152.

Bardaji, L., Pérez-Martínez, I., Rodríguez-Moreno, L., Rodríguez-Palenzuela, P., Sundin, G.W., Ramos, C., and Murillo, J. (2011). Sequence and role in virulence of the three plasmid complement of the model tumor-inducing bacterium *Pseudomonas savastanoi* pv. *savastanoi* NCPPB 3335. PLoS ONE *6*, e25705.

von Bargen, K., and Haas, A. (2009). Molecular and infection biology of the horse pathogen *Rhodococcus equi*. FEMS Microbiol. Rev. *33*, 870–891.

Barry, S.M., Kers, J.A., Johnson, E.G., Song, L., Aston, P.R., Patel, B., Krasnoff, S.B., Crane, B.R., Gibson, D.M., Loria, R., *et al.* (2012). Cytochrome P450–catalyzed L-tryptophan nitration in thaxtomin phytotoxin biosynthesis. Nat. Chem. Biol. *8*, 814–816.

Bart, R., Cohn, M., Kassen, A., McCallum, E.J., Shybut, M., Petriello, A., Krasileva, K., Dahlbeck, D., Medina, C., Alicai, T., *et al.* (2012). High-throughput genomic sequencing of cassava bacterial blight strains identifies conserved effectors to target for durable resistance. Proc. Natl. Acad. Sci. U.S.A. *109*, E1972–E1979.

Bauer, D.W., Wei, Z.M., Beer, S.V., and Collmer, A. (1995). *Erwinia chrysanthemi* harpinEch: an elicitor of the hypersensitive response that contributes to soft-rot pathogenesis. Mol. Plant Microbe Interact. *8*, 484–491.

Bednarek, P. (2012). Chemical warfare or modulators of defence responses – the function of secondary metabolites in plant immunity. Curr. Opin. Plant Biol. *15*, 407–414.

Bentley, S.D., and Parkhill, J. (2004). Comparative genomic structure of prokaryotes. Annu. Rev. Genet. *38*, 771–792.

Bentley, S.D., Corton, C., Brown, S.E., Barron, A., Clark, L., Doggett, J., Harris, B., Ormond, D., Quail, M.A., May, G., *et al.* (2008). Genome of the actinomycete plant pathogen *Clavibacter michiganensis* subsp. *sepedonicus* suggests recent niche adaptation. J. Bacteriol. *190*, 2150–2160.

Bignell, D.R.D., Huguet-Tapia, J.C., Joshi, M.V., Pettis, G.S., and Loria, R. (2010a). What does it take to be a plant pathogen: genomic insights from *Streptomyces* species. Antonie Van Leeuwenhoek *98*, 179–194.

Bignell, D.R.D., Seipke, R.F., Huguet-Tapia, J.C., Chambers, A.H., Parry, R.J., and Loria, R. (2010b). *Streptomyces scabies* 87–22 contains a coronafacic acid-like biosynthetic cluster that contributes to plant–microbe interactions. Mol. Plant Microbe Interact. 23, 161–175.

Bischoff, V., Cookson, S.J., Wu, S., and Scheible, W.-R. (2009). Thaxtomin A affects CESA-complex density, expression of cell wall genes, cell wall composition, and causes ectopic lignification in *Arabidopsis thaliana* seedlings. J. Exp. Bot. *60*, 955–965.

Boller, T., and He, S.Y. (2009). Innate immunity in plants: an arms race between pattern recognition receptors in plants and effectors in microbial pathogens. Science *324*, 742–744.

Bouarab, K., Melton, R., Peart, J., Baulcombe, D., and Osbourn, A. (2002). A saponin-detoxifying enzyme mediates suppression of plant defences. Nature *418*, 889–892.

Brodhagen, M., Henkels, M.D., and Loper, J.E. (2004). Positive autoregulation and signaling properties of pyoluteorin, an antibiotic produced by the biological control organism *Pseudomonas fluorescens* Pf-5. Appl. Environ. Microbiol. *70*, 1758–1766.

de Bruijn, I., de Kock, M.J.D., Yang, M., de Waard, P., van Beek, T.A., and Raaijmakers, J.M. (2007). Genome-based discovery, structure prediction and functional analysis of cyclic lipopeptide antibiotics in *Pseudomonas* species. Mol. Microbiol. *63*, 417–428.

Bukhalid, R.A., and Loria, R. (1997). Cloning and expression of a gene from *Streptomyces scabies* encoding a putative pathogenicity factor. J. Bacteriol. *179*, 7776–7783.

Bulgarelli, D., Rott, M., Schlaeppi, K., Ver Loren van Themaat, E., Ahmadinejad, N., Assenza, F., Rauf, P., Huettel, B., Reinhardt, R., Schmelzer, E., *et al.* (2012). Revealing structure and assembly cues for Arabidopsis root-inhabiting bacterial microbiota. Nature *488*, 91–95.

Bulgarelli, D., Schlaeppi, K., Spaepen, S., Ver Loren van Themaat, E., and Schulze-Lefert, P. (2013). Structure and functions of the bacterial microbiota of plants. Annu. Rev. Plant Biol. *64*, 807–838.

Burger, A., Gräfen, I., Engemann, J., Niermann, E., Pieper, M., Kirchner, O., Gartemann, K.-H., and Eichenlaub, R. (2005). Identification of homologues to the pathogenicity factor Pat-1, a putative serine protease of *Clavibacter michiganensis* subsp. *michiganensis*. Microbiol. Res. *160*, 417–427.

Burts, M.L., Williams, W.A., DeBord, K., and Missiakas, D.M. (2005). EsxA and EsxB are secreted by an ESAT-6-like system that is required for the pathogenesis of *Staphylococcus aureus* infections. Proc. Natl. Acad. Sci. U.S.A. *102*, 1169–1174.

Cai, R., Lewis, J., Yan, S., Liu, H., Clarke, C.R., Campanile, F., Almeida, N.F., Studholme, D.J., Lindeberg, M., Schneider, D., *et al.* (2011). The plant pathogen *Pseudomonas syringae* pv. tomato is genetically monomorphic and under strong selection to evade tomato immunity. PLoS Pathog. *7*, e1002130.

Cezairliyan, B., Vinayavekhin, N., Grenfell-Lee, D., Yuen, G.J., Saghatelian, A., and Ausubel, F.M. (2013). Identification of *Pseudomonas aeruginosa* phenazines that kill *Caenorhabditis elegans*. PLoS Pathog. *9*, e1003101.

Challis, G.L., and Hopwood, D.A. (2003). Synergy and contingency as driving forces for the evolution of multiple secondary metabolite production by *Streptomyces* species. Proc. Natl. Acad. Sci. U.S.A. *100* (Suppl. 2), 14555–14561.

Chalupowicz, L., Zellermann, E.-M., Fluegel, M., Dror, O., Eichenlaub, R., Gartemann, K.-H., Savidor, A., Sessa, G., Iraki, N., Barash, I., *et al.* (2012). Colonization and movement of GFP-labeled *Clavibacter michiganensis* subsp. *michiganensis* during tomato infection. Phytopathology *102*, 23–31.

Chang, J.H., Urbach, J.M., Law, T.F., Arnold, L.W., Hu, A., Gombar, S., Grant, S.R., Ausubel, F.M., and Dangl, J.L. (2005). A high-throughput, near-saturating screen for type III effector genes from *Pseudomonas syringae*. Proc. Natl. Acad. Sci. U.S.A. *102*, 2549–2554.

Chater, K.F., Biró, S., Lee, K.J., Palmer, T., and Schrempf, H. (2010). The complex extracellular biology of *Streptomyces*. FEMS Microbiol. Rev. *34*, 171–198.

Chen, Z., Agnew, J.L., Cohen, J.D., He, P., Shan, L., Sheen, J., and Kunkel, B.N. (2007). *Pseudomonas syringae* type III effector AvrRpt2 alters *Arabidopsis thaliana* auxin physiology. Proc. Natl. Acad. Sci. U.S.A. *104*, 20131–20136.

Choi, J., Huh, S.U., Kojima, M., Sakakibara, H., Paek, K.-H., and Hwang, I. (2010). The cytokinin-activated transcription factor ARR2 promotes plant immunity via TGA3/NPR1-dependent salicylic acid signaling in *Arabidopsis*. Dev. Cell *19*, 284–295.

Chung, E.-H., da Cunha, L., Wu, A.-J., Gao, Z., Cherkis, K., Afzal, A.J., Mackey, D., and Dangl, J.L. (2011). Specific threonine phosphorylation of a host target by two unrelated type III effectors activates a host innate immune receptor in plants. Cell Host Microbe *9*, 125–136.

Cianciotto, N.P. (2005). Type II secretion: a protein secretion system for all seasons. Trends Microbiol. *13*, 581–588.

Cornelis, G.R. (2006). The type III secretion injectisome. Nat. Rev. Microbiol. *4*, 811–825.

Cornelis, K., Ritsema, T., Nijsse, J., Holsters, M., Goethals, K., and Jaziri, M. (2001). The plant pathogen *Rhodococcus fascians* colonizes the exterior and interior of the aerial parts of plants. Mol. Plant Microbe Interact. *14*, 599–608.

Cornelis, K., Maes, T., Jaziri, M., Holsters, M., and Goethals, K. (2002). Virulence genes of the phytopathogen *Rhodococcus fascians* show specific spatial and temporal expression patterns during plant infection. Mol. Plant Microbe Interact. *15*, 398–403.

Coulthurst, S.J., Barnard, A.M.L., and Salmond, G.P.C. (2005). Regulation and biosynthesis of carbapenem antibiotics in bacteria. Nat. Rev. Micro. *3*, 295–306.
Creason, A.L., Vandeputte, O.M., Savory, E.A., Davis II, E.W., Pubnam, M.L., Hu, E., Swader-Hines, D., Mol, A., Baucher, M., Prinsen, E., *et al.*, (2014). Analysis of genome sequences from plant pathogenic *Rhodococcus* reveals genetic novelties in virulence Loci. PLoS ONE *9*, e01996.
Crespi, M., Messens, E., Caplan, A.B., Van Montagu, M., and Desomer, J. (1992). Fasciation induction by the phytopathogen *Rhodococcus fascians* depends upon a linear plasmid encoding a cytokinin synthase gene. EMBO J. *11*, 795–804.
Crespi, M., Vereecke, D., Temmerman, W., Van Montagu, M., and Desomer, J. (1994). The fas operon of *Rhodococcus fascians* encodes new genes required for efficient fasciation of host plants. J. Bacteriol. *176*, 2492–2501.
Darriba, D., Taboada, G.L., Doallo, R., and Posada, D. (2011). ProtTest 3: fast selection of best-fit models of protein evolution. Bioinformatics *27*, 1164–1165.
Da Silva, A., Ferro, J., Reinach, F., Farah, C., Furlan, L., Quaggio, R., Monteiro-Vitorello, C., Van Sluys, M., Almeida, N., and Alves, L. (2002). Comparison of the genomes of two *Xanthomonas* pathogens with differing host specificities. Nature *417*, 459–463.
Davis, M.J., Gillaspie, A.G., Harris, R.W., and Lawson, R.H. (1980). Ratoon stunting disease of sugarcane: isolation of the causal bacterium. Science *210*, 1365–1367.
Delepelaire, P. (2004). Type I secretion in Gram-negative bacteria. Biochim. Biophys. Acta *1694*, 149–161.
Depuydt, S., Trenkamp, S., Fernie, A.R., Elftieh, S., Renou, J.-P., Vuylsteke, M., Holsters, M., and Vereecke, D. (2009). An integrated genomics approach to define niche establishment by *Rhodococcus fascians*. Plant Physiol. *149*, 1366–1386.
Desomer, J., Dhaese, P., and Van Montagu, M. (1988). Conjugative transfer of cadmium resistance plasmids in *Rhodococcus fascians* strains. J. Bacteriol. *170*, 2401–2405.
Djonović, S., Urbach, J.M., Drenkard, E., Bush, J., Feinbaum, R., Ausubel, J.L., Traficante, D., Risech, M., Kocks, C., Fischbach, M.A., *et al.* (2013). Trehalose biosynthesis promotes *Pseudomonas aeruginosa* pathogenicity in plants. PLoS Pathog. *9*, e1003217.
Doolittle, W.F., and Zhaxybayeva, O. (2009). On the origin of prokaryotic species. Genome Res. *19*, 744–756.
Dorofeeva, L.V., Evtushenko, L.I., Krausova, V.I., Karpov, A.V., Subbotin, S.A., and Tiedje, J.M. (2002). *Rathayibacter caricis* sp. nov. and *Rathayibacter festucae* sp. nov., isolated from the phyllosphere of *Carex* sp. and the leaf gall induced by the nematode *Anguina graminis* on *Festuca rubra* L., respectively. Int. J. Syst. Bacteriol. *52*, 1917–1923.
Dou, D., and Zhou, J.-M. (2012). Phytopathogen effectors subverting host immunity: different foes, similar battleground. Cell Host Microbe *12*, 484–495.
Dreier, J., Meletzus, D., and Eichenlaub, R. (1997). Characterization of the plasmid encoded virulence region pat-1 of phytopathogenic *Clavibacter michiganensis* subsp. *michiganensis*. Mol. Plant Microbe Interact. *10*, 195–206.
Duerkop, B.A., Varga, J., Chandler, J.R., Peterson, S.B., Herman, J.P., Churchill, M.E.A., Parsek, M.R., Nierman, W.C., and Greenberg, E.P. (2009). Quorum-sensing control of antibiotic synthesis in *Burkholderia thailandensis*. J. Bacteriol. *191*, 3909–3918.
Duval, I., and Beaudoin, N. (2009). Transcriptional profiling in response to inhibition of cellulose synthesis by thaxtomin A and isoxaben in *Arabidopsis thaliana* suspension cells. Plant Cell Rep. *28*, 811–830.
Duval, I., Brochu, V., Simard, M., Beaulieu, C., and Beaudoin, N. (2005). Thaxtomin A induces programmed cell death in *Arabidopsis thaliana* suspension-cultured cells. Planta *222*, 820–831.
Eason, J., Morris, R., and Jameson, P. (1996). The relationship between virulence and cytokinin production by *Rhodococcus fascians* (Tilford 1936) Goodfellow 1984. Plant Pathol. *45*, 323–331.
Ehrlich, G.D., Hiller, N.L., and Hu, F. (2008). What makes pathogens pathogenic. Genome Biol. *9*, 225.
Eichenlaub, R., and Gartemann, K.-H. (2011). The *Clavibacter michiganensis* subspecies: molecular investigation of Gram-positive bacterial plant pathogens. Annu. Rev. Phytopatho. *49*, 445–464.
Einset, J.W., and Skoog, F.K. (1977). Isolation and identification of ribosyl – zeatin from transfer RNA of *Corynebacterium fascians*. Biochem. Biophys. Res. Commun. *79*, 1117–1121.
Esko, J.D., Doering, T.L., and Raetz, C.R. (2009). Eubacteria and Archaea. In Essentials of Glycobiology, 2nd edn, A. Varki, R.D. Cummings, J.D. Esko, H.H. Freeze, P. Stanley, C.R. Bertozzi, G.W. Hart, and E. ME, eds. (Cold Spring Harbor Laboratory Press, Cold Spring Harbor, NY, USA), pp. 575–586.
Feinbaum, R.L., Urbach, J.M., Liberati, N.T., Djonović, S., Adonizio, A., Carvunis, A.-R., and Ausubel, F.M. (2012). Genome-wide identification of *Pseudomonas aeruginosa* virulence-related genes using a *Caenorhabditis elegans* infection model. PLoS Pathog. *8*, e1002813.

Feng, F., and Zhou, J.-M. (2012). Plant–bacterial pathogen interactions mediated by type III effectors. Curr. Opin. Plant Biol. *15*, 469–476.

Flärdh, K., and Buttner, M.J. (2009). Streptomyces morphogenetics: dissecting differentiation in a filamentous bacterium. Nat. Rev. Micro. *7*, 36–49.

Francis, I., Holsters, M., and Vereecke, D. (2010). The Gram-positive side of plant–microbe interactions. Environ. Microbiol. *12*, 1–12.

Francis, I., De Keyser, A., De Backer, P., Simón-Mateo, C., Kalkus, J., Pertry, I., Ardiles-Diaz, W., De Rycke, R., Vandeputte, O.M., Jaziri, El, M., *et al.* (2012). pFiD188, the linear virulence plasmid of *Rhodococcus fascians* D188. Mol. Plant Microbe Interact. *25*, 637–647.

Frébort, I., Kowalska, M., Hluska, T., Frébortová, J., and Galuszka, P. (2011). Evolution of cytokinin biosynthesis and degradation. J. Exp. Bot. *62*, 2431–2452.

Frederick, R.D., Ahmad, M., Majerczak, D.R., Arroyo-Rodríguez, A.S., Manulis, S., and Coplin, D.L. (2001). Genetic organization of the *Pantoea stewartii* subsp. *stewartii hrp* gene cluster and sequence analysis of the *hrpA, hrpC, hrpN,* and *wtsE* operons. Mol. Plant Microbe Interact. *14*, 1213–1222.

Frugier, F., Kosuta, S., Murray, J.D., Crespi, M., and Szczyglowski, K. (2008). Cytokinin: secret agent of symbiosis. Trends Plant Sci. *13*, 115–120.

Fyans, J.K., Bignell, D., Loria, R., Toth, I., and Palmer, T. (2013). The ESX/type VII secretion system modulates development, but not virulence, of the plant pathogen *Streptomyces scabies*. Mol. Plant Pathol. *14*, 119–130.

Gallegos, M.T., Schleif, R., Bairoch, A., Hofmann, K., and Ramos, J.L. (1997). Arac/XylS family of transcriptional regulators. Microbiol. Mol. Biol. Rev. *61*, 393–410.

Gao, B., and Gupta, R.S. (2012). Phylogenetic framework and molecular signatures for the main clades of the phylum Actinobacteria. Microbiol. Mol. Biol. Rev. *76*, 66–112.

Gartemann, K.-H., Abt, B., Bekel, T., Burger, A., Engemann, J., Flügel, M., Gaigalat, L., Goesmann, A., Gräfen, I., Kalinowski, J., *et al.* (2008). The genome sequence of the tomato-pathogenic actinomycete *Clavibacter michiganensis* subsp. *michiganensis* NCPPB382 reveals a large island involved in pathogenicity. J. Bacteriol. *190*, 2138–2149.

Garufi, G., Butler, E., and Missiakas, D. (2008). ESAT-6-like protein secretion in *Bacillus anthracis*. J. Bacteriol. *190*, 7004–7011.

Genin, S., and Boucher, C. (2004). Lessons learned from the genome analysis of *Ralstonia solanacearum*. Annu. Rev. Phytopathol. *42*, 107–134.

Glazebrook, J. (2005). Contrasting mechanisms of defense against biotrophic and necrotrophic pathogens. Annu. Rev. Phytopathol. *43*, 205–227.

Goh, E.-B., Yim, G., Tsui, W., McClure, J., Surette, M.G., and Davies, J. (2002). Transcriptional modulation of bacterial gene expression by subinhibitory concentrations of antibiotics. Proc. Natl. Acad. Sci. U.S.A. *99*, 17025–17030.

González-Lamothe, R., Mitchell, G., Gattuso, M., Diarra, M.S., Malouin, F., and Bouarab, K. (2009). Plant antimicrobial agents and their effects on plant and human pathogens. Int. J. Mol. Sci. *10*, 3400–3419.

Gross, H., and Loper, J.E. (2009). Genomics of secondary metabolite production by *Pseudomonas* spp. Nat. Prod. Rep. *26*, 1408–1446.

Guan, D., Grau, B.L., Clark, C.A., Taylor, C.M., Loria, R., and Pettis, G.S. (2012). Evidence that thaxtomin C is a pathogenicity determinant of *Streptomyces ipomoeae*, the causative agent of *Streptomyces* soil rot disease of sweet potato. Mol. Plant Microbe Interact. *25*, 393–401.

Gust, A.A., Biswas, R., Lenz, H.D., Rauhut, T., Ranf, S., Kemmerling, B., Götz, F., Glawischnig, E., Lee, J., Felix, G., *et al.* (2007). Bacteria-derived peptidoglycans constitute pathogen-associated molecular patterns triggering innate immunity in *Arabidopsis*. J Biol. Chem. *282*, 32338–32348.

Guttman, D.S., Vinatzer, B.A., Sarkar, S.F., Ranall, M.V., Kettler, G., and Greenberg, J.T. (2002). A functional screen for the type III (Hrp) secretome of the plant pathogen *Pseudomonas syringae*. Science *295*, 1722–1726.

Harveson, R.M., and Schwartz, H.F. (2007). Bacterial diseases of dry edible beans in the central high plains. Online. Plant Health Progress. doi: 10.1094/PHP-2007-0125-01-DG.

Healy, F.G., Wach, M., Krasnoff, S.B., Gibson, D.M., and Loria, R. (2000). The *txtAB* genes of the plant pathogen *Streptomyces acidiscabies* encode a peptide synthetase required for phytotoxin thaxtomin A production and pathogenicity. Mol. Microbiol. *38*, 794–804.

Healy, F.G., Krasnoff, S.B., Wach, M., Gibson, D.M., and Loria, R. (2002). Involvement of a cytochrome P450 monooxygenase in thaxtomin A biosynthesis by *Streptomyces acidiscabies*. J. Bacteriol. *184*, 2019–2029.

Helgeson, J.P., and Leonard, N.J. (1966). Cytokinins: identification of compounds isolated from *Corynebacterium fascians*. Proc. Natl. Acad. Sci. U.S.A. *56*, 60–63.
Henderson, I.R., Navarro-Garcia, F., Desvaux, M., Fernandez, R.C., and Ala'Aldeen, D. (2004). Type V protein secretion pathway: the autotransporter story. Microbiol. Mol. Biol. Rev. *68*, 692–744.
Hocher, V., Alloisio, N., Auguy, F., Fournier, P., Doumas, P., Pujic, P., Gherbi, H., Queiroux, C., Da Silva, C., Wincker, P., *et al.* (2011). Transcriptomics of actinorhizal symbioses reveals homologs of the whole common symbiotic signaling cascade. Plant Physiol. *156*, 700–711.
Holeva, M.C., Bell, K.S., Hyman, L.J., Avrova, A.O., Whisson, S.C., Birch, P.R.J., and Toth, I.K. (2004). Use of a pooled transposon mutation grid to demonstrate roles in disease development for *Erwinia carotovora* subsp. *atroseptica* putative type III secreted effector (DspE/A) and helper (HrpN) proteins. Mol. Plant Microbe Interact. *17*, 943–950.
Hoy, J.W., Grisham, M.P., and Damann, K.E. (1999). Spread and increase of ratoon stunting disease of sugarcane and comparison of disease detection methods. Mol. Plant Microbe Interact. *83*, 1170–1175.
Hsu, F.-F., Soehl, K., Turk, J., and Haas, A. (2011). Characterization of mycolic acids from the pathogen *Rhodococcus equi* by tandem mass spectrometry with electrospray ionization. Anal. Biochem. *409*, 112–122.
Huguet-Tapia, J.C., and Loria, R. (2012). Draft genome sequence of *Streptomyces acidiscabies* 84–104, an emergent plant pathogen. J. Bacteriol. *194*, 1847.
Huguet-Tapia, J.C., Badger, J.H., Loria, R., and Pettis, G.S. (2011). *Streptomyces turgidiscabies* Car8 contains a modular pathogenicity island that shares virulence genes with other actinobacterial plant pathogens. Plasmid *65*, 118–124.
Irazoqui, J.E., Troemel, E.R., Feinbaum, R.L., Luhachack, L.G., Cezairliyan, B.O., and Ausubel, F.M. (2010). Distinct pathogenesis and host responses during infection of *C. elegans* by *P. aeruginosa* and *S. aureus*. PLoS Pathog. *6*, e1000982.
Jahr, H., Dreier, J., Meletzus, D., Bahro, R., and Eichenlaub, R. (2000). The endo-beta-1,4-glucanase CelA of *Clavibacter michiganensis* subsp. *michiganensis* is a pathogenicity determinant required for induction of bacterial wilt of tomato. Mol. Plant Microbe Interact. *13*, 703–714.
Jha, G., Rajeshwari, R., and Sonti, R.V. (2007). Functional interplay between two *Xanthomonas oryzae* pv. *oryzae* secretion systems in modulating virulence on rice. Mol. Plant Microbe Interact. *20*, 31–40.
Johnson, E.G., Krasnoff, S.B., Bignell, D.R.D., Chung, W.-C., Tao, T., Parry, R.J., Loria, R., and Gibson, D.M. (2009). 4-Nitrotryptophan is a substrate for the non-ribosomal peptide synthetase TxtB in the thaxtomin A biosynthetic pathway. Mol. Microbiol. 73, 409–418.
Jones, J.D.G., and Dangl, J.L. (2006). The plant immune system. Nature *444*, 323–329.
Joshi, M.V., and Loria, R. (2007). *Streptomyces turgidiscabies* possesses a functional cytokinin biosynthetic pathway and produces leafy galls. Mol. Plant Microbe Interact. *20*, 751–758.
Joshi, M.V., Bignell, D.R.D., Johnson, E.G., Sparks, J.P., Gibson, D.M., and Loria, R. (2007a). The AraC/XylS regulator TxtR modulates thaxtomin biosynthesis and virulence in *Streptomyces scabies*. Mol. Microbiol. *66*, 633–642.
Joshi, M., Rong, X., Moll, S., Kers, J., Franco, C., and Loria, R. (2007b). *Streptomyces turgidiscabies* secretes a novel virulence protein, Nec1, which facilitates infection. Mol. Plant Microbe Interact. *20*, 599–608.
Joshi, M.V., Mann, S.G., Antelmann, H., Widdick, D.A., Fyans, J.K., Chandra, G., Hutchings, M.I., Toth, I., Hecker, M., Loria, R., *et al.* (2010). The twin arginine protein transport pathway exports multiple virulence proteins in the plant pathogen *Streptomyces scabies*. Mol. Microbiol. 77, 252–271.
Kakimoto, T. (2003). Biosynthesis of cytokinins. J. Plant Res. *116*, 233–239.
Karlin, S., Campbell, A.M., and Mrázek, J. (1998). Comparative DNA analysis across diverse genomes. Annu. Rev. Genet. *32*, 185–225.
Katagiri, F., and Tsuda, K. (2010). Understanding the plant immune system. Mol. Plant Microbe Interact. *23*, 1531–1536.
Katoh, K., and Standley, D.M. (2013). MAFFT multiple sequence alignment software version 7: improvements in performance and usability. Mol. Biol. Evol. *30*, 772–780.
Katsir, L., Schilmiller, A.L., Staswick, P.E., He, S.Y., and Howe, G.A. (2008). COI1 is a critical component of a receptor for jasmonate and the bacterial virulence factor coronatine. Proc. Natl. Acad. Sci. U.S.A. *105*, 7100–7105.
Kaup, O., Gräfen, I., Zellermann, E.-M., Eichenlaub, R., and Gartemann, K.-H. (2005). Identification of a tomatinase in the tomato-pathogenic actinomycete *Clavibacter michiganensis* subsp. *michiganensis* NCPPB382. Mol. Plant Microbe Interact. *18*, 1090–1098.

Kers, J.A., Wach, M.J., Krasnoff, S.B., Widom, J., Cameron, K.D., Bukhalid, R.A., Gibson, D.M., Crane, B.R., and Loria, R. (2004). Nitration of a peptide phytotoxin by bacterial nitric oxide synthase. Nature *429*, 79–82.

Kers, J.A., Cameron, K.D., Joshi, M.V., Bukhalid, R.A., Morello, J.E., Wach, M.J., Gibson, D.M., and Loria, R. (2005). A large, mobile pathogenicity island confers plant pathogenicity on *Streptomyces* species. Mol. Microbiol. *55*, 1025–1033.

Kidarsa, T.A., Goebel, N.C., Zabriskie, T.M., and Loper, J.E. (2011). Phloroglucinol mediates cross-talk between the pyoluteorin and 2,4-diacetylphloroglucinol biosynthetic pathways in *Pseudomonas fluorescens* Pf-5. Mol. Microbiol. *81*, 395–414.

King, R.R., and Calhoun, L.A. (2009). The thaxtomin phytotoxins: sources, synthesis, biosynthesis, biotransformation and biological activity. Phytochemistry *70*, 833–841.

Klämbt, D., Thies, G., and Skoog, F. (1966). Isolation of cytokinins from *Corynebacterium fascians*. Proc. Natl. Acad. Sci. U.S.A. *56*, 52–59.

Koenig, R.L., Morris, R.O., and Polacco, J.C. (2002). tRNA is the source of low-level trans-zeatin production in *Methylobacterium* spp. J. Bacteriol. *184*, 1832–1842.

Koonin, E.V. (2009). Evolution of genome architecture. Int. J. Biochem. Cell Biol. *41*, 298–306.

Krzywinski, M., Schein, J., Birol, I., Connors, J., Gascoyne, R., Horsman, D., Jones, S.J., and Marra, M.A. (2009). Circos: an information aesthetic for comparative genomics. Genome Res. *19*, 1639–1645.

Kurakawa, T., Ueda, N., Maekawa, M., Kobayashi, K., Kojima, M., Nagato, Y., Sakakibara, H., and Kyozuka, J. (2007). Direct control of shoot meristem activity by a cytokinin-activating enzyme. Nature *445*, 652–655.

Larkin, M.J., Kulakov, L.A., and Allen, C.C.R. (2005). Biodegradation and *Rhodococcus* – masters of catabolic versatility. Curr. Opin. Biotechnol. *16*, 282–290.

Lerat, S., Simao-Beaunoir, A.-M., and Beaulieu, C. (2009). Genetic and physiological determinants of *Streptomyces scabies* pathogenicity. Mol. Plant Pathol. *10*, 579–585.

Lerat, S., Simao-Beaunoir, A.-M., Wu, R., Beaudoin, N., and Beaulieu, C. (2010). Involvement of the plant polymer Suberin and the disaccharide cellobiose in triggering thaxtomin A biosynthesis, a phytotoxin produced by the pathogenic agent *Streptomyces scabies*. Phytopathology *100*, 91–96.

Letek, M., González, P., Macarthur, I., Rodríguez, H., Freeman, T.C., Valero-Rello, A., Blanco, M., Buckley, T., Cherevach, I., Fahey, R., *et al.* (2010). The genome of a pathogenic rhodococcus: cooptive virulence underpinned by key gene acquisitions. PLoS Genet. *6*, e1001145.

Linares, J.F., Gustafsson, I., Baquero, F., and Martinez, J.L. (2006). Antibiotics as intermicrobial signaling agents instead of weapons. Proc. Natl. Acad. Sci. U.S.A. *103*, 19484–19489.

Liu, H., Coulthurst, S.J., Pritchard, L., Hedley, P.E., Ravensdale, M., Humphris, S., Burr, T., Takle, G., Brurberg, M.-B., Birch, P.R.J., *et al.* (2008). Quorum sensing coordinates brute force and stealth modes of infection in the plant pathogen *Pectobacterium atrosepticum*. PLoS Pathog. *4*, e1000093.

Liu, J., Elmore, J.M., Lin, Z.-J.D., and Coaker, G. (2011). A receptor-like cytoplasmic kinase phosphorylates the host target RIN4, leading to the activation of a plant innate immune receptor. Cell Host Microbe *9*, 137–146.

Lorenz, M.C., and Fink, G.R. (2002). Life and death in a macrophage: role of the glyoxylate cycle in virulence. Eukaryotic Cell *1*, 657–662.

Loria, R., Kers, J., and Joshi, M. (2006). Evolution of plant pathogenicity in *Streptomyces*. Annu. Rev. Phytopathol. *44*, 469–487.

Loria, R., Bignell, D.R.D., Moll, S., Huguet-Tapia, J.C., Joshi, M.V., Johnson, E.G., Seipke, R.F., and Gibson, D.M. (2008). Thaxtomin biosynthesis: the path to plant pathogenicity in the genus *Streptomyces*. Antonie Van Leeuwenhoek *94*, 3–10.

Lundberg, D.S., Lebeis, S.L., Paredes, S.H., Yourstone, S., Gehring, J., Malfatti, S., Tremblay, J., Engelbrektson, A., Kunin, V., del Rio, T.G., *et al.* (2012). Defining the core *Arabidopsis thaliana* root microbiome. Nature *488*, 86–90.

McCann, H.C., Nahal, H., Thakur, S., and Guttman, D.S. (2012). Identification of innate immunity elicitors using molecular signatures of natural selection. Proc. Natl. Acad. Sci. U.S.A. *109*, 4215–4220.

McDowell, J.M., and Simon, S.Y. (2006). Recent insights into *R* gene evolution. Mol. Plant Pathol. *7*, 437–448.

McGowan, S.J., Barnard, A.M.L., Bosgelmez, G., Sebaihia, M., Simpson, N.J.L., Thomson, N.R., Todd, D.E., Welch, M., Whitehead, N.A., and Salmond, G.P.C. (2005). Carbapenem antibiotic biosynthesis in *Erwinia carotovora* is regulated by physiological and genetic factors modulating the quorum sensing-dependent control pathway. Mol. Microbiol. *55*, 526–545.

McKenney, P.T., Driks, A., and Eichenberger, P. (2013). The *Bacillus subtilis* endospore: assembly and functions of the multilayered coat. Nat. Rev. Microbiol. *11*, 33–44.

Mackey, D., Belkhadir, Y., Alonso, J.M., Ecker, J.R., and Dangl, J.L. (2003). Arabidopsis RIN4 is a target of the type III virulence effector AvrRpt2 and modulates RPS2-mediated resistance. Cell *112*, 379–389.

McNabe, M., Tennant, R., Danelishvili, L., Young, L., and Bermudez, L.E. (2011). *Mycobacterium avium* ssp. *hominissuis* biofilm is composed of distinct phenotypes and influenced by the presence of antimicrobials. Clin. Microbiol. Infect. *17*, 697–703.

McNally, R.R., Toth, I.K., Cock, P.J.A., Pritchard, L., Hedley, P.E., Morris, J.A., Zhao, Y., and Sundin, G.W. (2012). Genetic characterization of the HrpL regulon of the fire blight pathogen *Erwinia amylovora* reveals novel virulence factors. Mol. Plant Pathol. *13*, 160–173.

Maes, T., Vereecke, D., Ritsema, T., Cornelis, K., Thu, H.N., Van Montagu, M., Holsters, M., and Goethals, K. (2001). The *att* locus of *Rhodococcus fascians* strain D188 is essential for full virulence on tobacco through the production of an autoregulatory compound. Mol. Microbiol. *42*, 13–28.

Mansfield, J., Genin, S., Magori, S., Citovsky, V., Sriariyanum, M., Ronald, P., Dow, M., Verdier, V., Beer, S.V., Machado, M.A., *et al.* (2012). Top 10 plant-pathogenic bacteria in molecular plant pathology. Mol. Plant Pathol *13*, 614–629.

Manulis, S., Haviv-Chesner, A., Brandl, M.T., Lindow, S.E., and Barash, I. (1998). Differential involvement of indole-3-acetic acid biosynthetic pathways in pathogenicity and epiphytic fitness of *Erwinia herbicola* pv. *gypsophilae*. Mol. Plant Microbe Interact. *11*, 634–642.

Matsubara, S., Armstrong, D.J., and Skoog, F. (1968). Cytokinins in tRNA of *Corynebacterium fascians*. Plant Physiol. *43*, 451–453.

Maurelli, A.T., Fernández, R.E., Bloch, C.A., Rode, C.K., and Fasano, A. (1998). 'Black holes' and bacterial pathogenicity: a large genomic deletion that enhances the virulence of *Shigella* spp. and enteroinvasive *Escherichia coli*. Proc. Natl. Acad. Sci. U.S.A. *95*, 3943–3948.

Maurhofer, M., Baehler, E., Notz, R., Martinez, V., and Keel, C. (2004). Cross talk between 2,4-diacetylphloroglucinol-producing biocontrol pseudomonads on wheat roots. Appl. Environ. Microbiol. *70*, 1990–1998.

Medini, D., Serruto, D., Parkhill, J., Relman, D.A., Donati, C., Moxon, R., Falkow, S., and Rappuoli, R. (2008). Microbiology in the post-genomic era. Nat. Rev. Microbiol. *6*, 419–430.

Meletzus, D., Bermphol, A., Dreier, J., and Eichenlaub, R. (1993). Evidence for plasmid-encoded virulence factors in the phytopathogenic bacterium *Clavibacter michiganensis* subsp. *michiganensis* NCPPB382. J. Bacteriol. *175*, 2131–2136.

Mittler, R., Vanderauwera, S., Gollery, M., and Van Breusegem, F. (2004). Reactive oxygen gene network of plants. Trends Plant Sci. *9*, 490–498.

Miyata, S., Casey, M., Frank, D.W., Ausubel, F.M., and Drenkard, E. (2003). Use of the *Galleria mellonella* caterpillar as a model host to study the role of the type III secretion system in *Pseudomonas aeruginosa* pathogenesis. Infect. Immun. *71*, 2404–2413.

Monteiro-Vitorello, C.B., Camargo, L.E.A., Van Sluys, M.A., Kitajima, J.P., Truffi, D., do Amaral, A.M., Harakava, R., de Oliveira, J.C.F., Wood, D., de Oliveira, M.C., *et al.* (2004). The genome sequence of the Gram-positive sugarcane pathogen *Leifsonia xyli* subsp. *xyli*. Mol. Plant Microbe Interact. *17*, 827–836.

Mukaihara, T., Tamura, N., and Iwabuchi, M. (2010). Genome-wide identification of a large repertoire of *Ralstonia solanacearum* type III effector proteins by a new functional screen. Mol. Plant Microbe Interact. *23*, 251–262.

Mukhtar, M.S., Carvunis, A.-R., Dreze, M., Epple, P., Steinbrenner, J., Moore, J., Tasan, M., Galli, M., Hao, T., Nishimura, M.T., *et al.* (2011). Independently evolved virulence effectors converge onto hubs in a plant immune system network. Science *333*, 596–601.

Mur, L.A.J., Kenton, P., Lloyd, A.J., Ougham, H., and Prats, E. (2008). The hypersensitive response; the centenary is upon us but how much do we know? J. Exp. Bot. *59*, 501–520.

Murai, N., Skoog, F., Doyle, M.E., and Hanson, R.S. (1980). Relationships between cytokinin production, presence of plasmids, and fasciation caused by strains of *Corynebacterium fascians*. Proc. Natl. Acad. Sci. U.S.A. *77*, 619–623.

Naseem, M., and Dandekar, T. (2012). The role of auxin-cytokinin antagonism in plant–pathogen interactions. PLoS Pathog. *8*, e1003026.

Naseem, M., Philippi, N., Hussain, A., Wangorsch, G., Ahmed, N., and Dandekar, T. (2012). Integrated systems view on networking by hormones in *Arabidopsis* immunity reveals multiple crosstalk for cytokinin. Plant Cell *24*, 1793–1814.

Natale, P., Brüser, T., and Driessen, A.J.M. (2008). Sec- and Tat-mediated protein secretion across the bacterial cytoplasmic membrane – distinct translocases and mechanisms. Biochim. Biophys. Acta *1778*, 1735–1756.

Navarro, L., Zipfel, C., Rowland, O., Keller, I., Robatzek, S., Boller, T., and Jones, J.D.G. (2004). The transcriptional innate immune response to flg22. Interplay and overlap with Avr gene-dependent defense responses and bacterial pathogenesis. Plant Physiol. *135*, 1113–1128.

Navarro, L., Dunoyer, P., Jay, F., Arnold, B., Dharmasiri, N., Estelle, M., Voinnet, O., and Jones, J.D.G. (2006). A plant miRNA contributes to antibacterial resistance by repressing auxin signaling. Science *312*, 436–439.

Nicaise, V., Roux, M., and Zipfel, C. (2009). Recent advances in PAMP-triggered immunity against bacteria: pattern recognition receptors watch over and raise the alarm. Plant Physiol. *150*, 1638–1647.

Niederweis, M., Danilchanka, O., Huff, J., Hoffmann, C., and Engelhardt, H. (2010). *Mycobacterial* outer membranes: in search of proteins. Trends Microbiol. *18*, 109–116.

Nikolaeva, E., Park, S., Kang, S.T., and Kim, S. (2009). Ratios of cells with and without virulence genes in *Rhodococcus fascians* populations correlate with degrees of symptom development. Plant Dis. 1–8.

Nissinen, R., Lai, F.M., Laine, M.J., Bauer, P.J., Reilley, A.A., Li, X., De Boer, S.H., Ishimaru, C.A., and Metzler, M.C. (1997). *Clavibacter michiganensis* subsp. *Sepedonicus* elicits a hypersensitive response in tobacco and secretes hypersensitive response-inducing protein(s). Phytopathology *87*, 678–684.

Nissinen, R., Xia, Y., Mattinen, L., Ishimaru, C.A., Knudson, D.L., Knudson, S.E., Metzler, M., and Pirhonen, M. (2009). The putative secreted serine protease Chp-7 is required for full virulence and induction of a nonhost hypersensitive response by *Clavibacter michiganensis* subsp. *sepedonicus*. Mol. Plant Microbe Interact. *22*, 809–819.

Nizet, V., and Esko, J.D. (2009). Bacterial and viral infections. In Essentials of Glycobiology, A. Varki, R.D. Cummings, J.D. Esko, H.H. Freeze, P. Stanley, C.R. Bertozzi, G.W. Hart, and M.E. Etzler, eds (Cold Spring Harbor Laboratory Press, Cold Spring Harbor, NY, USA).

O'Brien, J., and Wright, G.D. (2011). An ecological perspective of microbial secondary metabolism. Curr. Opin. Biotechnol. *22*, 552–558.

Ochman, H., and Davalos, L.M. (2006). The nature and dynamics of bacterial genomes. Science *311*, 1730–1733.

Ortíz-Castro, R., Valencia-Cantero, E., and López-Bucio, J. (2008). Plant growth promotion by *Bacillus megaterium* involves cytokinin signaling. Plant Signal Behav. *3*, 263–265.

Pallai, R., Hynes, R.K., Verma, B., and Nelson, L.M. (2012). Phytohormone production and colonization of canola (*Brassica napus* L.) roots by *Pseudomonas fluorescens* 6–8 under gnotobiotic conditions. Can. J. Microbiol. *58*, 170–178.

Pareja-Jaime, Y., Roncero, M.I.G., and Ruiz-Roldan, M.C. (2008). Tomatinase from *Fusarium oxysporum* f. sp. *lycopersici* is required for full virulence on tomato plants. Mol. Plant Microbe Interact. *21*, 728–736.

Parkhill, J., Sebaihia, M., Preston, A., Murphy, L.D., Thomson, N., Harris, D.E., Holden, M.T.G., Churcher, C.M., Bentley, S.D., Mungall, K.L., *et al.* (2003). Comparative analysis of the genome sequences of *Bordetella pertussis*, *Bordetella parapertussis* and *Bordetella bronchiseptica*. Nat. Genet. 35, 32–40.

Pauwels, L., and Goossens, A. (2011). The JAZ proteins: a crucial interface in the Jasmonate signaling cascade. Plant Cell 23, 3089–3100.

Pawlowski, K., and Demchenko, K.N. (2012). The diversity of actinorhizal symbiosis. Protoplasma *249*, 967–979.

Peck, M.W., Stringer, S.C., and Carter, A.T. (2011). *Clostridium botulinum* in the post-genomic era. Food Microbiol. *28*, 183–191.

Pertry, I., Václavíková, K., Depuydt, S., Galuszka, P., Spíchal, L., Temmerman, W., Stes, E., Schmülling, T., Kakimoto, T., Van Montagu, M.C.E., *et al.* (2009). Identification of *Rhodococcus fascians* cytokinins and their modus operandi to reshape the plant. Proc. Natl. Acad. Sci. U.S.A. *106*, 929–934.

Pertry, I., Václavíková, K., Gemrotová, M., Spíchal, L., Galuszka, P., Depuydt, S., Temmerman, W., Stes, E., De Keyser, A., Riefler, M., *et al.* (2010). *Rhodococcus fascians* impacts plant development through the dynamic fas-mediated production of a cytokinin mix. Mol. Plant Microbe Interact. *23*, 1164–1174.

Petnicki-Ocwieja, T., Schneider, D.J., Tam, V.C., Chancey, S.T., Shan, L., Jamir, Y., Schechter, L.M., Janes, M.D., Buell, C.R., Tang, X., *et al.* (2002). Genomewide identification of proteins secreted by the Hrp type III protein secretion system of *Pseudomonas syringae* pv. tomato DC3000. Proc. Natl. Acad. Sci. U.S.A. *99*, 7652–7657.

Phizicky, E.M., and Hopper, A.K. (2010). tRNA biology charges to the front. Genes Dev. *24*, 1832–1860.

Plotnikova, J.M., Rahme, L.G., and Ausubel, F.M. (2000). Pathogenesis of the human opportunistic pathogen *Pseudomonas aeruginosa* PA14 in *Arabidopsis*. Plant Physiol. *124*, 1766–1774.

Poele, te, E.M., Bolhuis, H., and Dijkhuizen, L. (2008). Actinomycete integrative and conjugative elements. Antonie Van Leeuwenhoek *94*, 127–143.

Prithiviraj, B., Bais, H.P., Jha, A.K., and Vivanco, J.M. (2005). *Staphylococcus aureus* pathogenicity on *Arabidopsis thaliana* is mediated either by a direct effect of salicylic acid on the pathogen or by SA-dependent, NPR1-independent host responses. Plant J. *42*, 417–432.

Putnam, M.L., and Miller, M.L. (2007). *Rhodococcus fascians* in herbaceous perennials. Plant Dis. *91*, 1064–1076.

Pym, A.S., Brodin, P., Majlessi, L., Brosch, R., Demangel, C., Williams, A., Griffiths, K.E., Marchal, G., Leclerc, C., and Cole, S.T. (2003). Recombinant BCG exporting ESAT-6 confers enhanced protection against tuberculosis. Nat. Med. *9*, 533–539.

Rajaonson, S., Vandeputte, O.M., Vereecke, D., Kiendrebeogo, M., Ralambofetra, E., Stévigny, C., Duez, P., Rabemanantsoa, C., Mol, A., Diallo, B., *et al.* (2011). Virulence quenching with a prenylated isoflavanone renders the Malagasy legume *Dalbergia pervillei* resistant to *Rhodococcus fascians*. Environ. Microbiol. *13*, 1236–1252.

Rangaswamy, V., Jiralerspong, S., Parry, R., and Bender, C.L. (1998). Biosynthesis of the *Pseudomonas* polyketide coronafacic acid requires monofunctional and multifunctional polyketide synthase proteins. Proc. Natl. Acad. Sci. U.S.A. *95*, 15469–15474.

Rathbone, M.P., and Hall, R.H. (1972). Concerning the presence of the cytokinin, N 6-(Δ2-isopentnyl) adenine, in cultures of *Corynebacterium fascians*. Planta *108*, 93–102.

Ray, A., Cot, M., Puzo, G., Gilleron, M., and Nigou, J. (2013). Bacterial cell wall macroamphiphiles: pathogen-/microbe-associated molecular patterns detected by mammalian innate immune system. Biochimie *95*, 33–42.

Robert-Seilaniantz, A., Grant, M., and Jones, J.D.G. (2011a). Hormone crosstalk in plant disease and defense: more than just JASMONATE-SALICYLATE antagonism. Annu. Rev. Phytopathol. *49*, 317–343.

Robert-Seilaniantz, A., Maclean, D., Jikumaru, Y., Hill, L., Yamaguchi, S., Kamiya, Y., and Jones, J.D.G. (2011b). The microRNA miR393 re-directs secondary metabolite biosynthesis away from camalexin and towards glucosinolates. Plant J. *67*, 218–231.

Roden, J.A., Belt, B., Ross, J.B., Tachibana, T., Vargas, J., and Mudgett, M.B. (2004). A genetic screen to isolate type III effectors translocated into pepper cells during *Xanthomonas* infection. Proc. Natl. Acad. Sci. U.S.A. *101*, 16624–16629.

Romero, D., Traxler, M.F., López, D., and Kolter, R. (2011). Antibiotics as signal molecules. Chem. Rev. *111*, 5492–5505.

Ryan, R.P., and Dow, J.M. (2008). Diffusible signals and interspecies communication in bacteria. Microbiology *154*, 1845–1858.

Savidor, A., Teper, D., Gartemann, K.-H., Eichenlaub, R., Chalupowicz, L., Manulis-Sasson, S., Barash, I., Tews, H., Mayer, K., Giannone, R.J., *et al.* (2012). The *Clavibacter michiganensis* subsp. *michiganensis*-tomato interactome reveals the perception of pathogen by the host and suggests mechanisms of infection. J. Proteome Res. *11*, 736–750.

Scarbrough, E., Armstrong, D.J., Skoog, F., Frihart, C.R., and Leonard, N.J. (1973). Isolation of cis-Zeatin from *Corynebacterium fascians* Cultures. Proc. Natl. Acad. Sci. U.S.A. *70*, 3825–3829.

Scheible, W.-R., Fry, B., Kochevenko, A., Schindelasch, D., Zimmerli, L., Somerville, S., Loria, R., and Somerville, C.R. (2003). An *Arabidopsis* mutant resistant to thaxtomin A, a cellulose synthesis inhibitor from *Streptomyces* species. Plant Cell *15*, 1781–1794.

Schwessinger, B., and Ronald, P.C. (2012). Plant innate immunity: perception of conserved microbial signatures. Annu. Rev. Plant Biol. *63*, 451–482.

Seipke, R.F., and Loria, R. (2008). *Streptomyces scabies* 87–22 possesses a functional tomatinase. J. Bacteriol. *190*, 7684–7692.

Seipke, R.F., Kaltenpoth, M., and Hutchings, M.I. (2012). *Streptomyces* as symbionts: an emerging and widespread theme? FEMS Microbiol. Rev. *36*, 862–876.

Shao, F., Golstein, C., Ade, J., Stoutemyer, M., Dixon, J.E., and Innes, R.W. (2003). Cleavage of *Arabidopsis* PBS1 by a bacterial type III effector. Science *301*, 1230–1233.

Sisto, A., Cipriani, M.G., and Morea, M. (2004). Knot formation caused by *Pseudomonas syringae* subsp. *savastanoi* on olive plants is *hrp*-dependent. Phytopathology *94*, 484–489.

Snowdon, A.L. (2010). Post-harvest Diseases and Disorders of Fruits and Vegetables, Volume 2: Vegetables (Manson Publishing Ltd, London, UK).

Stamatakis, A. (2006). RAxML-VI-HPC: maximum likelihood-based phylogenetic analyses with thousands of taxa and mixed models. Bioinformatics *22*, 2688–2690.

Stange, R., Jeffares, D., Young, C., Scott, D., Eason, J., and Jameson, P. (1996). PCR amplification of the *fas-1* gene for the detection of virulent strains of *Rhodococcus fascians*. Plant Pathol. *45*, 407–417.

Stanley, S.A., Raghavan, S., Hwang, W.W., and Cox, J.S. (2003). Acute infection and macrophage subversion by *Mycobacterium tuberculosis* require a specialized secretion system. Proc. Natl. Acad. Sci. U.S.A. *100*, 13001–13006.

Stavrinides, J., McCann, H.C., and Guttman, D.S. (2008). Host-pathogen interplay and the evolution of bacterial effectors. Cell Microbiol. *10*, 285–292.

Stes, E., Vandeputte, O.M., Jaziri, El, M., Holsters, M., and Vereecke, D. (2011). A successful bacterial coup d'état: how *Rhodococcus fascians* redirects plant development. Annu. Rev. Phytopathol. *49*, 69–86.

Strøm, A.R., and Kaasen, I. (1993). Trehalose metabolism in *Escherichia coli*: stress protection and stress regulation of gene expression. Mol. Microbiol. *8*, 205–210.

Studholme, D.J., Glover, R.H., and Boonham, N. (2011). Application of high-throughput DNA sequencing in phytopathology. Annu. Rev. Phytopathol. *49*, 87–105.

Sun, Q., Sun, Y., Walker, M.A., and Labavitch, J.M. (2013). Vascular occlusions in grapevines with Pierce's disease make disease symptom development worse. Plant Physiol. *161*, 1529–1541.

Szczesny, R., Jordan, M., Schramm, C., Schulz, S., Cogez, V., Bonas, U., and Büttner, D. (2010). Functional characterization of the Xcs and Xps type II secretion systems from the plant pathogenic bacterium *Xanthomonas campestris* pv *vesicatoria*. New Phytol. *187*, 983–1002.

Takai, S., Hines, S.A., Sekizaki, T., Nicholson, V.M., Alperin, D.A., Osaki, M., Takamatsu, D., Nakamura, M., Suzuki, K., Ogino, N., *et al.* (2000). DNA sequence and comparison of virulence plasmids from *Rhodococcus equi* ATCC 33701 and 103. Infect. Immun. *68*, 6840–6847.

Takasuka, T.E., Book, A.J., Lewin, G.R., Currie, C.R., and Fox, B.G. (2013). Aerobic deconstruction of cellulosic biomass by an insect-associated *Streptomyces*. Sci. Rep. 3, 1030.

Tao, Y., Xie, Z., Chen, W., Glazebrook, J., Chang, H.-S., Han, B., Zhu, T., Zou, G., and Katagiri, F. (2003). Quantitative nature of *Arabidopsis* responses during compatible and incompatible interactions with the bacterial pathogen *Pseudomonas syringae*. Plant Cell *15*, 317–330.

Thimann, K.V., and Sachs, T. (1966). The role of cytokinins in the 'fasciation' disease caused by *Corynebacterium fascians*. Am. J. Bot. 731–739.

Thomas, C.M., and Nielsen, K.M. (2005). Mechanisms of, and barriers to, horizontal gene transfer between bacteria. Nat Rev. Microbiol. *3*, 711–721.

Torres, M.A., and Dangl, J.L. (2005). Functions of the respiratory burst oxidase in biotic interactions, abiotic stress and development. Curr. Opin. Plant Biol. *8*, 397–403.

Toth, I.K., and Birch, P.R.J. (2005). Rotting softly and stealthily. Curr. Opin. Plant Biol. *8*, 424–429.

Troemel, E.R., Chu, S.W., Reinke, V., Lee, S.S., Ausubel, F.M., and Kim, D.H. (2006). p38 MAPK regulates expression of immune response genes and contributes to longevity in *C. elegans*. PLoS Genet. *2*, e183.

Underhill, D.M., Ozinsky, A., Smith, K.D., and Aderem, A. (1999). Toll-like receptor-2 mediates mycobacteria-induced proinflammatory signaling in macrophages. Proc. Natl. Acad. Sci. U.S.A. *96*, 14459–14463.

Underwood, W. (2012). The plant cell wall: a dynamic barrier against pathogen invasion. Front. Plant Sci. 3, 85.

Valls, M., Genin, S., and Boucher, C. (2006). Integrated regulation of the type III secretion system and other virulence determinants in *Ralstonia solanacearum*. PLoS Pathog. *2*, e82.

Vandeputte, O., Oden, S., Mol, A., Vereecke, D., Goethals, K., Jaziri, El, M., and Prinsen, E. (2005). Biosynthesis of auxin by the Gram-positive phytopathogen *Rhodococcus fascians* is controlled by compounds specific to infected plant tissues. Appl. Environ.Microbiol. *71*, 1169–1177.

Vereecke, D., Messens, E., Klarskov, K., De Bruyn, A., Van Montagu, M., and Goethals, K. (1997). Patterns of phenolic compounds in leafy galls of tobacco. Planta *201*, 342–348.

Vereecke, D., Cornelis, K., Temmerman, W., Jaziri, M., Van Montagu, M., Holsters, M., and Goethals, K. (2002). Chromosomal locus that affects pathogenicity of *Rhodococcus fascians*. J. Bacteriol. *184*, 1112–1120.

Walters, D.R., McRoberts, N., and Fitt, B.D.L. (2008). Are green islands red herrings? Significance of green islands in plant interactions with pathogens and pests. Biol. Rev. Camb. Philos. Soc. *83*, 79–102.

Wang, L., Rong, W., and He, C. (2008). Two *Xanthomonas* extracellular polygalacturonases, PghAxc and PghBxc, are regulated by type III secretion regulators HrpX and HrpG and are required for virulence. Mol. Plant Microbe Interact. *21*, 555–563.

Wanner, L.A. (2006). A survey of genetic variation in streptomyces isolates causing potato common scab in the United States. Phytopathology *96*, 1363–1371.

Willems, R.J.L., Hanage, W.P., Bessen, D.E., and Feil, E.J. (2011). Population biology of Gram-positive pathogens: high-risk clones for dissemination of antibiotic resistance. FEMS Microbiol. Rev. *35*, 872–900.

Williams, M.A., Schmidt, R.L., and Lenz, L.L. (2012). Early events regulating immunity and pathogenesis during *Listeria monocytogenes* infection. Trends Immunol. *33*, 488–495.

Wozniak, R.A.F., and Waldor, M.K. (2010). Integrative and conjugative elements: mosaic mobile genetic elements enabling dynamic lateral gene flow. Nat. Rev. Microbiol. *8*, 552–563.

Yam, K.C., Okamoto, S., Roberts, J.N., and Eltis, L.D. (2011). Adventures in *Rhodococcus* – from steroids to explosives. Can. J. Microbiol. *57*, 155–168.

Zeidler, D., Zähringer, U., Gerber, I., Dubery, I., Hartung, T., Bors, W., Hutzler, P., and Durner, J. (2004). Innate immunity in *Arabidopsis thaliana*: lipopolysaccharides activate nitric oxide synthase (NOS) and induce defense genes. Proc. Natl. Acad. Sci. U.S.A. *101*, 15811–15816.

Zgurskaya, H.I., Evtushenko, L.I., Akimov, V.N., and Kalakoutskii, L.V. (1993). *Rathayibacter* gen. nov., including the species *Rathayibacter rathayi* comb. nov., *Rathayibacter tritici* comb. nov., *Rathayibacter iranicus* comb. nov., and six strains from annual grasses. Int. J. Syst. Bacteriol. *43*, 143–149.

The Molecular Interactions Between Human-pathogenic Bacteria and Plants

6

Nicola J. Holden, Ashleigh Holmes, Yannick Rossez and Robert W. Jackson

Abstract

Human-pathogenic bacteria that are able to persist and proliferate outside animal hosts can interact directly with plants to exploit them as alternative hosts. The interactions are complex involving adaptive processes for both the bacterium and plant. Experimental evidence has been increasing over recent time and shows functional roles for various bacterial factors, including those required for adherence and metabolism. Furthermore, plant hosts are able to respond to bacterial colonization and mount an effective response, although some pathogens appear to be able to overcome these responses. As with interactions between human-pathogenic bacteria and animal hosts, or with phytopathogens and plant hosts, there are important specificities that influence the outcome of these interactions. Our current understanding of the relationships between bacteria and plant hosts is on the verge of identifying the true nature of these interactions, which will in the longer term enable targeted solutions to increase the safety of fresh produce.

Introduction

Food-borne human-pathogenic bacteria are transmitted to their susceptible human host by a dizzying array of vehicles. For bacteria that are transmitted through the faecal–oral pathway, there have been reports of just about any means of transmission including dust particles, flying insects, wild-life and plants. All these vehicles have played a role in the complex web of transmission from food source to the consumers' plate. However, on closer inspection one finds that for many of these vehicles, transmission from one part of the food web to another is not a passive and inert process, but involves two-way interactions between the transient host and food-borne pathogen.

Although plants as host or vectors for human pathogens have only relatively recently received increased attention from the scientific community and popular press, there is little doubt that fresh produce-associated infections and outbreaks are not particularly novel. Early reports of infections and outbreaks come from contaminated bean sprouts in the 1970s (Portnoy *et al.*, 1976) and from home-preserved vegetables (Gangaros *et al.*, 1971; Terranova *et al.*, 1978). A combination of changes such as those in epidemiological analysis, outbreak reporting, microbiological detection, changes in agricultural practices and scale of fresh food distribution have all resulted in a relatively recently increased awareness of the role of fruit and vegetables in food-borne infections. This coupled with some very large-scale and high profile outbreaks (CDC, 2006; Frank *et al.*, 2011; Michino *et al.*, 1999) have put fresh produce as a new strategic priority at the level of the growers, producers, retailers and now even governments.

There is a growing body of scientific evidence investigating different aspects of the association between food-borne pathogens and plants and it has become clear that there are quite defined and specific interactions that enable many of the food-borne pathogens to colonize plants and utilize them as alternative hosts. Understandably, the work has focused on the food-borne pathogens that give rise to the majority of reported produce-associated outbreaks, mainly toxigenic *Escherichia coli* and non-typhoidal *Salmonella enterica*. The primary reservoir for this group of pathogens is mainly farm animals. Outbreaks of *Listeria monocytogenes*, *Bacillus* and *Clostridium* species have also occurred, although these pathogens are normally associated with the wider environment and soil. This review will draw together the evidence to support the notion of a two-way interaction between food-borne pathogens and their plant hosts and discuss the implications for understanding the 'big picture' biology underlying transmission of pathogenic bacteria through the food-web. Enteric viruses are another serious source of outbreaks, accounting for almost as many infections as the enteric bacteria. However, as viruses (and other enteric parasites such *Crytosporidium*) require mammalian hosts for replication, they will not be considered in this review.

At this point it is important to make the distinction between colonization of plants that are growing and are considered 'pre-harvest' and those that have been harvested and can be contaminated during the processing, packaging and transportation stages, i.e. post-harvest colonization. The main difference is in the ability of the plant host to respond to the bacterial colonization and mount a defence response, which in turn the bacteria must be able to adapt to, or be cleared by the response. There are other important differences such as the availability of nutrients through plant cell damage or leakage; temperature differences in the field compared to containers/factory/transportation vehicles; or humidity of the plant tissue, e.g. in packaged goods (Choi *et al.*, 2011). All of these differences will impact bacterial and plant physiology and therefore, the success or failure of colonization of plant tissue by the pathogens. In order to fully appreciate the molecular interactions between the microbe and its host, the physiological status of both partners must be taken into consideration, and with that in mind, this review will focus on pre-harvest colonization of plant tissue, but will use post-harvest examples where necessary.

Bacterial factors

Ability to persist and grow on plant hosts

Factors that determine the success of plant colonization by human-pathogenic bacteria are derived from the plant, bacteria (Fig. 6.1) and to some extent, the environmental conditions. To be able to successfully colonize plant hosts, human-pathogenic bacteria must be able to reach the plant host, metabolize and sequester any available nutrients, and maintain a foothold in a preferred habitat. They must be able to respond to any changes in the host defence response and finally have the ability to disseminate to other, suitable hosts. Plant factors (considered later) centre around the plant defence response, which extends to the production of secondary metabolites and physical structures. Bacterial factors are underpinned by the timely and appropriate expression of genetic determinants. There is growing

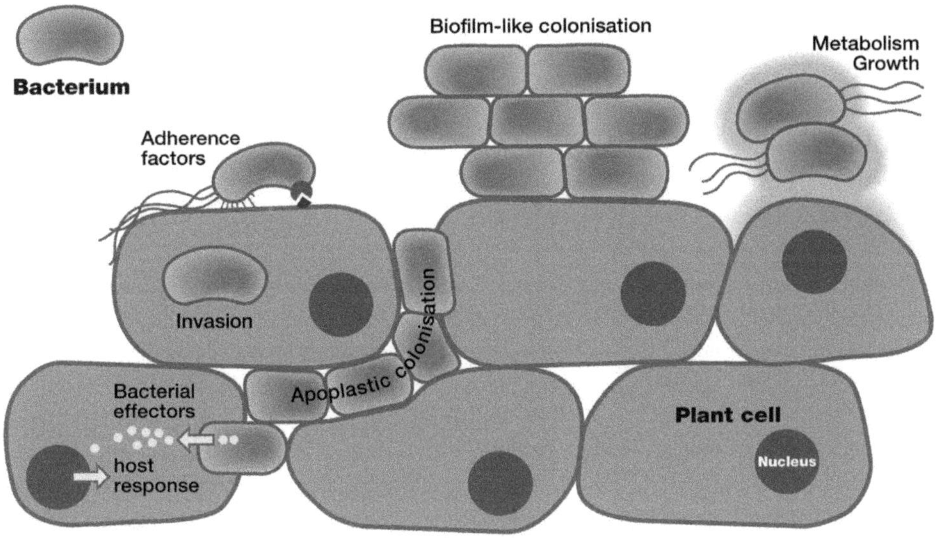

Figure 6.1 Cartoon showing the possible interactions between human-pathogenic bacteria (magenta) and plant hosts (plant cells in green). A colour version of this figure is located in the plate section at the back of the book.

evidence to support the hypothesis that human pathogens are not merely attaching to plant tissues and are actually metabolizing and, in some cases multiplying, gaining access to the intercellular spaces and even invading inside plant cells (Carey *et al.*, 2009; Gu *et al.*, 2011; Hou *et al.*, 2012; Klerks *et al.*, 2007b; Kroupitski *et al.*, 2009b; Kutter *et al.*, 2006; Kyle *et al.*, 2010; Schikora *et al.*, 2008). There is also evidence of differential colonization performances by different pathogens (Jeter and Matthysse, 2005; Klerks *et al.*, 2007a; Parker *et al.*, 2012; Patel and Sharma, 2010; Quilliam *et al.*, 2012b) in different plant niches (Quilliam *et al.*, 2012b; Semenov *et al.*, 2010) and evidence for host-specific effects (Alegre *et al.*, 2010; Berger *et al.*, 2011; Quilliam *et al.*, 2012a). Evidence is emerging to show evolution of a pathogen with increased colonization fitness in different animal and plant niches after passage through new environments (Oliveira *et al.*, 2011). Finally, environmental factors including the physiochemical environment present challenges and it might have originally been expected that the plant microenvironment would be more desolate and harsh to enteric bacteria. However, some of these pathogens appear to be well adapted to some of these conditions: for example, acidic fruits were considered to be 'safe' from contamination with human pathogens, but recent evidence shows that *E. coli*, *Salmonella* and *Listeria* can all replicate or persist on peach tissue (Alegre *et al.*, 2010; Eblen *et al.*, 2004) and probably do so through the activation of stress-response systems that can enhance survival in low pH conditions (Bergholz *et al.*, 2009). Human pathogens also have to cope with the 'local' microbial community, which may provide opportunities or barriers to colonization. For example, *Salmonella* appears to grow better in soft rot tomato fruits colonized by the phytopathogen *Pectobacterium* compared to fruits that lack *Pectobacterium* suggesting some kind of synergistic effect (Noel *et al.*, 2010b). In another interesting study, the presence of *Salmonella* in tomato flowers appears to cause a shift in the microbial populations of the resulting fruit to exclude or favour different bacterial types (Shi *et al.*, 2009). It will be fascinating to determine whether this might represent a paradigm comparable to *Salmonella* overcoming intestinal colonization resistance in the animal gut, by triggering resistance mechanisms in the host to alter the microbial composition and open up a 'clean' niche for pathogen colonization (Stecher *et al.*, 2007).

The physiological status of the bacteria changes in association with plant hosts where the bacteria are able to enter a less metabolically active state, termed viable but non-culturable (VBNC), from which they can be resuscitated. This has important implications for researchers trying to quantify the event of bacterial colonization. While many studies have used microbial counts to enumerate bacteria, molecular techniques have also been used, e.g. Ibekwe *et al.* (2009), which may be important in quantification of low levels of bacteria, or for those that have entered a VBNC state. One study has shown that although the culturability of *E. coli* O157:H7 declines on lettuce plants, the bacteria are still viable (Dinu and Bach, 2011). Fluorescently tagged *E. coli* O157:H7 inoculated onto the leaves of lettuce plants decreased to undetectable levels after six days, as determined by agar plate counts, when the plants were maintained at 8°C, which is close to the minimum growth temperature for the bacteria. However, microscopy analysis showed the presence of viable bacteria on the leaf surface (the fluorescent tag had been shown to act as a surrogate for viability and provided similar viable counts to commercial Live/Dead kits) and viable counts remained around 6 $\log_{10}$ for several days after.

Once the human pathogens have come into contact with plants and entered the plant tissue, it is clear that, under certain experimental conditions, the pathogens can survive and, in some cases, grow. In the latter case, this is likely to be occurring through metabolic activity supporting pathogen replication. Somewhat surprisingly, there is only meagre evidence to support this hypothesis. The clearest evidence to date is provided by a series of transcriptomic studies using *in vivo* expression technology (IVET) or microarrays that have sought to examine bacterial gene expression during plant colonization. Noel *et al.* used IVET to screen for *Salmonella* genes expressed within red tomato fruits (Noel *et al.*, 2010a). Of the 51 inducible genes identified from a limited screen, 13 are involved in general metabolism. Curiously, one gene, *fadH*, was more highly expressed in green immature fruits than ripened red fruits. Further experimentation found that linoleic acid, which is present in higher quantities in green fruit, specifically up-regulated *fadH*. These observations potentially indicate a specific adaptive mechanism for early colonization of tomato fruits, possibly using linoleic acid as a carbon source. Other genes identified by IVET were postulated to be involved in biofilm formation, regulation and stress responses such as cold and acid shock; similar observations were seen in three studies examining *E. coli* O157 gene expression during interaction with lettuce leaves or exposure to lettuce leaf lysates, or apple juice (Bergholz *et al.*, 2009; Fink *et al.*, 2012; Kyle *et al.*, 2010). These studies provided greater insight to the bacterial adaptation. Kyle *et al.* found that *E. coli* colonizing injured lettuce tissue activated transport systems for uptake of carbohydrates commonly found in plants, but also found that the pathogen activated several virulence genes including type III secretion genes on various LEE pathogenicity islands and the SPI-1 pathogenicity island (Kyle *et al.*, 2010). Accompanying this were genes encoding oxidative and osmotic stress tolerance systems, systems also found to be turned on within apple juice (Bergholz *et al.*, 2009). Bergholz and colleagues also found a number of energy and general metabolism systems, and type III secretion systems, were up-regulated, reinforcing the idea that human pathogens are possibly suppressing resistance mechanisms, using the plant as a host for resources and channelling them into bacterial replication. These observations contrast partially with those of Fink *et al.* who found that *E. coli* O157 down-regulated metabolism genes when colonizing lettuce leaf surfaces, but it did activate stress response systems and sessility (biofilm) systems (Fink *et al.*, 2012). These studies actually provide a very similar picture to that seen with bacterial plant pathogens: they remain epiphytic with virulence and metabolism systems ramped down until such time they can reach the intercellular spaces (e.g. via wounds) whereby they activate gene systems to enable the bacterium to cope with the change in environment and suppress plant resistance. Whether the latter actually happens for all human pathogen–plant interactions is a key question. Of note, Bergholz *et al.* found that in *E. coli* O157 a large number of hypothetical genes were up-regulated, many of which are found in *E. coli* K-12, while several are O157-specific (Bergholz *et al.*, 2007). This underlines a major gap in our knowledge of how these bacteria respond to plant compounds and thus requires a deeper investigative approach.

The role of adherence factors in plant–microbe interactions

The ability of bacteria to adhere to a substrate is an important and necessary step to facilitate persistence and proliferation. Adherence can occur through a variety of mechanisms, from simple cell surface or organelle charge through to specific protein–protein interactions. Adherence and subsequent cell proliferation can take the form of discrete colonies, or can become more complex biofilm communities. Organelles and structures present on the bacterial cell surface are able to mediate interactions with abiotic and biotic surfaces and play key roles in the success of the bacteria within a given niche. While some factors interact directly with host surface proteins or surface moieties, others play a role in bacteria–bacteria interactions, for example in the development of biofilm communities. Many studies have focused on the role of adherence factors that mediate bacteria–host interactions, not least because of their clear phenotype and tractability. The role of adherence factors of human pathogens has been well characterized in mammalian hosts, where many of them are considered to act as virulence factors because of their contribution to host colonization. While adherence factors tend to be subject to tight regulation and respond to environmental cues, several have also been shown to play a role in adherence to plant tissue (Table 6.1).

Fimbrial and afimbrial adherence factors

Gram-negative bacteria encode several fimbrial gene clusters that belong to the chaperone-usher family. The name derives from the functional role of the chaperone and usher proteins, which are required for secretion and assembly of the structural components across the bacterial envelope. The chaperone interacts with the fimbrillin subunit in the periplasm and facilitates addition to the growing fimbrial filament via the usher outer membrane protein (reviewed in Nuccio and Baumler, 2007). The presence of many fimbrial clusters within a bacterial genome provokes interest in their function: some pathogens can encode as many as 16 fimbrial clusters. However, their analysis *in vitro* can be challenging as many are not expressed under conditions tested. *E. coli* K-12 encodes 12 fimbrial clusters and analysis of six of them demonstrated that they were poorly expressed yet functional *in vitro* (Korea *et al.*, 2010). Similarly, *E. coli* O157:H7 encodes 16 fimbrial clusters, half of which are conserved or partially represented in *E. coli* K-12 and the others are specific to pathogenic strains (Low *et al.*, 2006). Many of these fimbriae are expressed optimally at 37°C and have been characterized in mammalian host colonization. An interesting observation was

Table 6.1 Bacterial adherence factors that have a potential or confirmed role *in planta*, encoded either by human pathogens, or by plant pathogens where homologues exist in human pathogens

Adherence factor	Bacteria spp.	Plant species used for interaction assays	References
Ecp/MAT	*Escherichia coli* O157:H7	Spinach leaves	Saldana *et al*. (2011)
Type I fimbriae	*Klebsiella pneumoniae*; *Enterobacter agglomerans*; *Erwinia amylovora*	Grass roots; Gala apple tree	Gerlach *et al*. (1989), Haahtela and Korhonen (1985), Haahtela *et al*. (1985), Koczan *et al*. (2011)
Type III fimbriae	*K. pneumoniae*	Grass and cereal roots	Duguid (1959), Haahtela and Korhonen (1985), Haahtela *et al*. (1985)
Type IV fimbriae	*Ralstonia solanacearum*; *Azoarcus* species	Potato plants; grass roots and fungus mycelium	Dorr *et al*. (1998), Wairuri *et al*. (2012)
Cellulose	*Salmonella enterica*; *E. coli* O157:H7; *Agrobacterium tumefaciens*; *Rhizobium leguminosarum* bv. *trifolii*	Alfalfa sprouts; tomato roots; *Arabidopsis thaliana* and red clover	Ausmees *et al*. (1999b), Barak *et al*. (2007), Matthysse *et al*. (2008), Matthysse and McMahan (1998)
Curli	*S. enterica* sv. Newport; *E. coli* O157:H7	Alfalfa sprouts; baby spinach leaves	Barak *et al*. (2005), Jeter and Matthysse (2005), Saldana *et al*. (2011), Torres *et al*. (2005)
XadA1 and XadB	*Xanthomonas*	Rice leaves	Das *et al*. (2009)
OmpA	*E. coli* O157:H7	Alfalfa sprouts and seed coats	Torres *et al*. (2005)
Flagella	*Listeria monocytogenes*; *S. enterica* sv. Senftenberg phase-1 flagella; *Pseudomonas fluorescens*; *Pseudomonas syringae*; *Azospirillum* spp.; enterotoxigenic *E. coli*; enteropathogenic *E. coli*	Alfalfa, broccoli, radish sprouts; potato roots; bean seedlings; wheat, rocket, spinach, lettuce and basil leaves	Allerberger and Sessitsch (2009), Berger *et al*. (2009), Croes *et al*. (1993), Deweger *et al*. (1987), Gorski *et al*. (2009), Haefele and Lindow (1987), Shaw *et al*. (2011)
T3SS	*Listeria monocytogenes*; *S. enterica*; *E. coli* O157:H7	Cabbage tissue; stomata on spinach rocket, lettuce	Ells and Truelstrup Hansen (2006), Shaw *et al*. (2008)

that some of the fimbrial clusters were optimally expressed at lower temperatures (28°C), indicating a potential role outside the animal host (Low *et al.*, 2006). The *yag* cluster is conserved in *E. coli* strains and was found to be expressed by meningitis-associated *E. coli* at 20°C in LB broth, thus renamed MAT (meningitis and temperature) fimbriae (Pouttu *et al.*, 2001). Subsequent work has shown that these fimbriae are expressed in tissue-culture media (DMEM) at 26°C and 37°C, and are in fact widespread in pathogenic and non-pathogenic *E. coli*, leading to the name *E. coli* common pilus (ECP) (Rendon *et al.*, 2007). Consequently, ECP has been shown to be involved in adherence to mammalian epithelial cells for different pathovars of *E. coli* including EHEC (enterohaemorrhagic) (Rendon *et al.*, 2007), EPEC (enteropathogenic) (Saldana *et al.*, 2009), ETEC (entertoxigenic) (Blackburn *et al.*, 2009) and EAEC (enteroaggregative) (Avelino *et al.*, 2010). ECP also plays a role in

E. coli O157:H7 adherence to baby spinach leaves, as ECP deficient strains show an 83% reduction in adherence (Saldana *et al.*, 2011).

Type I fimbriae are well described in *Enterobacteriaceae* and are characterized by mannose-sensitive agglutination of red blood cells. The function of type I fimbriae has been characterized showing a role for adherence of *Klebsiella* and *E. coli* to epithelial cells of the urinary tract (Connell *et al.*, 1996; Struve *et al.*, 2008). It has also been demonstrated that the type I fimbriae of plant-associated isolates of *Klebsiella pneumoniae* and *Enterobacter agglomerans* mediated adherence to grass roots (Gerlach *et al.*, 1989; Haahtela and Korhonen, 1985; Haahtela *et al.*, 1985). Recently, a putative type I fimbrial cluster of *Erwinia amylovora* has been shown to function in initial attachment to surfaces, which in turn impacts virulence, as mutants are less virulent than wild-type (Koczan *et al.*, 2011). Virulence in this study was measured by lesion formation on pears and the progression of disease in apple shoots. Bioinformatics analysis shows that this cluster is in fact more closely related to another member of the chaperone-usher family, K88, than to type 1. Expression of type 1 fimbriae has been shown to be controlled by the metabolic sensor CRP-cAMP and is up-regulated in stressed conditions, i.e. low nutrients (Muller *et al.*, 2009). A transcriptomic study of *E. coli* O157:H7 in response to lettuce lysates reported an up-regulation of the main structural sub-unit of type I fimbriae under the conditions tested (Kyle *et al.*, 2010). However, the relevance of this expression is unknown as the strain of *E. coli* O157 used contains a 16-bp deletion in the type 1 promoter, rendering its expression minimal or off (Roe *et al.*, 2001). Type 1 fimbriae from different members of the *Enterobacteriacae* exhibit a wide diversity in amino acid identity for both the main structural subunit (FimA) and the adhesin (FimH) (Dwyer *et al.*, 2011; Sokurenko *et al.*, 1997; Stahlhut *et al.*, 2009). They are known to show different specificities for mannose-containing substrates, as demonstrated in an elegant study using chimeric FimA–FimH hybrids from *E. coli, K. pneumoniae* and *S. enterica* (Duncan *et al.*, 2005). Furthermore, some coding sequences annotated as type 1 fimbrial adhesins do not share homology with the mannose-binding domain in *E. coli* FimH, e.g. *Pantoea ananatis*, CDS PAJ_0941 (termed '*fimH*').

Type III fimbriae are characterized by a haemagglutination phenotype which is mannose resistant and dependent upon pre-treatment of erythrocytes with tannins (Duguid, 1959; Korhonen *et al.*, 1983). *K. pneumoniae* express Type III fimbriae, which appear to facilitate adherence to plant roots (Duguid, 1959; Haahtela and Korhonen, 1985; Korhonen *et al.*, 1983). A significant role in adherence to grass and cereal roots was demonstrated for Type III fimbriae with *Klebsiella* strains expressing both fimbrial types but only showed a reduction in adherence in the presence of antibodies specific to Type III and not to Type I fimbriae (Haahtela *et al.*, 1985). This in turn is supported by studies showing that the virulence of *K. pneumoniae* in urinary tract infections is not dependent upon Type III expression but Type I is essential (Struve *et al.*, 2009). This demonstrates that different fimbrial types are expressed and have functions in various environments and habitats. The phytobacterium *Pseudomonas fluorescens* expresses fimbriae with mannose resistance agglutination phenotype, which mediates adherence to maize roots (Vesper, 1987). The haemagglutination phenotype was partially inhibited by galactose and by galacturonic acid, suggesting functional recognition of these carbohydrates.

Type IV fimbriae (or pili) are expressed by many Gram-negative bacteria and are involved in a variety of phenotypes: biofilm formation (O'Toole and Kolter, 1998), twitching motility (Bradley, 1980), transformation competence (Fussenegger *et al.*, 1997), and bacteriophage infection (Bradley, 1974). An essential role in adhesion of pathogenic bacteria to their host cells, e.g. for initial attachment of EPEC to brush border cells (Tobe and Sasakawa, 2002), has also been shown. Moreover, type IV fimbriae appear to play a role in the pathogenicity of *Ralstonia solanacearum*, a soil-borne Gram-negative bacterium that causes bacterial wilt disease in more than 200 plant species. Mutation of type IV pili resulted in an impaired ability to cause wilting of potato plants (Wairuri *et al.*, 2012). The Gram-negative endophytic bacteria, *Azoarcus* spp. occurs in association with grasses and interacts with rhizosphere fungi. Type IV-dependent adherence was demonstrated for the interactions of a nitrogen-fixing *Azoarcus* species with plant roots and fungal mycelia, both hosts for this bacterium (Dorr *et al.*, 1998). This broad host range indicates that there may be common molecular target recognized by these pili.

E. coli O157 has been shown to attach and form a strong biofilm on a number of different produce surfaces (Boyer *et al.*, 2007; Ryu and Beuchat, 2004; Silagyi *et al.*, 2009). Cellulose is an important component of the biofilm matrix, along with curli, and cellulose synthase genes are widespread among the *Enterobacteriaceae*. The phytobacterium *Agrobacterium tumefaciens* (Matthysse and McMahan, 1998) and *Rhizobium leguminosarum* bv. *trifolii* (Ausmees *et al.*, 1999) have been shown to synthesize and use cellulose for adherence to plant tissues; cellulose production by *P. fluorescens* is a key factor for root and shoot colonization (Gal *et al.*, 2003). Cellulose also plays a role in adherence of human pathogens to plant tissue. Cellulose synthase mutants of *Salmonella enterica* (Barak *et al.*, 2007) and *E. coli* O157:H7 (Matthysse *et al.*, 2008) adhered less well to alfalfa sprouts compared to wild-type strains. Complementation of *E. coli* K-12 with cellulose synthase (*celA*) from *A. tumefaciens* resulted in an increase in adherence to alfalfa sprouts (Matthysse *et al.*, 2008). Both cellulose and curli play a role in the transfer of *S. enterica* from contaminated irrigation water to parsley (Lapidot and Yaron, 2009), but not for initial attachment (Lapidot *et al.*, 2006).

Curli are long, thin, curved fibres, which are expressed in response to low temperature, low oxygen, low osmolarity and nutrient limitation (Barnhart and Chapman, 2006). Curli are an important component of bacterial biofilms, where their expression is often co-regulated with cellulose production (Barnhart and Chapman, 2006). Recently, it has been described that the curli expression status (positive or negative) of *E. coli* O157:H7 can determine its survival during particular environmental stresses (Carter *et al.*, 2011). It was observed that curli-producing *E. coli* O157:H7 bacteria can survive nutrient poor conditions, whereas curli-deficient cells are acid tolerant. A homogeneous curli-negative *E. coli* O157:H7 population grown in nutrient limiting conditions becomes heterogeneous following prolonged exposure to acidified media, containing a mixed population of curli-positive and curli-negative cells (Carter *et al.*, 2011).

Curli are conserved in *Salmonella* and *E. coli*, indicating a shared function (Romling *et al.*, 1998). Several lines of evidence suggest that this common function could be in the interactions with plant tissues. Curli was identified as a factor required for adherence of *S. enterica* serovar Newport to alfalfa sprouts, using a transposon mutant screen (Barak *et al.*, 2005; Patel *et al.*, 2010). Introduction of the curli operon from *E. coli* O157:H7 into K-12 conferred the ability to bind to alfalfa sprouts but the reciprocal deletion of *csgA*, which encodes the main structural subunit, from *E. coli* O157:H7 did not significantly affect binding, indicating that other adherence factors were also important and that a functional redundancy exists (Jeter and Matthysse, 2005; Torres *et al.*, 2005). Curli are also involved in *E. coli* O157:H7 adherence to baby spinach leaves, demonstrated using immunofluorescence microscopy, which showed a reduction in bacterial numbers of a curli-deficient strain compared to its isogenic parent (Saldana *et al.*, 2011). However, bacteria deficient in curli can still attach to spinach leaves, confirming previous studies that adherence is a multifactorial process. Macarisin *et al.* reported that curli and cellulose are not required for loose attachment, but that curli expression is important for strong attachment of *E. coli* O157:H7 to spinach leaves (Macarisin *et al.*, 2012). Expression of curli was determined to be optimal at 22°C and the incubation conditions used for the bacteria prior to plant interactions (at 37°C) would not have been conducive to fibre production. Therefore, the observed increase in bacterial numbers 24 hours after inoculation of spinach leaf surfaces may be as a result of increased expression of curli at 22°C.

In Gram-negative bacteria, the outer membrane is an asymmetric bilayer with phospholipids in the inner monolayer and the lipopolysaccharide (LPS) in the outer monolayer. Outer membrane proteins play key roles in the structural integrity of the outer membrane, constituting a semipermeable barrier, as membrane-bound enzymes or components of signal transduction cascades. Some outer membrane proteins are implicated in bacterial adhesion and are correlated with the bacterial virulence in animal or plant hosts, for example: PagC of *S. enterica* serovar Typhimurium (Miller *et al.*, 1992), HecA and TolC of the phytopathogen *Dickeya dadantii* (Barabote *et al.*, 2003; Rojas *et al.*, 2002) and OmpA from *E. coli*. OmpA was reported to mediate invasion of brain microvasculature endothelial cells for *E. coli* K-1 (Weiser and Gotschlich, 1991) with GlcNAcβ1-4GlcNAc glycan epitopes as the molecular target (Prasadarao *et al.*, 1996). One study has implicated *E. coli* O157:H7 OmpA in plant tissue adhesion, showing that the *E. coli* O157:H7 *ompA* mutant strain was unable to bind to alfalfa sprouts and seed coats (Torres *et al.*, 2005). Interestingly, the aforementioned glycan epitope is represented in plant tissue and more particularly on the plant *N*-glycans (Lerouge *et al.*, 1998).

There has been a lot of interest in non-fimbrial adhesins, designated as oligomeric coiled-coil adhesins, because of their wide occurrence and crucial role in host cell attachment of a broad range of hosts from mammals to plants. All members of this protein family are trimeric autotransporter adhesins (TAAs) and have a tripartite structure consisting of a stalk and membrane anchor, conserved in all TAAs, and the head domain of which is likely the adhesive part (Linke *et al.*, 2006). The prototypical TAA is *Yersinia* adhesin A (YadA) from enteropathogenic *Yersinia enterocolitica* and *Yersinia pseudotuberculosis* (Bolin *et al.*, 1982). The proteins form a capsule-like structure of trimeric 'lollipop'-shaped surface projections. After the initial step of invasion of the intestinal mucosa, YadA is the predominant adhesin in infected tissue and mediates adherence to the tissue (Hoiczyk *et al.*, 2000). Recently TAA homologues have been identified in other pathogens, including *Bartonella* (Riess *et al.*, 2004), *Neisseria* spp. (Comanducci *et al.*, 2002), *Moraxella* (Lafontaine *et al.*, 2000) and *Xanthomonas* (da Silva *et al.*, 2002). Functional data have been obtained for TAAs encoded by the rice pathogen *Xanthomonas oryzae* pv. *oryzae*, called XadA1 (*Xanthomonas* adhesin-like protein A) and its paralogue XadB. These adhesins have been shown to play an important role in the early stages of plant infection, by a reduction of approximately 50% of the *xad* mutant in comparison with the wild type (Das *et al.*, 2009; Ray *et al.*, 2002). Furthermore *xadA1* was identified in all sequenced *Xanthomonas campestris* and *Xanthomonas axonopodis* pathovars, the causal agents of black rot in crucifers and other plants (pepper, tomato and citrus) respectively (Bogdanove *et al.*, 2011; Thieme *et al.*, 2005). Whereas *xadB* is restricted to *X. oryzae* strains, *X. axonopodis* strains encode *xadA2*, another paralogue. The observed differences in the adhesin repertoires of *Xanthomonas* with the evidence of sequence variations among them, may contribute to host and tissue specificity (Mhedbi-Hajri *et al.*, 2011) or, if the adhesins are recognized as PAMPs, be a result of protein evolution driven by selection against recognition by the plant host (McCann *et al.*, 2012). Moreover, genome sequences of other phytopathogenic bacteria like *Xyllela fastidiosa* and *Ralstonia solanacearum* have revealed genes for TAA family members (Salanoubat *et al.*, 2002; Simpson *et al.*, 2000). The TAA family members are expressed by other Gram-negative pathogenic human bacteria such as *E. coli* O157:H7 (Totsika *et al.*, 2012) and *S. enterica* serovar Typhimurium (Raghunathan *et al.*, 2011). The expression of EhaG and SadA, *E. coli* 0157:H7 and *Salmonella* TAA respectively, resulted in cell aggregation, biofilm formation and increased adhesion to human tissues. The role of TAA proteins expressed by human-pathogenic bacteria in the plant adhesion process has yet to be elucidated.

The role of flagella and the T3SS in adherence to plant tissue

The ability of bacteria to swim and swarm is largely driven by the action of flagella in liquid or on semi-solid surfaces. Flagellar biosynthesis and regulation is complex and often co-ordinated to be expressed in the correct environmental conditions (reviewed in Chilcott and Hughes, 2000). Flagella-driven motility contributes to chemotaxis, which has been shown in different *S. enterica* serovars as a response to plant root exudates (Klerks *et al.*, 2007a). The flagellum can also function as an adhesin, however the attribution of adherence via the flagella in the absence of motility is difficult to ascertain. Flagella-mediated adherence to plant tissues has previously been described for phytobacteria such *as P. fluorescens* to potato roots (Deweger *et al.*, 1987), *P. syringae* to bean seedlings (Haefele and Lindow, 1987) and nitrogen-fixing bacteria such as *Azospirillum* to wheat (Croes *et al.*, 1993). In the case of human pathogens, flagella have been shown to play a role in adherence to plant tissues by *S. enterica* serovars, pathogenic *E. coli* and *Listeria monocytogenes*. Flagella-mediated attachment to basil leaf epidermis appears to be serovar specific as *S. enterica* serovar Senftenberg phase-1 flagella mutants were shown to be significantly reduced in attachment compared to an isogenic parent, a phenotype not observed with a *S. enterica* serovar Typhimurium *fliC* mutant (Berger *et al.*, 2009). Gorski and colleagues tested the role of *Listeria monocytogenes* flagellin (*flaA*) and the flagellar motor (*motAB*) in adherence to alfalfa, broccoli and radish sprouts (Gorski *et al.*, 2009). They observed that deficiency in motility, Δ*motAB*, did not affect adherence compared to the parent strain, but although *flaA* contributed to adherence to sprouts, the extent depended upon the bacterial strain and sprout species tested, indicating that other bacterial factors have a role. *E. coli* flagella serotypes H6 and H7 have been characterized for adherence to mammalian cell lines *in vitro* and been shown to interact with mucins (Erdem *et al.*, 2007). Flagella produced by *E. coli* O157:H7 mediated attachment to baby spinach leaves, as shown by immunofluorescence microscopy (Saldana *et al.*, 2011). Expression of flagella was shown to be heterogenous *in planta* so that not all of the population was flagellate. Furthermore, adherent bacteria were be observed for the *fliC* deficient strain, albeit in fewer numbers, suggesting a combinatorial effect from multiple adherence factors. A *fliC* mutant of enterotoxigenic *E. coli* was found to be significantly less adherent to rocket, spinach, lettuce and basil leaves (Shaw *et al.*, 2011). Flagella-mediated adherence to plant tissue for EPEC has been shown to confer tropism for stomata, although experimentation with EPEC *fliC* mutants showed that flagella were not essential for adherence to epidermal leaf cells (Berger *et al.*, 2009).

In Gram-negative bacteria, flagella and the type III secretion system (T3SS) are evolutionarily conserved in their overall structure and assembly, for example in the mechanism of secretion and of the main structural subunits and assembly of the polymeric organelle (Cornelis, 2006). They both consist of an ATP-dependent basal apparatus spanning the cytoplasmic membrane and a hollow needle-like complex spanning the periplasm and outer membrane. This structure allows the translocation of the proteins which form the filament structure that spans outwith the bacterium and for the T3SS that mediates attachment with the host cell. This conservation may extend to functionality since it has been proposed that the flagella apparatus may be able to secrete proteins other than those involved in biogenesis. Flagella-mediated secretion has been demonstrated for a *Yersinia enterocolitica* phospholipase (Young and Young, 2002; Young *et al.*, 1999) as well as for a *Campylobacter jejuni* adhesin, FlaC, which has been shown to facilitate binding to Hep-2 cells (Song *et al.*, 2004). However, secretion using this mechanism of an adhesin for bacterial adherence to plant tissue has yet to be shown.

Type III secretion systems are well described for both phytopathogens and human pathogens (Abby and Rocha, 2012; Hueck, 1998). The T3SS apparatus in EPEC and EHEC differs from the *Yersinia* and *Salmonella* T3SS in that it has a filamentous extension from the needle EscF structure to 'reach' the host cell (Wilson *et al.*, 2001). This translocon filament is composed of EspA (Ebel *et al.*, 1996; Knutton *et al.*, 1998) and has functional homology to flagellin, as it has been shown that the EspA filament extends in a similar mechanism to that of the flagella (Delahay *et al.*, 2005). The main structural difference between plant and animal T3SS is the length of the translocon or pilus, for example a *Pseudomonas syringae* Hrp filament is ~2–6 μm in length compared to EspA from EHEC which is ~600 nm (Cornelis, 2006). This greater pilus length in phytopathogens facilitates traversing the plant cell wall for translocation of effector proteins into the cell. The T3SS is also strictly regulated, and often in co-ordination with the prior down-regulation of flagellar expression to facilitate intimate attachment to mammalian epithelial cells (Iyoda *et al.*, 2006; Soscia *et al.*, 2007). Expression of this secretion apparatus is also temperature dependent, where optimum expression in *Salmonella* and *E. coli* is at 37°C whereas for the phytopathogens *Pseudomonas syringae* and *Erwinia* is at 20°C. The growth conditions of the bacteria prior to assessment of plant tissue attachment can have an effect on binding, most likely due to the optimum expression of adherence factors. This has been shown in studies with *Listeria monocytogenes* where bacteria grown at 37°C did not adhere as strongly to cabbage tissue as those grown at 22°C or 10°C (Ells and Truelstrup Hansen, 2006). Greater numbers of *S. enterica* serovar Senftenberg were recovered from basil leaves when infected from a culture grown at 20°C compared to 37°C (Berger *et al.*, 2009). EspA from *E. coli* O157:H7 adhered to stomata on spinach and lettuce leaves, but this was only observed from adherence assays incubated at 37°C, not 20°C, for 1 hour (Shaw *et al.*, 2008).

Plant factors

Location of human-pathogenic bacteria on plants

Bacteria have the ability to colonize almost every tissue type of plants and can be isolated from every surface: above and below ground, external and internal. The majority of research has understandably focused on colonization of above-ground plant tissue by human-pathogenic bacteria, since this is the tissue that is most likely to be used for human consumption. However, leafy tissue has a lower threshold for bacterial colonization in general, compared to roots. This is down to several reasons, including low water availability, lower nutrient availability and exposure to UV light (Brandl, 2006). Furthermore, nutrients are not evenly distributed but tend to be present in 'oases' leading to patchy colonization by plant-associated bacteria. Nevertheless, the leaves and fruit are important sources of contamination by human-pathogenic bacteria. Colonization studies of the leaves of produce plants by EHEC have shown that the number of culturable bacteria declines over time, e.g. for *E. coli* O157:H7 on lettuce plants (Cooley *et al.*, 2006; Oliveira *et al.*, 2012); and for *E. coli* O157:H7 and non-pathogenic *E. coli* on spinach plants (Gutiérrez-Rodríguez *et al.*, 2012; Patel *et al.*, 2010). Market-place surveys and epidemiological studies from outbreaks have shown that human-pathogenic bacteria are also present in fruiting tissue of plants (Eblen *et al.*, 2004; Wells and Butterfield, 1997), especially for *S. enterica* serovars (Hanning *et al.*, 2009). Pre-harvest studies investigating colonization of tomatoes by *S. enterica* serovars have shown that the bacteria can colonize all plant parts from the roots, stems, leaves, flowers and fruit (Barak and Liang, 2008; Guo *et al.*, 2001). On tomato leaves, the trichcomes appear to be a preferred location of colonization (Barak *et al.*, 2011). A similar preference was observed for *Staphylococcus aureus* on *A. thaliana* leaves (Prithiviraj *et al.*, 2005).

Human-pathogenic bacteria are also able to enter internal locations in plants, which has important public health consequences as these populations of bacteria are protected from conventional sanitation procedures used in processing. Inoculation of tomato leaves with *S. enterica* serovar Typhimurium showed that bacterial cells had ingressed into the leaves, moved to the mid-rib vein and could sometimes be detected in the leaf xylem (Gu *et al.*, 2011). Tomato fruit collected from the colonized plants revealed that the bacteria were located within the fruit pulp, clearly demonstrated with the use of a fluorescently marked strain (Gu *et al.*, 2011). Several studies have shown internalization of EHEC in plant tissue. The bacteria are able to take advantage of natural openings in the plant tissue, for example stomatal cavities or emergence of lateral roots (Dong *et al.*, 2003; Warriner *et al.*, 2003). These locations are relatively protected, for example from physical or chemical damage. However, there are some reports that show an endophytic-like behaviour of human pathogens, and as such are able to enter deeper within plant tissue (Deering *et al.*, 2012; Hirneisen *et al.*, 2012), where they are considered to be truly internalized (as defined by Rosenleuth and colleagues, (Rosenblueth and Martinez-Romero, 2006)). For example, EHEC has the ability to localize between cells of living *Medicago* plants (Dong *et al.*, 2003), while *S. enterica* Typhimurium localized to the mesophyll and even systemic tissue of tomato leaves (Gu *et al.*, 2011). These studies indicate that human-pathogenic enterobacteria have the ability to enter plant tissue perhaps using similar mechanisms as plant-associated bacteria, an ability not shared by some non-pathogenic members of the *E. coli* species, e.g. *E. coli* K-12 isolates (Dong *et al.*, 2003). Phytopathogens have specific adaptations, however, enabling more extensive colonization which can eventually result in symptomatic disease, for example colonization of the xylem of orchard plants by the enteric phytopathogen *Erwinia amylovora* (Malnoy *et al.*, 2012).

Bacteria gain access into internal plant tissue through natural openings, and these are quite different above and below ground. Work on leafy tissue has shown that human-pathogenic bacteria are able to access internal tissue through open stomatal pores (Berger *et al.*, 2009; Golberg *et al.*, 2011; Gomes *et al.*, 2009; Saldana *et al.*, 2011), much like phytopathogens, where they can potentially gain access to deeper tissue. However, the guard cells are known to play a role in defence, closing on PAMP perception of invading microbes (Melotto *et al.*, 2006; Zhang *et al.*, 2009). This feature has been manipulated by phytopathogenic bacteria, as elegantly demonstrated using epidermal peels of *A. thaliana* infected with *Pseudomonas syringae*, which is able to prevent closure of the stomatal aperture by interfering with PAMP signalling. In contrast, *E. coli* O157:H7 lacks this ability and its presence induced *A. thaliana* stomata to remain closed for the 8-hour duration of the experiment (Melotto *et al.*, 2006). Differences between human pathogens have also been demonstrated in their ability of penetrate plant tissue via stomatal pores, where *S. enterica* serovar Typhimurium appeared to induced a weaker PTI response coupled with stomatal closure than *E. coli* O157:H7, which presumably accounts for the observed differences in their accumulation in leaf tissue (Roy *et al.*, 2013). Work by Kroupitski and colleagues has shown that entry of *S. enterica* serovar Typhimurium into lettuce leaf stomata was induced by light, with significantly lower levels of internalized bacteria under dark conditions when stomata were closed (Kroupitski *et al.*, 2009a). Furthermore, *S. enterica* was induced to move towards open stomata, also under light conditions. Remarkably, this mimics the phenotype observed for *P. syringae*, induced to move towards open *A. thaliana* stomata (Melotto *et al.*, 2006). Entry of human-pathogenic bacteria into root systems also relies on natural openings and these frequently occur where lateral roots emerge (Cooley *et al.*, 2003; Dong *et al.*, 2003; Klerks *et al.*, 2007b), a similar tactic to that employed by other endophytic bacteria (James *et al.*, 2002; Mano and Morisaki, 2008).

The basis for bacterial entry into plant tissue has not been fully elucidated, although there does appear to be a role for flagella-driven motility (Cooley *et al.*, 2003; Kroupitski *et al.*, 2009a). Not all endophytic bacteria encode flagella, e.g. *K. pneumoniae*, which suggests that other factors are important, which may be bacterial mechanisms such as twitching motility, or plant-derived factors. Studies using surrogates for microbes, such as fluorescent microspheres, have been initiated to investigate plant-factors that facilitate microbial association with or internalization into plants (Solomon and Matthews, 2005). In one case microspheres were found to accumulate in the roots of onion plants over several days (Chancellor *et al.*, 2006), suggesting a passive up-take process occurred. Another study investigating short-term adherence to post-harvest lettuce leaves found no difference in the number of associated *E. coli* O157:H7 and fluorescent microspheres (Solomon and Matthews, 2006). Furthermore, the findings were the same for live or dead bacteria, suggesting that bacterial factors did not play a role in interaction with the plant tissue. These studies highlight the requirement for detailed molecular investigation of the processes involved.

EHEC has been shown to persist for extended periods in soil and the rhizosphere of plants. Studies using spiked manure or compost have found that *E. coli* O157:H7 persisted in soil used for either lettuce or parsley propagation for more than five months (Islam *et al.*, 2004). Parsley plants appeared to present a more favourable environment for the pathogen as it persisted for a greater period on parsley compared to lettuce plants, and was present in the soil covered by the parsley plants for a longer time than in the soil covered by lettuce plants. Inoculum level is also important, as demonstrated for both *E. coli* O157:H7 and for *S. enterica* serovar Typhimurium in the bulk soil and rhizosphere of cabbage (Ongeng *et al.*, 2011) and on spinach (Gutiérrez-Rodríguez *et al.*, 2012). Inoculation with a high density of bacteria (7 log cfu/g) resulted in a 4 to 5-fold increase in persistence, measured by the time to the limit of detection in cabbage rhizosphere and bulk soil, compared to a lower dose (4 log cfu/g), which is greater than that expected from inoculation level alone (Ongeng *et al.*, 2011). There may be various explanations for this effect, including quorum sensing: a communication strategy employed by bacteria based on secretion and perception of diffusible molecules that has profound implications on population-level gene expression. Gene expression of enteric phytopathogens is known to be regulated *in planta* in a quorum-sensing dependent manner (Liu *et al.*, 2008; Pollumaa *et al.*, 2012) and similar regulation may occur for human pathogens in association with plants. Colonization and persistence of human pathogens on the leaves of plants tend to be orders of magnitude lower than seen on the roots and rhizosphere on a total population basis (Brandl *et al.*, 2004; Patel *et al.*, 2010; Quilliam *et al.*, 2012b), although the bacteria tend to be clustered into colonies of varying densities. Comparison of persistence of bacteria on different plant species raises the issue of plant species-specific differences, not least the role of plant root exudates in colonization by human pathogens (Hartmann *et al.*, 2009). Root exudates are involved in a wide range of functions and are able to mediate interactions between the plants and rhizosphere-associated bacteria (Bais *et al.*, 2006; De Hoff *et al.*, 2009). Some exudates have been shown to mimic quorum sensing molecules, including those commonly employed by bacteria, *N*-acyl homoserine lactones (AHLs) (Bauer and Mathesius, 2004), whilst sensing of bacterially derived AHL molecules has also been demonstrated (Schuhegger *et al.*, 2006). AHL mimics secreted by plant roots appear to be relatively widespread, although there does seem to be some prevalence in nodulating plants (Bais *et al.*, 2006). The influence of exudates on the colonization of human-pathogenic bacteria has yet to be fully explored, although there is evidence for chemotactic response of *S. enterica* to exudates (Klerks *et al.*, 2007a) as well as a swarming response (Barak *et al.*, 2009). In addition, differential gene expression of *E. coli* has been demonstrated for bacteria recovered from the rhizosphere of lettuce (Hou *et al.*, 2012).

Plant immunity

Plant hosts are able to respond to bacterial colonization through recognition of conserved molecules, to control the extent of microbial growth at the plant's extent. Overt phytopathogens have evolved mechanisms to overcome the response and enter into the archetypal arms-race with their susceptible hosts (Boller and He, 2009). The first layer of responsive plant defence is through recognition of conserved microbial molecules termed pathogen- or microbe-associated molecular patterns (PAMPs, MAMPs) by recognition receptors (PRRs) on the plant's cell surface, which results in induction of PAMP-triggered immunity (PTI). Bacterial factors that trigger a PTI response include surface factors such as flagella and peptidoglycan, as well as abundant cytosolic proteins, elongation factor Tu (EF-Tu) and cold shock protein A (CspA) (Segonzac and Zipfel, 2011). A recent study using a bioinformatics approach identified a much larger potential repertoire of PAMPs, on the basis of positive selection within coding regions (McCann *et al.*, 2012). This approach applied to the 'core genome set' derived from six Gram-negative phytopathogens revealed that as much as 35% of the 'core genome' contained at least one positively selected residue within a coding sequence. Functional validation for 10 candidate peptides showed significant virulence suppression activity occurred with eight of them in *A. thaliana* (McCann *et al.*, 2012). Phytopathogens are able to suppress PTI though delivery of effector proteins, via type three secretion systems (T3SS) that inject the bacterial proteins directly into the plant cell. Effectors suppress or modulate the plant host through molecular mimicry and/or inhibition of eukaryotic cellular processes (Jones and Dangl, 2006). Resistant hosts are able to counter the effect of bacterial effectors via specific interaction with *R* gene proteins, most frequently nucleotide binding, leucine rich repeat proteins (NB-LRR), which induces a second arm of defence termed effector-triggered immunity (ETI). Often, a hypersensitive-response (HR) is triggered, in which cells at the site of infection undergo apoptosis in an attempt to halt pathogen spread. Some effectors can suppress this layer of defence, leading to effector-triggered suppression (ETS) (Jones and Dangl, 2006).

Effector-driven immunity is an excellent example of co-evolution between pathogen and host, and occurs on a much more rapid evolutionary time-scale than PTI. Given the specific relationships involved, it is intriguing that some human pathogens appear to be able to also interact with plant defences. Infiltration of *E. coli* O157:H7 into *Arabidopsis thaliana* leaves did not result in bacterial growth or any symptomatic disease, but a PTI was induced (Thilmony *et al.*, 2006). Furthermore, whole transcriptome analysis revealed a remarkable overlap between the genes induced by the human pathogen compared with those induced by a T3SS mutant of *Pseudomonas syringae* (*hrp*-). It is of note that perception of flagellate bacteria for either pathogen did not result in major transcriptional changes. Infiltration with *E. coli* O157:H7 *filC*- did not result in differential expression of any genes compared to the WT, and for the same mutant of *P. syringae*, only 20 genes were induced > two-fold. This suggests that for infiltrated bacteria, flagellin plays only a minor role in PTI, although it was noted that perhaps the response is different for bacteria present on leaf surfaces. Studies where *A. thaliana* were inoculated with EHEC on the surface, by leaf dipping, showed that EHEC surface factors like flagella, LPS and curli, all acted as PAMPS (Seo and Matthews, 2012). Mutation of the surface factors led to an increase in the bacterial counts five days post infection, and expression of *A. thaliana* BGL2 (β-1,3-glucanase, a PR2 gene regulated by salicylic acid (SA) and known as a molecular marker of systemic acquired resistance) decreased more than two fold following challenge with the surface factor mutants (Seo and Matthews, 2012). Studies with *A. thaliana* infiltrated with *S. enterica* serovar Typhimurium support this concept as plants defective in flagellin perception still exhibited activation of MPK6, a marker of PTI (Schikora *et al.*, 2008), although in *N. benthamiana* flagellin recognition appears to be via FliC and not the alternative phase 2 subunit, FljB (Meng *et al.*, 2013). This also may explain, in part, the microbe–plant relationship demonstrated by some endophytic bacteria that lack both flagella and a T3SS, such as *K. pneumoniae* (Dong *et al.*, 2001).

In contrast to what was found for *E. coli* O157:H7, inoculation of *S. enterica* serovar Typhimurium on *A. thaliana* leaves did result in symptomatic disease, in the form of chlorosis and wilting (Schikora *et al.*, 2008). Symptomatic disease of *A. thaliana* was also observed following colonization by the Gram-positive pathogen *Staphylococcus aureus* (Prithiviraj *et al.*, 2005). Furthermore, chlorosis occurred even when the plants were inoculated at the roots, from spiked compost, suggesting that this pathogen is able to translocate from the roots. Inoculation with *S. enterica* serovar Typhimurium resulted in a strong PTI response, including expression of PR genes belonging to both the ethylene (ET)- and jasmonic acid (JA)-dependent pathways (Schikora *et al.*, 2008). Subsequent work showed that as for *E. coli* O157:H7, there was a strong overlap between the *A. thaliana* genes expressed following infiltration with *S.* Typhimurium as with *P. syringae*. What is remarkable is that there appeared to be a specific response to *S.* Typhimurium, and there was some evidence to suggest that the human pathogen was able to suppress PTI in a T3SS-dependent manner (Schikora *et al.*, 2011). *S.* Typhimurium encodes two T3SS, located on two pathogenicity islands, SPI-1 and SPI-2, respectively. During colonization of mammalian hosts, the SPI-1 T3SS is involved in cellular invasion: effector genes are translocated into the host cell, such as a macrophage, where they are able to induce actin filament rearrangements, which in turn cause membrane ruffling and engulf the bacterial cell within a membrane-bound vacuole (reviewed in Srikanth *et al.*, 2011). Once the bacteria are intracellular, the second SPI-2 T3SS is induced, where effectors are required to maintain the vacuole and prevent the toxic effects of pH changes and production of ROS within the host cell (Ruiz-Albert *et al.*, 2002). Intact T3SS of both types appeared to be necessary for complete growth within *A. thaliana* leaves, and in particular, T3SS SPI-1 appeared to be required to suppress plant host defences (Schikora *et al.*, 2011; Shirron and Yaron, 2011). Furthermore, infiltration of *A. thaliana* with *S. enterica* serovar Typhimurium containing mutations in genes of T3SS-SPI-1 and SPI-2 resulted in decreased bacterial colonization within the infiltrated leaf, suggesting a positive role for both T3SS in colonization (Schikora *et al.*, 2011). The lower levels of colonization exhibited by the mutants were coupled with more obvious HR-like symptoms, indicative of a role of *S. enterica* effectors in suppression of the HR. However, the picture is not yet completely clear, as different mutations in either a structural component of the T3SS-SPI-1 apparatus, or a T3SS-SPI-1 effector was found to result in an increase in the level of bacterial colonization of alfalfa root (Iniguez *et al.*, 2005), suggesting in this case, perception of the T3SS and induction of defence. In support of plants mounting a HR due to effector recognition, the type III effector protein of *S. enterica*, SseF, was shown to trigger a HR in *Nicotiana benthamiana* when expressed by *Agrobacterium tumefaciens* or secreted from *Xanthomonas campestris* pv. *vesicatoria* (Ustun *et al.*, 2012). This highlights the plant's potential to recognize the human pathogen effector via an R protein. Whether this is due to the R protein recognizing SseF through co-adaptation or is because of perception of a common domain shared with phytopathogen effectors remains to be seen. These discrepancies suggest specific interactions that are dependent on plant species and tissue type. However, the application of novel modelling methods to identify putative interactor proteins should help uncover host targets and resistance mechanisms (Schleker *et al.*, 2012). Furthermore, they also point to the need for a better understanding of bacteria gene expression *in planta*. Although intriguing, it is perhaps not altogether surprising that a member of the *Enterobacteriaceae* that encounters a huge variety of environments is well equipped to adapt to those situations.

There is also evidence for a role in suppression of plant defence by virulence factors expressed from other pathogens. *Staphylococcus aureus* mutants lacking virulence factors known to play a role in symptomatic infection of mammals, exhibited ~50% reduced colonization levels compared to the isogenic parental bacteria, coupled with significantly lower levels of plant mortality (Prithiviraj *et al.*, 2005). Mammalian virulence factors of *Pseudomonas aeruginosa* also appear to contribute to colonization of *A. thaliana* and control of symptomatic disease (Plotnikova *et al.*, 2000). Subsequent studies benefited from a screen of *P. aeruginosa* transposon mutants *in planta*, to find bacterial factors required for colonization and elicitation of disease (Rahme *et al.*, 2000). The mutants were screened in plant, nematode and mouse infection models and identified 15 genes that shared virulence functionality across different kingdoms.

Further experiments have revealed important differences between the response elicited by the human pathogen and the plant endophyte. Inoculation of *A. thaliana npr1* (non-expressor of pathogenesis related protein 1) and *nahG* (encodes salicylate hydroxylase) mutants with *S. enterica* variants suggested that the T3SS-SPI1 of *S. enterica* induced both the SA-mediated and SA-independent response pathways (Iniguez *et al.*, 2005), supported by studies that showed increased colonization of *S. enterica* serovar Typhimurium in *A. thaliana coi1–16* (coronatine insensitive 1) and *ein2–1* (ethylene insensitive 2) (Schikora *et al.*, 2008). A similar situation occurred with *S. aureus*, where mutation of *A. thaliana npr1* and *nahG* resulted in increased levels of bacterial colonization at the roots (Prithiviraj *et al.*, 2005). In contrast, the response to an endophytic *K. pneumoniae* isolate induced only SA-independent responses (Iniguez *et al.*, 2005). Furthermore, PR1 expression in *A. thaliana* was induced in the presence of wild type *S. enterica*, but not for *K. pneumoniae* or for the T3SS-SPI1 *S. enterica* mutant (Iniguez *et al.*, 2005). These experiments were amongst the first to demonstrate specific host-defence responses of plants to *S. enterica* (later corroborated with whole transcriptome analysis (Schikora *et al.*, 2011)). In addition, they add an important insight into the colonization strategy employed by the endophyte *K. pneumoniae*.

The ecological perspective

Molecular analysis of the two-way cross-talk between microbes and their plant hosts is clearly vital to fully understand the nature of the relationships. However, this two-way cross-talk is unlikely to occur in isolation under 'real world' conditions that exist in agriculture. Therefore, it is vital that both areas of molecular and ecological microbiology are considered together to provide a more realistic over-view. Plants are complex organisms and every niche associated with the plant host encompasses a wide variety of physiochemical environments. On the one hand, plant propagation conditions may be extremely conducive to bacterial growth, for example the relatively high temperature and humidity conditions used for sprouted seeds, while on the other hand other crops and growth regimes may present far more challenging conditions. Due consideration also needs to be given to the different microenvironments inhabited by human-pathogenic bacteria; the availability of nutrients and water; physical structures that may influence interactions; plant immunity, which may well be 'primed' by the presence of other microbes; and the interactions with other microbial species, whether positive or negative. The work showing stomatal responses to human-pathogenic bacteria highlights the need to carry out plant–microbe interaction experiments under conditions that permit realistic interactions. They also emphasize the need to differentiate between observations made with living plants compared to post-harvest, processed plant material. The presence of other microbial species has been shown to influence the colonization potential by human-pathogenic bacteria, in both negative and positive ways (Cooley *et al.*, 2003; Hora *et al.*, 2005). Our work on three-week colonization of barley plants has revealed that EHEC colonizes the roots of the plants to relatively high levels (~5 $\log_{10}$ cfu/g), similar to that seen for *E. coli* isolated from barley roots (unpublished). Furthermore, colonization to these levels occurs in the presence of native microflora present in un-sterilized plant growth compost. This work together with similar undertaken elsewhere, shows that *E. coli* is able to compete with native microbes and successfully colonize plants. Ecological surveys have revealed that *E. coli* persists in so-called 'environmental' habitats including beaches (Walk *et al.*, 2007), watersheds (Lyautey *et al.*, 2010), alpine grasslands (Texier *et al.*, 2008) and soil (Ishii *et al.*, 2009), leading to designation of 'naturalized' isolates that are able to persist and proliferate in the absence of mammalian hosts (Ishii and Sadowsky, 2008). The concept reinforces the view that many human pathogens have a great capacity to adapt to different environments and challenges our dogma that those, especially classed as zoonoses (i.e. animal associated and causing disease in humans) are only ever present outwith mammalian or animal hosts for a limited time. Genomic comparisons between human-pathogenic bacteria and plant-associated bacteria have revealed striking levels of homology exist for some regions of the genome, especially in regions of horizontal transfer, raising the possibility of ancestral commonality (Holden, 2010; Toth *et al.*, 2006). It is perhaps not surprising therefore, to find that some human-pathogenic bacteria are able to adapt to plant hosts.

Conclusions

In conclusion, a complex series of molecular interactions takes place between human-pathogenic bacteria and their plant hosts. The interaction is very much a two-way process resulting in a progressive molecular dialogue which underpins the outcome of colonization. The interaction is also a dynamic one, with each partner adapting as the interaction proceeds. There are clear parallels with the interactions that occur between human-pathogenic bacteria and human hosts and with phytopathogens and plant hosts. In fact, the same basic steps of colonization hold true whichever kingdom the host belongs to. This has been exemplified in the mechanisms of bacterial adherence and immune host responses. What is intriguing are the specific differences that relate to bacterial adaptation to alternative hosts, including plants.

Future directions

Evidence gathered on the interactions between human-pathogenic bacteria and plant hosts consolidates the view that plants do not simply act as inert vectors to transfer bacteria into the food chain, but instead can act as alternative hosts. Our challenge now is to fully understand the nature of the bacteria–plant interactions, giving due consideration to the experimental conditions and how they may influence the outcome. This covers a range of experiments from molecular analysis, for example the determination of gene expression of both plant and bacterial partners during different points of colonization, through to microbiological data gathered from realistic agricultural settings. The increasing world-wide demand for provision of food has put unprecedented pressure on agriculture and food producers. A greater reliance on scaling-up of production systems, distribution and imported produce comes with additional challenges, not least in ensuring high levels of food safety. Ultimately, the research needs to provide evidence that can be confidently applied to preventing, or at least controlling transmission of human pathogens through the food chain.

Acknowledgements

NH is supported by the Scottish Government Rural and Environment Science and Analytical Services Division; AH is supported by the Biotechnology and Biological Sciences Research Council; YR is supported by the Leverhulme Trust; RJ is partially supported by the Leverhulme Trust and the Biotechnology and Biological Sciences Research Council. Thanks go to Cavan Convery for graphical assistance with Fig. 6.1.

Web resources

Website name and short description (host site)	Web address (url)
Genomics and molecular biology databases	
CGSG: the *E. coli* genetic stock centre (Yale University, USA)	http://cgsc2.biology.yale.edu/index.php
*Echo*Base: an integrated post-genomic database for *E. coli* (University of York, UK)	http://www.york.ac.uk/res/thomas/index.cfm
EcoGene: a multiple tool website for interrogating *E. coli* experimental gene expression data in the EcoGene database (University of Miami School of Medicine, USA)	http://www.ecogene.org/
Microbial Genomes: genome resources and annotation tools (National Library of Medicine, Bethesda, USA)	http://www.ncbi.nlm.nih.gov/genomes/MICROBES/microbial_taxtree.html
SGSC: the *Salmonella* genetic stock centre (University of Calgary, Canada)	http://people.ucalgary.ca/~kesander/index.html
T3SE database: a type 3 secretion effector database (National University of Singapore, Singapore)	http://effectors.bic.nus.edu.sg/T3SEdb/index.php
The Genome Gateway: a gateway to genome information and resources (Horizon Press)	http://www.horizonpress.com/gateway/genome.html
xbase: a genome resource for comparative bacterial genomics (University of Birmingham, UK)	http://www.xbase.ac.uk/colibase/
Sub-species typing	
CRISPRs database (Institut de Génétique et Microbiologie, Orsay, France)	http://crispr.u-psud.fr/
MLST databases (University of Oxford, UK)	http://pubmlst.org/databases.shtml
MLVA database (National Institute for Public Health and the Environment (RIVM), the Netherlands)	http://www.mlva.net/default.asp
PulseNet (Centers for Disease Control, USA)	http://www.cdc.gov/pulsenet/
Public health	
A-Z index for foodborne illness (Centers for Disease Control, USA)	http://www.cdc.gov/foodsafety/diseases/
ECDC: European Centre for Disease Prevention and Control (European Union)	http://ecdc.europa.eu/en/Pages/home.aspx
ProMED-mail: program for monitoring emerging diseases (Harvard School of Public Health, USA (run as part of the International Society for Infectious Diseases))	http://www.promedmail.org/index.html
UK infectious diseases epidemiological data (Health Protection Agency, UK)	http://www.hpa.org.uk/Topics/InfectiousDiseases/
World-wide data on disease statistics (World Health Organization)	http://www.who.int/research/en/

References

Abby, S.S., and Rocha, E.P.C. (2012). The non-flagellar type III secretion system evolved from the bacterial flagellum and diversified into host-cell adapted systems. PLoS Genet *8*, e1002983.

Alegre, I., Abadias, M., Anguera, M., Usall, J., and Vinas, I. (2010). Fate of *Escherichia coli* O157:H7, *Salmonella* and *Listeria* innocua on minimally-processed peaches under different storage conditions. Food Microbiol. *27*, 862–868.

Allerberger, F., and Sessitsch, A. (2009). Incidence and microbiology of salad-borne disease. CAB Rev.: Perspect. Agric. Vet. Sci. Nutr. Nat. Res. *4*, 1–13.

Ausmees, N., Jonsson, H., Hoglund, S., Ljunggren, H., and Lindberg, M. (1999). Structural and putative regulatory genes involved in cellulose synthesis in *Rhizobium leguminosarum* bv. *trifolii*. Microbiology *145*, 1253–1262.

Avelino, F., Saldana, Z., Islam, S., Monteiro-Neto, V., Dall'Agnol, M., Eslava, C.A., and Giron, J.A. (2010). The majority of enteroaggregative *Escherichia coli* strains produce the *E. coli* common pilus when adhering to cultured epithelial cells. Int. J. Med. Microbiol. *300*, 440–448.

Bais, H.P., Weir, T.L., Perry, L.G., Gilroy, S., and Vivanco, J.M. (2006). The role of root exudates in rhizosphere interactions with plants and other organisms. Annu. Rev. Plant Biol. *57*, 233–266.

Barabote, R.D., Johnson, O.L., Zetina, E., San Francisco, S.K., Fralick, J.A., and San Francisco, M.J.D. (2003). *Erwinia chrysanthemi tolC* is involved in resistance to antimicrobial plant chemicals and is essential for phytopathogenesis. J. Bacteriol. *185*, 5772–5778.

Barak, J.D., and Liang, A.S. (2008). Role of soil, crop debris, and a plant pathogen in *Salmonella enterica* contamination of tomato plants. PLoS ONE 3, e1657.

Barak, J.D., Gorski, L., Naraghi-Arani, P., and Charkowski, A.O. (2005). *Salmonella enterica* virulence genes are required for bacterial attachment to plant tissue. Appl. Environ. Microbiol. *71*, 5685–5691.

Barak, J.D., Jahn, C.E., Gibson, D.L., and Charkowski, A.O. (2007). The role of cellulose and O-antigen capsule in the colonization of plants by *Salmonella enterica*. Mol. Plant Microbe Interact. *20*, 1083–1091.

Barak, J.D., Gorski, L., Liang, A.S., and Narm, K.-E. (2009). Previously uncharacterized *Salmonella enterica* genes required for swarming play a role in seedling colonization. Microbiology *155*, 3701–3709.

Barak, J.D., Kramer, L.C., and Hao, L.-Y. (2011). Colonization of tomato plants by *Salmonella enterica* is cultivar dependent, and type 1 trichomes are preferred colonization sites. Appl. Environ. Microbiol. *77*, 498–504.

Barnhart, M.M., and Chapman, M.R. (2006). Curli biogenesis and function. Annu. Rev. Microbiol. *60*, 131–147.

Bauer, W.D., and Mathesius, U. (2004). Plant responses to bacterial quorum sensing signals. Curr. Opin. Plant Biol. *7*, 429–433.

Berger, C.N., Shaw, R.K., Ruiz-Perez, F., Nataro, J.P., Henderson, I.R., Pallen, M.J., and Frankel, G. (2009). Interaction of enteroaggregative *Escherichia coli* with salad leaves. Environ. Microbiol. Rep. *1*, 234–239.

Berger, C.N., Brown, D.J., Shaw, R.K., Minuzzi, F., Feys, B., and Frankel, G. (2011). *Salmonella enterica* strains belonging to O serogroup 1,3,19 induce chlorosis and wilting of *Arabidopsis thaliana* leaves. Environ. Microbiol. *13*, 1299–1308.

Bergholz, T., Wick, L., Qi, W., Riordan, J., Ouellette, L., and Whittam, T. (2007). Global transcriptional response of *Escherichia coli* O157:H7 to growth transitions in glucose minimal medium. BMC Microbiol. *7*, 97.

Bergholz, T.M., Vanaja, S.K., and Whittam, T.S. (2009). Gene expression induced in *Escherichia coli* O157:H7 upon exposure to model apple juice. Appl. Environ. Microbiol. *75*, 3542–3553.

Blackburn, D., Husband, A., Saldana, Z., Nada, R.A., Klena, J., Qadri, F., and Giron, J.A. (2009). Distribution of the *Escherichia coli* common pilus among diverse strains of human enterotoxigenic *E. coli*. J. Clin. Microbiol. *47*, 1781–1784.

Bogdanove, A.J., Koebnik, R., Lu, H., Furutani, A., Angiuoli, S.V., Patil, P.B., Van Sluys, M.A., Ryan, R.P., Meyer, D.F., Han, S.W., *et al.* (2011). Two new complete genome sequences offer insight into host and tissue specificity of plant pathogenic *Xanthomonas* spp. J. Bacteriol. *193*, 5450–5464.

Bolin, I., Norlander, L., and Wolf-Watz, H. (1982). Temperature-inducible outer membrane protein of *Yersinia pseudotuberculosis* and *Yersinia enterocolitica* is associated with the virulence plasmid. Infect. Immun. *37*, 506–512.

Boller, T., and He, S.Y. (2009). Innate immunity in plants: an arms race between pattern recognition receptors in plants and effectors in microbial pathogens. Science *324*, 742–744.

Boyer, R.R., Sumner, S.S., Williams, R.C., Pierson, M.D., Popham, D.L., and Kniel, K.E. (2007). Influence of curli expression by *Escherichia coli* O157:H7 on the cell's overall hydrophobicity, charge, and ability to attach to lettuce. J. Food Prot. *70*, 1339–1345.

Bradley, D.E. (1974). The adsorption of *Pseudomonas aeruginosa* pilus-dependent bacteriophages to a host mutant with nonretractile pili. Virology *58*, 149–163.

Bradley, D.E. (1980). A function of *Pseudomonas aeruginosa* PAO polar pili: twitching motility. Can. J. Microbiol. *26*, 146–154.

Brandl, M.T. (2006). Fitness of human enteric pathogens on plants and implications for food safety. Annu. Rev. Phytopathol. *44*, 367–392.

Brandl, M.T., Haxo, A.F., Bates, A.H., and Mandrell, R.E. (2004). Comparison of survival of *Campylobacter jejuni* in the phyllosphere with that in the rhizosphere of spinach and radish plants. Appl. Environ. Microbiol. *70*, 1182–1189.

Carey, C.M., Kostrzynska, M., and Thompson, S. (2009). *Escherichia coli* O157:H7 stress and virulence gene expression on Romaine lettuce using comparative real-time PCR. J. Microbiol. Methods *77*, 235–242.

Carter, M.Q., Brandl, M.T., Louie, J.W., Kyle, J.L., Carychao, D.K., Cooley, M.B., Parker, C.T., Bates, A.H., and Mandrell, R.E. (2011). Distinct acid resistance and survival fitness displayed by Curli variants of enterohemorrhagic *Escherichia coli* O157:H7. Appl. Environ. Microbiol. *77*, 3685–3695.

C.D.C. (Centers for Disease Control and Prevention) (2006). Ongoing multistate outbreak of *Escherichia coli* serotype O157:H7 infections associated with consumption of fresh spinach – United States, September 2006. Morb. Mortal. Wkly Rep. *55*, 1045–1046.

Chancellor, D.D., Tyagi, S., Bazaco, M.C., Bacvinskas, S., Chancellor, M.B., Dato, V.M., and de Miguel, F. (2006). Green onions: potential mechanism for hepatitis A contamination. J. Food Prot. *69*, 1468–1472.

Chilcott, G.S., and Hughes, K.T. (2000). Coupling of flagellar gene expression to flagellar assembly in *Salmonella enterica* serovar Typhimurium and *Escherichia coli*. Microbiol. Mol. Biol. Rev. *64*, 694–708.

Choi, S., Bang, J., Kim, H., Beuchat, L.R., and Ryu, J.H. (2011). Survival and colonization of *Escherichia coli* O157:H7 on spinach leaves as affected by inoculum level and carrier, temperature and relative humidity. J. Appl. Microbiol. *111*, 1465–1472.

Comanducci, M., Bambini, S., Brunelli, B., Adu-Bobie, J., Arico, B., Capecchi, B., Giuliani, M.M., Masignani, V., Santini, L., Savino, S., *et al.* (2002). NadA, a novel vaccine candidate of *Neisseria meningitidis*. J. Exp. Med. *195*, 1445–1454.

Connell, I., Agace, W., Klemm, P., Schembri, M., Marild, S., and Svanborg, C. (1996). Type 1 fimbrial expression enhances *Escherichia coli* virulence for the urinary tract. Proc. Natl. Acad. Sci. U.S.A. *93*, 9827–9832.

Cooley, M.B., Miller, W.G., and Mandrell, R.E. (2003). Colonization of *Arabidopsis thaliana* with *Salmonella enterica* and enterohemorrhagic *Escherichia coli* O157:H7 and competition by *Enterobacter asburiae*. Appl. Environ. Microbiol. *69*, 4915–4926.

Cooley, M.B., Chao, D., and Mandrell, R.E. (2006). *Escherichia coli* O157:H7 survival and growth on lettuce is altered by the presence of epiphytic bacteria. J. Food Prot. *69*, 2329–2335.

Cornelis, G.R. (2006). The type III secretion injectisome. Nat. Rev. Microbiol. *4*, 811–825.

Croes, C.L., Moens, S., Vanbastelaere, E., Vanderleyden, J., and Michiels, K.W. (1993). The polar flagellum mediates *Azospirillum brasilense* adsorption to wheat roots. J. Gen. Microbiol. *139*, 2261–2269.

Das, A., Rangaraj, N., and Sonti, R.V. (2009). Multiple adhesin-like functions of *Xanthomonas oryzae* pv. oryzae are involved in promoting leaf attachment, entry, and virulence on rice. Mol. Plant Microbe Interact. *22*, 73–85.

Deering, A.J., Mauer, L.J., and Pruitt, R.E. (2012). Internalization of *E. coli* O157:H7 and *Salmonella* spp. in plants: a review. Food Res. Int. *45*, 567–575.

De Hoff, P.L., Brill, L.M., and Hirsch, A.M. (2009). Plant lectins: the ties that bind in root symbiosis and plant defense. Mol. Genet. Genom. *282*, 1–15.

Delahay, R.M., Knutton, S., Shaw, R.K., Hartland, E.L., Pallen, M.J., and Frankel, G. (2005). The coiled-coil domain of EspA is essential for the assembly of the type III secretion translocon on the surface of enteropathogenic *Escherichia coli*. J. Biol. Chem. *280*, 19436–19436.

Deweger, L.A., Vandervlugt, C.I.M., Wijfjes, A.H.M., Bakker, P.A.H.M., Schippers, B., and Lugtenberg, B. (1987). Flagella of a plant-growth-stimulating *Pseudomonas fluorescens* strain are required for colonization of potato roots. J. Bacteriol. *169*, 2769–2773.

Dinu, L.-D., and Bach, S. (2011). Induction of viable but nonculturable *Escherichia coli* O157:H7 in the phyllosphere of lettuce: a food safety risk factor. Appl. Environ. Microbiol. *77*, 8295–8302.

Dong, Y., Glasner, J.D., Blattner, F.R., and Triplett, E.W. (2001). Genomic interspecies microarray hybridization: rapid discovery of three thousand genes in the maize endophyte, *Klebsiella pneumoniae* 342, by microarray hybridization with *Escherichia coli* K-12 open reading frames. Appl. Environ. Microbiol. *67*, 1911–1921.

Dong, Y., Iniguez, A.L., Ahmer, B.M., and Triplett, E.W. (2003). Kinetics and strain specificity of rhizosphere and endophytic colonization by enteric bacteria on seedlings of *Medicago sativa* and *Medicago truncatula*. Appl. Environ. Microbiol. *69*, 1783–1790.

Dorr, J., Hurek, T., and Reinhold-Hurek, B. (1998). Type IV pili are involved in plant–microbe and fungus–microbe interactions. Mol. Microbiol. *30*, 7–17.

Duguid, J.P. (1959). Fimbriae and adhesive properties in *Klebsiella* strains. J. Gen. Microbiol. *21*, 271–286.

Duncan, M.J., Mann, E.L., Cohen, M.S., Ofek, I., Sharon, N., and Abraham, S.N. (2005). The distinct binding specificities exhibited by enterobacterial type 1 fimbriae are determined by their fimbrial shafts. J. Biol. Chem. *280*, 37707–37716.

Dwyer, B.E., Newton, K.L., Kisiela, D., Sokurenko, E.V., and Clegg, S. (2011). Single nucleotide polymorphisms of *fimH* associated with adherence and biofilm formation by serovars of *Salmonella enterica*. Microbiology *157*, 3162–3171.

Ebel, F., Deibel, C., Kresse, A.U., Guzman, C.A., and Chakraborty, T. (1996). Temperature- and medium-dependent secretion of proteins by Shiga toxin-producing *Escherichia coli*. Infect. Immun. *64*, 4472–4479.

Eblen, B.S., Walderhaug, M.O., Edelson-Mammel, S., Chirtel, S.J., De Jesus, A., Merker, R.I., Buchanan, R.L., and Miller, A.J. (2004). Potential for internalization, growth, and survival of *Salmonella* and *Escherichia coli* O157:H7 in oranges. J. Food Prot. *67*, 1578–1584.

Ells, T.C., and Truelstrup Hansen, L. (2006). Strain and growth temperature influence *Listeria* spp. attachment to intact and cut cabbage. Int. J. Food Microbiol. *111*, 34–42.

Erdem, A.L., Avelino, F., Xicohtencatl-Cortes, J., and Giron, J.A. (2007). Host protein binding and adhesive properties of H6 and H7 flagella of attaching and effacing *Escherichia coli*. J. Bacteriol. *189*, 7426–7435.

Fink, R.C., Black, E.P., Hou, Z., Sugawara, M., Sadowsky, M.J., and Diez-Gonzalez, F. (2012). Transcriptional responses of *Escherichia coli* K-12 and O157:H7 associated with lettuce leaves. Appl. Environ. Microbiol. *78*, 1752–1764.

Frank, C., Werber, D., Cramer, J.P., Askar, M., Faber, M., an der Heiden, M., Bernard, H., Fruth, A., Prager, R., Spode, A., *et al.* (2011). Epidemic profile of Shiga-toxin-producing *Escherichia coli* O104:H4 outbreak in Germany. N. Engl. J. Med. *365*, 1771–1780.

Fussenegger, M., Rudel, T., Barten, R., Ryll, R., and Meyer, T.F. (1997). Transformation competence and type-4 pilus biogenesis in *Neisseria gonorrhoeae* – a review. Gene *192*, 125–134.

Gal, M., Preston, G.M., Massey, R.C., Spiers, A.J., and Rainey, P.B. (2003). Genes encoding a cellulosic polymer contribute toward the ecological success of *Pseudomonas fluorescens* SBW25 on plant surfaces. Mol. Ecol. *12*, 3109–3121.

Gangaros, E.J., Donadio, J.A., Armstrong, R.W., Meyer, K.F., Brachman, P.S., and Dowell, V.R. (1971). Botulism in the United States, 1899–1969. Am. J. Epidemiol. *93*, 93–101.

Gerlach, G.F., Clegg, S., and Allen, B.L. (1989). Identification and characterization of the genes encoding the type 3 and type 1 fimbrial adhesins of *Klebsiella pneumoniae*. J. Bacteriol. *171*, 1262–1270.

Golberg, D., Kroupitski, Y., Belausov, E., Pinto, R., and Sela, S. (2011). *Salmonella* Typhimurium internalization is variable in leafy vegetables and fresh herbs. Int. J. Food Microbiol. *145*, 250–257.

Gomes, C., Da Silva, P., Moreira, R.G., Castell-Perez, E., Ellis, E.A., and Pendleton, M. (2009). Understanding *E. coli* internalization in lettuce leaves for optimization of irradiation treatment. Int. J. Food Microbiol. *135*, 238–247.

Gorski, L., Duhé, J.M., and Flaherty, D. (2009). The use of flagella and motility for plant colonization and fitness by different strains of the foodborne pathogen *Listeria monocytogenes*. PLoS ONE *4*, e5142.

Gu, G., Hu, J., Cevallos-Cevallos, J.M., Richardson, S.M., Bartz, J.A., and van Bruggen, A.H.C. (2011). Internal colonization of *Salmonella enterica* serovar Typhimurium in tomato plants. PLoS ONE *6*, e27340.

Guo, X., Chen, J., Brackett, R.E., and Beuchat, L.R. (2001). Survival of salmonellae on and in tomato plants from the time of inoculation at flowering and early stages of fruit development through fruit ripening. Appl. Environ. Microbiol. *67*, 4760–4764.

Gutiérrez-Rodríguez, E., Gundersen, A., Sbodio, A.O., and Suslow, T.V. (2012). Variable agronomic practices, cultivar, strain source and initial contamination dose differentially affect survival of *Escherichia coli* on spinach. J. Appl. Microbiol. *112*, 109–118.

Haahtela, K., and Korhonen, T.K. (1985). *In vitro* adhesion of N(2)-fixing enteric bacteria to roots of grasses and cereals. Appl. Environ. Microbiol. *49*, 1186–1190.

Haahtela, K., Tarkka, E., and Korhonen, T.K. (1985). Type 1 fimbria-mediated adhesion of enteric bacteria to grass roots. Appl. Environ. Microbiol. *49*, 1182–1185.

Haefele, D.M., and Lindow, S.E. (1987). Flagellar motility confers epiphytic fitness advantages upon *Pseudomonas syringae*. Appl. Environ. Microbiol. *53*, 2528–2533.

Hanning, I.B., Nutt, J.D., and Ricke, S.C. (2009). Salmonellosis outbreaks in the United States due to fresh produce: sources and potential intervention measures. Foodborne Pathog. Dis. *6*, 635–648.

Hartmann, A., Schmid, M., van Tuinen, D., and Berg, G. (2009). Plant-driven selection of microbes. Plant Soil *321*, 235–257.

Hirneisen, K.A., Sharma, M., and Kniel, K.E. (2012). Human enteric pathogen internalization by root uptake into food crops. Foodborne Pathog. Dis. *9*, 396–405.

Hoiczyk, E., Roggenkamp, A., Reichenbecher, M., Lupas, A., and Heesemann, J. (2000). Structure and sequence analysis of *Yersinia* YadA and *Moraxella* UspAs reveal a novel class of adhesins. EMBO J. *19*, 5989–5999.

Holden, N.J. (2010). Plants as reservoirs for human enteric pathogens. CAB Rev.: Perspect. Agric. Vet. Sci. Nutr. Nat. Res. *5*, 11 pp.

Hora, R., Warriner, K., Shelp, B.J., and Griffiths, M.W. (2005). Internalization of *Escherichia coli* O157: H7 following biological and mechanical disruption of growing spinach plants. J. Food Prot. *68*, 2506–2509.

Hou, Z., Fink, R.C., Black, E., Sugawara, M., Zhang, Z., Diez-Gonzalez, F., and Sadowsky, M.J. (2012). Gene expression profiling of *Escherichia coli* in response to interactions with the lettuce rhizosphere. J. Appl. Microbiol. *113*, 1076–1086.

Hueck, C.J. (1998). Type III protein secretion systems in bacterial pathogens of animals and plants. Microbiol. Mol. Biol. Rev. *62*, 379–433.

Ibekwe, A.M., Grieve, C.M., Papiernik, S.K., and Yang, C.H. (2009). Persistence of *Escherichia coli* O157:H7 on the rhizosphere and phyllosphere of lettuce. Lett. Appl. Microbiol. *49*, 784–790.

Iniguez, A.L., Dong, Y., Carter, H.D., Ahmer, B.M.M., Stone, J.M., and Triplett, E.W. (2005). Regulation of enteric endophytic bacterial colonization by plant defenses. Mol. Plant Microbe Interact. *18*, 169–178.

Ishii, S., and Sadowsky, M.J. (2008). *Escherichia coli* in the environment: Implications for water quality and human health. Microbes Environ. *23*, 101–108.

Ishii, S., Yan, T., Vu, H., Hansen, D.L., Hicks, R.E., and Sadowsky, M.J. (2009). Factors controlling long-term survival and growth of naturalized *Escherichia coli* populations in temperate field soils. Microbes Environ. *25*, 8–14.

Islam, M., Doyle, M.P., Phatak, S.C., Millner, P., and Jiang, X. (2004). Persistence of enterohemorrhagic *Escherichia coli* O157:H7 in soil and on leaf lettuce and parsley grown in fields treated with contaminated manure composts or irrigation water. J. Food Prot. *67*, 1365–1370.

Iyoda, S., Koizumi, N., Satou, H., Lu, Y., Saitoh, T., Ohnishi, M., and Watanabe, H. (2006). The GrlR-GrlA regulatory system coordinately controls the expression of flagellar and LEE-encoded type III protein secretion systems in enterohemorrhagic *Escherichia coli*. J. Bacteriol. *188*, 5682–5692.

James, E.K., Gyaneshwar, P., Mathan, N., Barraquio, W.L., Reddy, P.M., Iannetta, P.P., Olivares, F.L., and Ladha, J.K. (2002). Infection and colonization of rice seedlings by the plant growth-promoting bacterium *Herbaspirillum seropedicae* Z67. Mol. Plant Microbe Interact. *15*, 894–906.

Jeter, C., and Matthysse, A.G. (2005). Characterization of the binding of diarrheagenic strains of *E. coli* to plant surfaces and the role of curli in the interaction of the bacteria with alfalfa sprouts. Mol. Plant Microbe Interact. *18*, 1235–1242.

Jones, J.D.G., and Dangl, J.L. (2006). The plant immune system. Nature *444*, 323–329.

Klerks, M.M., Franz, E., van Gent-Pelzer, M., Zijlstra, C., and van Bruggen, A.H. (2007a). Differential interaction of *Salmonella enterica* serovars with lettuce cultivars and plant–microbe factors influencing the colonization efficiency. ISME J. *1*, 620–631.

Klerks, M.M., van Gent-Pelzer, M., Franz, E., Zijlstra, C., and van Bruggen, A.H. (2007b). Physiological and molecular responses of *Lactuca sativa* to colonization by *Salmonella enterica* serovar Dublin. Appl. Environ. Microbiol. *73*, 4905–4914.

Knutton, S., Rosenshine, I., Pallen, M.J., Nisan, I., Neves, B.C., Bain, C., Wolff, C., Dougan, G., and Frankel, G. (1998). A novel EspA-associated surface organelle of enteropathogenic *Escherichia coli* involved in protein translocation into epithelial cells. EMBO J. *17*, 2166–2176.

Koczan, J.M., Lenneman, B.R., McGrath, M.J., and Sundin, G.W. (2011). Cell surface attachment structures contribute to biofilm formation and xylem colonization by *Erwinia amylovora*. Appl. Environ. Microbiol. *77*, 7031–7039.

Korea, C.-G., Badouraly, R., Prevost, M.-C., Ghigo, J.-M., and Beloin, C. (2010). *Escherichia coli* K-12 possesses multiple cryptic but functional chaperone–usher fimbriae with distinct surface specificities. Environ. Microbiol. *12*, 1957–1977.

Korhonen, T.K., Tarkka, E., Ranta, H., and Haahtela, K. (1983). Type 3 fimbriae of *Klebsiella* sp.: molecular characterization and role in bacterial adhesion to plant roots. J. Bacteriol. *155*, 860–865.

Kroupitski, Y., Golberg, D., Belausov, E., Pinto, R., Swartzberg, D., Granot, D., and Sela, S. (2009a). Internalization of *Salmonella enterica* in leaves is induced by light and involves chemotaxis and penetration through open stomata. Appl. Environ. Microbiol. *75*, 6076–6086.

Kroupitski, Y., Pinto, R., Brandl, M.T., Belausov, E., and Sela, S. (2009b). Interactions of *Salmonella enterica* with lettuce leaves. J. Appl. Microbiol. *106*, 1876–1885.

Kutter, S., Hartmann, A., and Schmid, M. (2006). Colonization of barley (*Hordeum vulgare*) with *Salmonella enterica* and *Listeria* spp. FEMS Microbiol. Ecol. *56*, 262–271.

Kyle, J.L., Parker, C.T., Goudeau, D., and Brandl, M.T. (2010). Transcriptome analysis of *Escherichia coli* O157:H7 exposed to lysates of lettuce leaves. Appl. Environ. Microbiol. *76*, 1375–1387.

Lafontaine, E.R., Cope, L.D., Aebi, C., Latimer, J.L., McCracken, G.H., Jr., and Hansen, E.J. (2000). The UspA1 protein and a second type of UspA2 protein mediate adherence of *Moraxella catarrhalis* to human epithelial cells *in vitro*. J. Bacteriol. *182*, 1364–1373.

Lapidot, A., and Yaron, S. (2009). Transfer of *Salmonella enterica* serovar Typhimurium from contaminated irrigation water to parsley is dependent on curli and cellulose, the biofilm matrix components. J. Food Prot. *72*, 618–623.

Lapidot, A., Romling, U., and Yaron, S. (2006). Biofilm formation and the survival of *Salmonella* Typhimurium on parsley. Int. J. Food Microbiol. *109*, 229–233.

Lerouge, P., Cabanes-Macheteau, M., Rayon, C., Fischette-Laine, A.C., Gomord, V., and Faye, L. (1998). *N*-glycoprotein biosynthesis in plants: recent developments and future trends. Plant Mol. Biol. *38*, 31–48.

Linke, D., Riess, T., Autenrieth, I.B., Lupas, A., and Kempf, V.A. (2006). Trimeric autotransporter adhesins: variable structure, common function. Trends Microbiol. *14*, 264–270.

Liu, H., Coulthurst, S.J., Pritchard, L., Hedley, P.E., Ravensdale, M., Humphris, S., Burr, T., Takle, G., Brurberg, M.B., Birch, P.R., *et al.* (2008). Quorum sensing coordinates brute force and stealth modes of infection in the plant pathogen *Pectobacterium atrosepticum*. PLoS Pathog. *4*, e1000093.

Low, A.S., Holden, N., Rosser, T., Roe, A.J., Constantinidou, C., Hobman, J.L., Smith, D.G., Low, J.C., and Gally, D.L. (2006). Analysis of fimbrial gene clusters and their expression in enterohaemorrhagic *Escherichia coli* O157:H7. Environ. Microbiol. *8*, 1033–1047.

Lyautey, E., Lu, Z., Lapen, D.R., Wilkes, G., Scott, A., Berkers, T., Edge, T.A., and Topp, E. (2010). Distribution and diversity of *Escherichia coli* populations in the south nation river drainage basin, eastern Ontario, Canada. Appl. Environ. Microbiol. *76*, 1486–1496.

Macarisin, D., Patel, J., Bauchan, G., Giron, J.A., and Sharma, V.K. (2012). Role of curli and cellulose expression in adherence of *Escherichia coli* O157:H7 to spinach leaves. Foodborne Pathog. Dis. *9*, 160–167.

McCann, H.C., Nahal, H., Thakur, S., and Guttman, D.S. (2012). Identification of innate immunity elicitors using molecular signatures of natural selection. Proc. Natl. Acad. Sci. U.S.A. *109*, 4215–4220.

Malnoy, M., Martens, S., Norelli, J.L., Barny, M.-A., Sundin, G.W., Smits, T.H.M., and Duffy, B. (2012). Fire blight: applied genomic insights of the pathogen and host. Annu. Rev. Phytopathol. *50*, 475–494.

Mano, H., and Morisaki, H. (2008). Endophytic bacteria in the rice plant. Microbes Environ. *23*, 109–117.

Matthysse, A.G., and McMahan, S. (1998). Root colonization by *Agrobacterium tumefaciens* is reduced in *cel*, *attB*, *attD*, and *attR* mutants. Appl. Environ. Microbiol. *64*, 2341–2345.

Matthysse, A.G., Deora, R., Mishra, M., and Torres, A.G. (2008). Polysaccharides cellulose, poly-β–1,6-*N*-acetylglucosamine, and colanic acid are required for optimal binding of *Escherichia coli* O157:H7 strains to alfalfa sprouts and K-12 strains to plastic but not for binding to epithelial cells. Appl. Environ. Microbiol. *74*, 2384–2390.

Melotto, M., Underwood, W., Koczan, J., Nomura, K., and He, S.Y. (2006). Plant stomata function in innate immunity against bacterial invasion. Cell *126*, 969–980.

Meng, F., Altier, C., and Martin, G.B. (2013). *Salmonella* colonization activates the plant immune system and benefits from association with plant-pathogenic bacteria. Environ. Microbiol. *15*, 2418–2430.

Mhedbi-Hajri, N., Darrasse, A., Pigne, S., Durand, K., Fouteau, S., Barbe, V., Manceau, C., Lemaire, C., and Jacques, M.A. (2011). Sensing and adhesion are adaptive functions in the plant pathogenic xanthomonads. BMC Evol. Biol. *11*, 67.

Michino, H., Araki, K., Minami, S., Takaya, S., Sakai, N., Miyazaki, M., Ono, A., and Yanagawa, H. (1999). Massive outbreak of *Escherichia coli* O157:H7 infection in schoolchildren in Sakai City, Japan, associated with consumption of white radish sprouts. Am. J. Epidemiol. *150*, 787–796.

Miller, V.L., Beer, K.B., Loomis, W.P., Olson, J.A., and Miller, S.I. (1992). An unusual *pagC*::Tn*phoA* mutation leads to an invasion-defective and virulence-defective phenotype in *Salmonellae*. Infect. Immun. *60*, 3763–3770.

Muller, C.M., Aberg, A., Straseviciene, J., Emody, L., Uhlin, B.E., and Balsalobre, C. (2009). Type 1 fimbriae, a colonization factor of uropathogenic *Escherichia coli*, are controlled by the metabolic sensor CRP-cAMP. PLoS Pathog. *5*, e1000303.

Noel, J.T., Arrach, N., Alagely, A., McClelland, M., and Teplitski, M. (2010a). Specific responses of *Salmonella enterica* to tomato varieties and fruit ripeness identified by *in vivo* expression technology. PLoS ONE *5*, e12406.

Noel, J.T., Joy, J., Smith, J.N., Fatica, M., Schneider, K.R., Ahmer, B.M., and Teplitski, M. (2010b). *Salmonella* SdiA recognizes N-acyl homoserine lactone signals from *Pectobacterium carotovorum in vitro*, but not in a bacterial soft rot. Mol. Plant Microbe Interact. *23*, 273–282.

Nuccio, S.P., and Baumler, A.J. (2007). Evolution of the chaperone/usher assembly pathway: fimbrial classification goes Greek. Microbiol. Mol. Biol. Rev. *71*, 551–575.

Oliveira, M., Wijnands, L., Abadias, M., Aarts, H., and Franz, E. (2011). Pathogenic potential of *Salmonella* Typhimurium DT104 following sequential passage through soil, packaged fresh-cut lettuce and a model gastrointestinal tract. Int. J. Food Microbiol. *148*, 149–155.

Oliveira, M., Viñas, I., Usall, J., Anguera, M., and Abadias, M. (2012). Presence and survival of *Escherichia coli* O157:H7 on lettuce leaves and in soil treated with contaminated compost and irrigation water. Int. J. Food Microbiol. *156*, 133–140.

Ongeng, D., Muyanja, C., Ryckeboer, J., Geeraerd, A.H., and Springael, D. (2011). Rhizosphere effect on survival of *Escherichia coli* O157:H7 and *Salmonella enterica* serovar Typhimurium in manure-amended soil during cabbage (*Brassica oleracea*) cultivation under tropical field conditions in Sub-Saharan Africa. Int. J. Food Microbiol. *149*, 133–142.

O'Toole, G.A., and Kolter, R. (1998). Flagellar and twitching motility are necessary for *Pseudomonas aeruginosa* biofilm development. Mol. Microbiol. *30*, 295–304.

Parker, C.T., Kyle, J.L., Huynh, S., Carter, M.Q., Brandl, M.T., and Mandrell, R.E. (2012). Distinct transcriptional profiles and phenotypes exhibited by *Escherichia coli* O157:H7 isolates related to the 2006 spinach-associated outbreak. Appl. Environ. Microbiol. *78*, 455–463.

Patel, J., and Sharma, M. (2010). Differences in attachment of *Salmonella enterica* serovars to cabbage and lettuce leaves. Int. J. Food Microbiol. *139*, 41–47.

Patel, J., Millner, P., Nou, X., and Sharma, M. (2010). Persistence of enterohaemorrhagic and nonpathogenic *E. coli* on spinach leaves and in rhizosphere soil. J. Appl. Microbiol. *108*, 1789–1796.

Plotnikova, J.M., Rahme, L.G., and Ausubel, F.M. (2000). Pathogenesis of the human opportunistic pathogen *Pseudomonas aeruginosa* PA14 in *Arabidopsis*. Plant Physiol. *124*, 1766–1774.

Pollumaa, L., Alamae, T., and Mae, A. (2012). Quorum sensing and expression of virulence in *Pectobacteria*. Sensors *12*, 3327–3349.

Portnoy, B.L., Goepfert, J.M., and Harmon, S.M. (1976). Outbreak of *Bacillus cereus* food poisoning resulting from contaminated vegetable sprouts. Am. J. Epidemiol. *103*, 589–594.

Pouttu, R., Westerlund-Wikstrom, B., Lang, H., Alsti, K., Virkola, R., Saarela, U., Siitonen, A., Kalkkinen, N., and Korhonen, T.K. (2001). matB, a common fimbrillin gene of *Escherichia coli*, expressed in a genetically conserved, virulent clonal group. J. Bacteriol. *183*, 4727–4736.

Prasadarao, N.V., Wass, C.A., and Kim, K.S. (1996). Endothelial cell GlcNAcβ1-4GlcNAc epitopes for outer membrane protein A enhance traversal of *Escherichia coli* across the blood–brain barrier. Infect. Immun. *64*, 154–160.

Prithiviraj, B., Bais, H.P., Jha, A.K., and Vivanco, J.M. (2005). *Staphylococcus aureus* pathogenicity on *Arabidopsis thaliana* is mediated either by a direct effect of salicylic acid on the pathogen or by SA-dependent, NPR1-independent host responses. Plant J. *42*, 417–432.

Quilliam, R.S., Chalmers, R.M., Williams, A.P., Chart, H., Willshaw, G.A., Kench, S.M., Edwards-Jones, G., Evans, J., Thomas, D.R., Salmon, R.L., *et al.* (2012a). Seroprevalence and risk factors associated with *Escherichia coli* O157 in a farming population. Zoonoses Pub. Health *59*, 83–88.

Quilliam, R.S., Williams, A.P., and Jones, D.L. (2012b). Lettuce cultivar mediates both phyllosphere and rhizosphere activity of *Escherichia coli* O157:H7. PLoS ONE *7*, e33842.

Raghunathan, D., Wells, T.J., Morris, F.C., Shaw, R.K., Bobat, S., Peters, S.E., Paterson, G.K., Jensen, K.T., Leyton, D.L., Blair, J.M., *et al.* (2011). SadA, a trimeric autotransporter from *Salmonella enterica* serovar

Typhimurium, can promote biofilm formation and provides limited protection against infection. Infect. Immun. *79*, 4342–4352.

Rahme, L.G., Ausubel, F.M., Cao, H., Drenkard, E., Goumnerov, B.C., Lau, G.W., Mahajan-Miklos, S., Plotnikova, J., Tan, M.W., Tsongalis, J., *et al.* (2000). Plants and animals share functionally common bacterial virulence factors. Proc. Natl. Acad. Sci. U.S.A. *97*, 8815–8821.

Ray, S.K., Rajeshwari, R., Sharma, Y., and Sonti, R.V. (2002). A high-molecular-weight outer membrane protein of *Xanthomonas oryzae* pv. *oryzae* exhibits similarity to non-fimbrial adhesins of animal pathogenic bacteria and is required for optimum virulence. Mol. Microbiol. *46*, 637–647.

Rendon, M.A., Saldana, Z., Erdem, A.L., Monteiro-Neto, V., Vazquez, A., Kaper, J.B., Puente, J.L., and Giron, J.A. (2007). Commensal and pathogenic *Escherichia coli* use a common pilus adherence factor for epithelial cell colonization. Proc. Natl. Acad. Sci. U.S.A. *104*, 10637–10642.

Riess, T., Andersson, S.G., Lupas, A., Schaller, M., Schafer, A., Kyme, P., Martin, J., Walzlein, J.H., Ehehalt, U., Lindroos, H., *et al.* (2004). *Bartonella* adhesin A mediates a proangiogenic host cell response. J. Exp. Med. *200*, 1267–1278.

Roe, A.J., Currie, C., Smith, D.G., and Gally, D.L. (2001). Analysis of type 1 fimbriae expression in verotoxigenic *Escherichia coli*: a comparison between serotypes O157 and O26. Microbiology *147*, 145–152.

Rojas, C.M., Ham, J.H., Deng, W.L., Doyle, J.J., and Collmer, A. (2002). HecA, a member of a class of adhesins produced by diverse pathogenic bacteria, contributes to the attachment, aggregation, epidermal cell killing, and virulence phenotypes of *Erwinia chrysanthemi* EC16 on *Nicotiana clevelandii* seedlings. Proc. Natl. Acad. Sci. U.S.A. *99*, 13142–13147.

Romling, U., Bian, Z., Hammar, M., Sierralta, W.D., and Normark, S. (1998). Curli fibers are highly conserved between *Salmonella typhimurium* and *Escherichia coli* with respect to operon structure and regulation. J. Bacteriol. *180*, 722–731.

Rosenblueth, M., and Martinez-Romero, E. (2006). Bacterial endophytes and their interactions with hosts. Mol. Plant Microbe Interact. *19*, 827–837.

Roy, D., Panchal, S., Rosa, B.A., and Melotto, M. (2013). *Escherichia coli* O157:H7 induces stronger plant immunity than *Salmonella enterica* Typhimurium SL1344. Phytopathology *103*, 326–332.

Ruiz-Albert, J., Yu, X.J., Beuzon, C.R., Blakey, A.N., Galyov, E.E., and Holden, D.W. (2002). Complementary activities of SseJ and SifA regulate dynamics of the *Salmonella typhimurium* vacuolar membrane. Mol. Microbiol. *44*, 645–661.

Ryu, J.H., and Beuchat, L.R. (2004). Factors affecting production of extracellular carbohydrate complexes by *Escherichia coli* O157: H7. Int. J. Food Microbiol. *95*, 189–204.

Salanoubat, M., Genin, S., Artiguenave, F., Gouzy, J., Mangenot, S., Arlat, M., Billault, A., Brottier, P., Camus, J.C., Cattolico, L., *et al.* (2002). Genome sequence of the plant pathogen *Ralstonia solanacearum*. Nature *415*, 497–502.

Saldana, Z., Erdem, A.L., Schuller, S., Okeke, I.N., Lucas, M., Sivananthan, A., Phillips, A.D., Kaper, J.B., Puente, J.L., and Giron, J.A. (2009). The *Escherichia coli* common pilus and the bundle-forming pilus act in concert during the formation of localized adherence by enteropathogenic *E. coli*. J. Bacteriol. *191*, 3451–3461.

Saldana, Z., Sanchez, E., Xicohtencatl-Cortes, J., Puente, J.L., and Giron, J.A. (2011). Surface structures involved in plant stomata and leaf colonization by Shiga-toxigenic *Escherichia coli* O157:H7. Front. Microbiol. *2*, 119.

Schikora, A., Carreri, A., Charpentier, E., and Hirt, H. (2008). The dark side of the salad: *Salmonella typhimurium* overcomes the innate immune response of *Arabidopsis thaliana* and shows an endopathogenic lifestyle. PLoS ONE *3*, e2279.

Schikora, A., Virlogeux-Payant, I., Bueso, E., Garcia, A.V., Nilau, T., Charrier, A., Pelletier, S., Menanteau, P., Baccarini, M., Velge, P., *et al.* (2011). Conservation of *Salmonella* infection mechanisms in plants and animals. PLoS ONE *6*, e24112.

Schleker, S., Garcia-Garcia, J., Klein-Seetharaman, J., and Oliva, B. (2012). Prediction and comparison of *Salmonella*-human and *Salmonella-Arabidopsis* interactomes. Chem. Biodivers. *9*, 991–1018.

Schuhegger, R., Ihring, A., Gantner, S., Bahnweg, G., Knappe, C., Vogg, G., Hutzler, P., Schmid, M., Van Breusegem, F., Eberl, L., *et al.* (2006). Induction of systemic resistance in tomato by N-acyl-L-homoserine lactone-producing rhizosphere bacteria. Plant Cell Environ. *29*, 909–918.

Segonzac, C., and Zipfel, C. (2011). Activation of plant pattern-recognition receptors by bacteria. Curr. Opin. Microbiol. *14*, 54–61.

Semenov, A.M., Kuprianov, A.A., and van Bruggen, A.H. (2010). Transfer of enteric pathogens to successive habitats as part of microbial cycles. Microb. Ecol. *60*, 239–249.

Seo, S., and Matthews, K.R. (2012). Influence of the plant defense response to *Escherichia coli* O157:H7 cell surface structures on survival of that enteric pathogen on plant surfaces. Appl. Environ. Microbiol. *78*, 5882–5889.

Shaw, R.K., Berger, C.N., Feys, B., Knutton, S., Pallen, M.J., and Frankel, G. (2008). Enterohemorrhagic *Escherichia coli* exploits EspA filaments for attachment to salad leaves. Appl. Environ. Microbiol. *74*, 2908–2914.

Shaw, R.K., Berger, C.N., Pallen, M.J., Sjoeling, A., and Frankel, G. (2011). Flagella mediate attachment of enterotoxigenic *Escherichia coli* to fresh salad leaves. Environ. Microbiol. Rep. *3*, 112–117.

Shi, X., Wu, Z., Namvar, A., Kostrzynska, M., Dunfield, K., and Warriner, K. (2009). Microbial population profiles of the microflora associated with pre- and postharvest tomatoes contaminated with *Salmonella typhimurium* or *Salmonella montevideo*. J. Appl. Microbiol. *107*, 329–338.

Shirron, N., and Yaron, S. (2011). Active suppression of early immune response in tobacco by the human pathogen *Salmonella* Typhimurium. PLoS ONE *6*, e18855.

Silagyi, K., Kim, S.H., Lo, Y.M., and Wei, C.I. (2009). Production of biofilm and quorum sensing by *Escherichia coli* O157:H7 and its transfer from contact surfaces to meat, poultry, ready-to-eat deli, and produce products. Food Microbiol. *26*, 514–519.

da Silva, A.C.R., Ferro, J.A., Reinach, F.C., Farah, C.S., Furlan, L.R., Quaggio, R.B., Monteiro-Vitorello, C.B., Sluys, M.A.V., Almeida, N.F., Alves, L.M.C., *et al.* (2002). Comparison of the genomes of two *Xanthomonas* pathogens with differing host specificities. Nature *417*, 459–463.

Simpson, A.J., Reinach, F.C., Arruda, P., Abreu, F.A., Acencio, M., Alvarenga, R., Alves, L.M., Araya, J.E., Baia, G.S., Baptista, C.S., *et al.* (2000). The genome sequence of the plant pathogen *Xylella fastidiosa*. The *Xylella fastidiosa* consortium of the organization for nucleotide sequencing and analysis. Nature *406*, 151–159.

Sokurenko, E.V., Chesnokova, V., Doyle, R.J., and Hasty, D.L. (1997). Diversity of the *Escherichia coli* type 1 fimbrial lectin. Differential binding to mannosides and uroepithelial cells. J. Biol. Chem. *272*, 17880–17886.

Solomon, E.B., and Matthews, K.R. (2005). Use of fluorescent microspheres as a tool to investigate bacterial interactions with growing plants. J. Food Prot. *68*, 870–873.

Solomon, E.B., and Matthews, K.R. (2006). Interaction of live and dead *Escherichia coli* O157: H7 and fluorescent microspheres with lettuce tissue suggests bacterial processes do not mediate adherence. Lett. Appl. Microbiol. *42*, 88–93.

Song, Y.C., Jin, S., Louie, H., Ng, D., Lau, R., Zhang, Y., Weerasekera, R., Al Rashid, S., Ward, L.A., Der, S.D., *et al.* (2004). FlaC, a protein of *Campylobacter jejuni* TGH9011 (ATCC43431) secreted through the flagellar apparatus, binds epithelial cells and influences cell invasion. Mol. Microbiol. *53*, 541–553.

Soscia, C., Hachani, A., Bernadac, A., Filloux, A., and Bleves, S. (2007). Cross talk between type III secretion and flagellar assembly systems in *Pseudomonas aeruginosa*. J. Bacteriol. *189*, 3124–3132.

Srikanth, C.V., Mercado-Lubo, R., Hallstrom, K., and McCormick, B.A. (2011). *Salmonella* effector proteins and host-cell responses. Cell. Mol. Life Sci. *68*, 3687–3697.

Stahlhut, S.G., Chattopadhyay, S., Struve, C., Weissman, S.J., Aprikian, P., Libby, S.J., Fang, F.C., Krogfelt, K.A., and Sokurenko, E.V. (2009). Population variability of the FimH type 1 fimbrial adhesin in *Klebsiella pneumoniae*. J. Bacteriol. *191*, 1941–1950.

Stecher, B., Robbiani, R., Walker, A.W., Westendorf, A.M., Barthel, M., Kremer, M., Chaffron, S., Macpherson, A.J., Buer, J., Parkhill, J., *et al.* (2007). *Salmonella enterica* serovar Typhimurium exploits inflammation to compete with the intestinal microbiota. PLoS Biol. *5*, e244.

Struve, C., Bojer, M., and Krogfelt, K.A. (2008). Characterization of *Klebsiella pneumoniae* type 1 fimbriae by detection of phase variation during colonization and infection and impact on virulence. Infect. Immun. *76*, 4055–4065.

Struve, C., Bojer, M., and Krogfelt, K.A. (2009). Identification of a conserved chromosomal region encoding *Klebsiella pneumoniae* type 1 and type 3 fimbriae and assessment of the role of fimbriae in pathogenicity. Infect. Immun. *77*, 5016–5024.

Terranova, W., Breman, J.G., Locey, R.P., and Speck, S. (1978). Botulism type-B: Epidemiological aspects of an extensive outbreak. Am. J. Epidemiol. *108*, 150–156.

Texier, S., Prigent-Combaret, C., Gourdon, M.H., Poirier, M.A., Faivre, P., Dorioz, J.M., Poulenard, J., Jocteur-Monrozier, L., Moenne-Loccoz, Y., and Trevisan, D. (2008). Persistence of culturable *Escherichia coli* fecal contaminants in dairy Alpine grassland soils. J. Environ. Qual. 37, 2299–2310.

Thieme, F., Koebnik, R., Bekel, T., Berger, C., Boch, J., Buttner, D., Caldana, C., Gaigalat, L., Goesmann, A., Kay, S., *et al.* (2005). Insights into genome plasticity and pathogenicity of the plant pathogenic bacterium *Xanthomonas campestris* pv. *vesicatoria* revealed by the complete genome sequence. J. Bacteriol. *187*, 7254–7266.

Thilmony, R., Underwood, W., and He, S.Y. (2006). Genome-wide transcriptional analysis of the *Arabidopsis thaliana* interaction with the plant pathogen *Pseudomonas syringae* pv. *tomato* DC3000 and the human pathogen *Escherichia coli* O157:H7. Plant J. *46*, 34–53.

Tobe, T., and Sasakawa, C. (2002). Species-specific cell adhesion of enteropathogenic *Escherichia coli* is mediated by type IV bundle-forming pili. Cell. Microbiol. *4*, 29–42.

Torres, A.G., Jeter, C., Langley, W., and Matthysse, A.G. (2005). Differential binding of *Escherichia coli* O157:H7 to alfalfa, human epithelial cells, and plastic is mediated by a variety of surface structures. Appl. Environ. Microbiol. *71*, 8008–8015.

Toth, I.K., Pritchard, L., and Birch, P.R. (2006). Comparative genomics reveals what makes an enterobacterial plant pathogen. Annu. Rev. Phytopathol. *44*, 305–336.

Totsika, M., Wells, T.J., Beloin, C., Valle, J., Allsopp, L.P., King, N.P., Ghigo, J.M., and Schembri, M.A. (2012). Molecular characterization of the EhaG and UpaG trimeric autotransporter proteins from pathogenic *Escherichia coli*. Appl. Environ. Microbiol. *78*, 2179–2189.

Ustun, S., Muller, P., Palmisano, R., Hensel, M., and Bornke, F. (2012). SseF, a type III effector protein from the mammalian pathogen *Salmonella enterica*, requires resistance-gene-mediated signalling to activate cell death in the model plant *Nicotiana benthamiana*. New Phytol. *194*, 1046–1060.

Vesper, S.J. (1987). Production of pili (fimbriae) by *Pseudomonas fluorescens* and correlation with attachment to corn roots. Appl. Environ. Microbiol. *53*, 1397–1405.

Wairuri, C.K., van der Waals, J.E., van Schalkwyk, A., and Theron, J. (2012). *Ralstonia solanacearum* needs Flp pili for virulence on potato. Mol. Plant Microbe Interact. *25*, 546–556.

Walk, S.T., Alm, E.W., Calhoun, L.M., Mladonicky, J.M., and Whittam, T.S. (2007). Genetic diversity and population structure of *Escherichia coli* isolated from freshwater beaches. Environ. Microbiol. *9*, 2274–2288.

Warriner, K., Spaniolas, S., Dickinson, M., Wright, C., and Waites, W.M. (2003). Internalization of bioluminescent *Escherichia coli* and *Salmonella* Montevideo in growing bean sprouts. J. Appl. Microbiol. *95*, 719–727.

Weiser, J.N., and Gotschlich, E.C. (1991). Outer membrane protein A (OmpA) contributes to serum resistance and pathogenicity of *Escherichia coli* K-1. Infect. Immun. *59*, 2252–2258.

Wells, J.M., and Butterfield, J.E. (1997). *Salmonella* contamination associated with bacterial soft rot of fresh fruits and vegetables in the marketplace. Plant Dis. *81*, 867–872.

Wilson, R.K., Shaw, R.K., Daniell, S., Knutton, S., and Frankel, G. (2001). Role of EscF, a putative needle complex protein, in the type III protein translocation system of enteropathogenic *Escherichia coli*. Cell. Microbiol. 3, 753–762.

Young, B.M., and Young, G.M. (2002). YplA is exported by the Ysc, Ysa, and flagellar type III secretion systems of *Yersinia enterocolitica*. J. Bacteriol. *184*, 1324–1334.

Young, G.M., Schmiel, D.H., and Miller, V.L. (1999). A new pathway for the secretion of virulence factors by bacteria: the flagellar export apparatus functions as a protein-secretion system. Proc. Natl. Acad. Sci. U.S.A. *96*, 6456–6461.

Zhang, H.J., Fang, Q., Zhang, Z.G., Wang, Y.C., and Zheng, X.B. (2009). The role of respiratory burst oxidase homologues in elicitor-induced stomatal closure and hypersensitive response in *Nicotiana benthamiana*. J. Exp. Bot. *60*, 3109–3122.

Recent Advances in *Pseudomonas* Biocontrol

7

Feyisara Eyiwumi Olorunleke, Nam Phuong Kieu and Monica Höfte

Abstract

Fluorescent pseudomonads have been intensively studied as biocontrol agents. These organisms exhibit an enormous metabolic versatility and especially isolates from the *Pseudomonas fluorescens* group produce a remarkable spectrum of secondary metabolites. Multilocus sequence analysis and phylogenomics have revealed the presence of at least five distinct subgroups in the *P. fluorescens* group with biocontrol isolates previously classified as *P. fluorescens* interspersed with strains classified in other species. Antibiotics such as phenazines, diacetylphloroglucinol, and hydrogen cyanide are produced by certain taxonomic groups within the genus *Pseudomonas* and appear to be ancestral. These compounds often have a physiological role for the producing strain, independent from their antibiotic activity. Other secondary metabolites are only found in certain *Pseudomonas* isolates and are apparently obtained by horizontal gene transfer. The recent boost in genome sequencing has revealed many orphan biosynthetic gene clusters in the genomes of *Pseudomonas* biocontrol strains. Genome mining has led to the discovery of many new antimicrobial compounds with a role in biocontrol of plant-pathogenic fungi, oomycetes and bacteria. Moreover, some *Pseudomonas* biocontrol strains produce potent insect toxins. The versatility of *Pseudomonas* biocontrol agents continues to surprise.

Fluorescent pseudomonads as biocontrol agents

Biopesticides based on the artificial introduction of living microorganisms into a plant system are an alternative or a supplementary way of reducing the use of chemical pesticides in agriculture. Biocontrol can be very useful when no alternative disease control strategies are available or in organic farming where the use of chemical pesticides is not allowed. There is also a genuine commercial interest in biopesticides because they can be used in rotation with chemical pesticides to reduce the development of pathogen resistance. The use of fluorescent *Pseudomonas* spp. as biocontrol agents has intensively been studied. They can be readily isolated from soil or the rhizosphere of plants, and are easily recognized on iron-limiting medium by the production of a yellow-green fluorescent pigment, called pyoverdine or pseudobactin, which acts as a siderophore or iron-chelating compound (Meyer, 2000). The bacteria exhibit an enormous metabolic versatility and some isolates produce a remarkable spectrum of secondary metabolites (Gross and Loper, 2009). Moreover, the modes of action are relatively easy to study in *Pseudomonas* spp. because these bacteria are amenable to mutation and modification using genetic tools (Chin-A-Woeng *et al.*, 2003).

The most commonly reported mechanisms of biocontrol by fluorescent *Pseudomonas* spp. include antibiosis (Gross and Loper, 2009; Haas and Defago, 2005), competition for nutrients and niches (Kamilova *et al.*, 2005), competition for iron mediated by siderophores (Leong, 1986), and induced systemic resistance (De Vleesschauwer and Hofte, 2009). Additionally, many rhizosphere fluorescent *Pseudomonas* spp. can promote plant growth in the absence of pathogens by increasing the availability and uptake of mineral nutrients via phosphate solubilizing enzymes or by enhancing root growth and morphology via the production of phytohormones. This subject has been recently reviewed by Lugtenberg and Kamilova (2009) and will not be discussed further. In this review we will focus on recent advances in the taxonomy of *Pseudomonas* biocontrol agents, novel insights in the evolution, role and function of the most important *Pseudomonas* secondary metabolites, and the discovery of new secondary metabolites by genome mining. Biocontrol in fluorescent pseudomonads has been intensively reviewed. An overview of interesting review papers about biocontrol of soilborne pathogens is given in Lugtenberg *et al.* (2013). The landmark paper by Haas and Defago (2005) gives an excellent introduction to *Pseudomonas* biocontrol and should be read by anyone interested in this subject. Other important reviews will be cited in the appropriate sections.

Taxonomy of *Pseudomonas* biocontrol agents

The genus *Pseudomonas* belongs to the class of the *γ-proteobacteria* within the phylum *Proteobacteria*. *Pseudomonas* is one of the most ubiquitous bacterial genera and different species have been isolated from very diverse ecological niches. Since its discovery the genus *Pseudomonas* has undergone numerous taxonomic changes. A detailed history of *Pseudomonas* taxonomy until 2009 is given by Peix *et al.* (2009). Currently, only the representatives of rRNA group I according to Palleroni *et al.* (1973) are included in the genus *Pseudomonas*. The number of *Pseudomonas* species is continuously increasing; up to now, more than 140 species have been validly described. Updates can be found on the DSMZ website (www.dsmz.de). Most *Pseudomonas* species are saprophytes that are commonly found in water and soil, but the genus includes also species that are associated with diseases in plants (e.g. *P. syringae*, *P. cichorii*, *P. viridiflava*, *P. corrugata*), mushrooms (e.g. *P. tolaasii*, *P. agarici*), humans and animals (e.g. *P. aeruginosa*) (Peix *et al.*, 2009). Within the *Pseudomonas* genus, two main lineages or intrageneric groups can be recognized, the *Pseudomonas aeruginosa* lineage (which includes the *P. aeruginosa*, *P. oleovorans* and *P. stutzeri* group) and the *P. fluorescens* lineage (Table 7.1). Based on phylogenetic analysis of 16S rRNA, *gyrB*, *rpoB* and *rpoD* sequences Mulet *et al.* (2010) recognized six groups within the *P. fluorescens* lineage: the *P. fluorescens*, *P. syringae*, *P. lutea*, *P. putida*, *P. anguilliseptica* and *P. straminea* group. The *P. fluorescens* group appears to be highly diverse and contains more than 50 named species divided over nine subgroups (Mulet *et al.*, 2010). Silby *et al.* (2009) were actually the first to show that the *P. fluorescens* group is highly diverse. They compared three *P. fluorescens* genomes and discovered that only 61% of genes were shared. Loper *et al.* (2012) conducted a comparative genomic analysis of 10 biocontrol strains within the *P. fluorescens* group and found that the core genome only represents 45–52% of the genome of any individual strain. Multilocus sequence analysis (MLSA) placed the strains in three main subclades (Table 7.1). Redondo-Nieto *et al.* (2013) analysed the genome sequence of the biocontrol strain *P. fluorescens* F113, compared it to 49 other complete and draft genomes of strains classified in

Table 7.1 Taxonomic position of some well-studied *Pseudomonas* biocontrol strains

Mulet *et al*. (2010)			Loper *et al*. (2009)	Redondo-Nieto *et al*. (2013)	Biocontrol strains[2]	Antibiotics[3]	Insect toxins	Reference
P. fluorescens lineage	*P. fluorescens* group[1]	*P. chlororaphis* subgroup	Subclade 1	Subgroup IV	*P. protegens* Pf-5	DAPG, PLT, PRN, RHI, HCN	FitD	Paulsen *et al*. (2005)
					P. protegens CHA0	DAPG, PLT, PRN, HCN	FitD	Haas and Defago (2005)
					Pseudomonas sp. CMR12a	PHZ, HCN	FitD	Perneel *et al*. (2007)
					Pseudomonas sp. CMR5c	PHZ, PLT, PRN, HCN	FitD	Perneel *et al*. (2007)
					***P. fluorescens* NZ17**	DAPG, HCN		Burlinson *et al*. (2013)
					P. chlororaphis PCL1391	PHZ, HCN	FitD	Chin-A-Woeng *et al*. (1998)
					P. chlororaphis 30-84	PHZ, PRN, HCN	FitD	Pierson and Thomashow (1992)
					P. chlororaphis 06	PHZ, PRN, HCN	FitD	Park *et al*. (2011)
					P. chlororaphis GP72	PHZ, PRN	FitD	Liu *et al*. (2007)
					P. chlororaphis PA23	PHZ, PRN		Selin *et al*. (2012)
		P. koreensis subgroup	Subclade 2	Subgroup II	*P. fluorescens* Pf0-1	HCN		Seaton *et al*. (2013)
		P. corrugata subgroup		Subgroup I	*P. fluorescens* F113	DAPG, HCN		Redondo-Nieto *et al*. (2013)
					P. brassicacearum Q8r1-96	DAPG, HCN		Loper *et al*. (2012)
					P. fluorescens Q2-87	DAPG, HCN		Loper *et al*. (2012)
					P. corrugata			

Table 7.1 Continued

Mulet *et al*. (2010)		Loper *et al*. (2009)	Redondo-Nieto *et al*. (2013)	Biocontrol strains[2]	Antibiotics[3]	Insect toxins	Reference
	P. fluorescens subgroup	Subclade 3	Subgroup V	*P. fluorescens* SBW25			Silby *et al*. (2009)
				P. fluorescens A506			Loper *et al*. (2012)
				P. fluorescens SS101			Loper *et al*. (2012)
				Pseudomonas BG33R			Loper *et al*. (2012)
				P. fluorescens 2-79	PCA		Mavrodi *et al*. (1998)
				P. synxantha	PCA		Parejko *et al*. (2013)
				P. orientalis	PCA		Parejko *et al*. (2013)
				P. aridus	PCA		Parejko *et al*. (2013)
				P. cerealis	PCA		Parejko *et al*. (2013)
				P. tolaasii			
P. syringae group				*P. syringae* 742RS			
				P. syringae ESC-10			Bull *et al*. (1997)
				P. syringae ESC-11			Bull *et al*. (1997)
				P. syringae			
				P. viridiflava			
				P. cichorii			

Mulet *et al*. (2010)		Loper *et al*. (2009)	Redondo-Nieto *et al*. (2013)	Biocontrol strains[2]	Antibiotics[3]	Insect toxins	Reference
	P. lutea group						
	P. putida group			*P. putida* 267			Kruijt *et al*. (2009)
				P. putida RW10S1	PRO		Li *et al*. (2011)
				P. putida PCL1445			Kuiper *et al*. (2004)
				P. entomophila	HCN		Ryall *et al*. (2009)
	P. anguilliseptica group						
	P. straminea group						
P. aeruginosa lineage	*P. aeruginosa* group			*P. aeruginosa* M18	PHZ, PLT, HCN		Ge *et al*. (2004)
				P. aeruginosa 7NSK2	PYO, HCN		De Vleesschauwer *et al*. (2006)
				P. aeruginosa PNA1	PHZ, HCN		Anjaiah *et al*. (1998)
				P. aeruginosa PaBP35	PHZ, HCN		Kumar *et al*. (2013)
	P. oleovorans group						
	P. stutzeri group						

[1]Within the *P. fluorenscens* group only subgroups with known biocontrol strains are shown.
[2]Plant or mushroom patrogens are indicated in bold.
[3]DAPG, di-acetylphloroglucinol; PLT, pyoluteorin; PRN, pyrrolnitrin; RHI, rhizoxin; HCN, hydrogen cyanide; PHZ, phenazines (PCA and PCN); PCA, phenazine-1-carboxylate; PRO, promysalin; PYO, pyocyanin.

the *Pseudomonas* genus and constructed a phylogenomic tree using the composition vector method. This analysis revealed the presence of at least five distinct subgroups in the *P. fluorescens* group with strains previously classified as *P. fluorescens* interspersed with strains classified in other species (Table 7.1). These studies show that the *P. fluorescens* group is a species complex that is clearly in need of revision.

Table 7.1 illustrates that biocontrol strains are commonly found within the *P. fluorescens* lineage. The *P. fluorescens* group shows a very strong strain-to-strain variation and only a subset of strains within the *P. fluorescens* group provides biocontrol. It should also be noted that *P. aeruginosa* isolates with strong biocontrol activities can be readily isolated from the (endo)rhizosphere of plants (Anjaiah *et al.*, 1998; Audenaert *et al.*, 2002; Kumar *et al.*, 2013). Moreover, various non-pathogenic isolates of *P. syringae* also show interesting biocontrol activities and are used to control postharvest pathogens on fruits (Janisiewicz and Korsten, 2002) and bacterial diseases caused by pathogenic *P. syringae* pathovars (Volksch and May, 2001; Wensing *et al.*, 2010). An overview of commercially available *Pseudomonas* biocontrol strains is given in Table 7.2.

How to characterize a new *Pseudomonas* isolate?

A commonly used tool to characterize and taxonomically identify *Pseudomonas* species is MLSA. MLSA is a rapid classification method for genotypic characterization using the sequences of multiple protein-encoding genes. The gene encoding 16S rRNA can be used to ascribe a strain to the genus *Pseudomonas,* but the resolution of 16S rRNA gene sequences at the intrageneric level is low (Mulet *et al.*, 2010). Other housekeeping genes such as *gyrB, rpoB, rpoD, atpD, carA* and *recA* have been studied. According to Mulet *et al.* (2010) the *rpoD* gene exhibits the highest number of polymorphic sites, followed by *gyrB* and *rpoB*. Mulet *et al.* (2010) suggest the following steps to infer whether a new isolate belongs to the genus *Pseudomonas* and to a precise species within the genus. The first step is the sequencing of the 16S rRNA gene with universal primers, which will tell whether the new isolate belongs to the genus *Pseudomonas*. A second step would be the sequencing of the *rpoD* and *gyrB* genes to allocate the new isolate to a specific group or subgroup. Analysis of the *rpoB* gene may be needed if a higher resolution is necessary. Multiple gene sequences from strains of *Pseudomonas* species are currently compiled in the PseudoMLSA database (Bennasar *et al.*, 2010) which is freely accessible. Users can upload their own sequence datasets to search against the PseudoMLSA sequence database.

Discovery, structure prediction and functional analysis of secondary metabolites in *Pseudomonas*

In recent years, the discovery, structure prediction and functional analysis of secondary metabolites in *Pseudomonas* spp. has found its roots in genome mining (de Bruijn *et al.*, 2007). Increased genome sequencing has led to increased identification of large orphan gene clusters for which the encoded metabolite is unknown. These clusters contain non-ribosomal peptide synthase (NRPS) and polyketide synthase (PKS) genes, which code for secondary metabolites such as lipopeptides and antibiotics. Genome mining for new NRPS-encoded products was accelerated with the development of the genomisotopic approach which combines genomic sequence analysis and isotope-guided fractionation in the isolation of natural products from orphan gene clusters (Gross *et al.*, 2007). In this

Table 7.2 Commercially available *Pseudomonas* biocontrol strains

Biocontrol strain	Commercial name	Usage	Crop/applications	Pathogen
Pseudomonas sp. DSMZ 13134	Proradix®	Fungicide, plant growth regulator	Barley, potatoes, greenhouse, laboratory, field	*Rhynchosporium secalis*, *Gaeumannomyces graminis*, *Rhizoctonia solani*, stem *Phytophthora*
P. chlororaphis MA34	Cedemon®	Fungicide	Seed-dressing in field grown cereals (barley, oats, triticale, wheat)	Leaf stripe (*Drechslera graminea*), covered smut in barley (*Ustilago hordeii*), glume blotch (*Stagonospora nodorum*) in wheat
P. fluorescens A506	Blightban®	Fungicide/ frost preventer	Foliar almond, peaches, apples, tomatoes, cherries, strawberries, potatoes	*Erwinia amylovora*, soil borne diseases caused by *Fusarium* and *Rhizoctonia* spp. and discoloration of pear and apple crops
P. aureofaciens strain Tx-1	Bioject Spot-Less	Fungicide	Lawn and grass management, recreational and ornamental lawns	Dollar spot (*Sclerotinia homeocarpa*) Anthracnose (*Colletotrichum* spp.), Pythium (*P. aphanidermatum*), pink snow mould (*Microdochium nivale*)
P. syringae strain 742RS	Frost ban	Fungicide/ frost preventer	Foliar almond, peaches, apples, tomatoes, cherries, strawberries, potatoes	
P. syringae strain	Snowmax®	Artificial snow making operations		
P. syringae strain ESC-10	Biosave 1000	Fungicide	Post harvested: apples, pears, lemons, sweet potatoes, potatoes	*Botrytis cinerea*, *Penicillium expansum*, *Penicillium digitatum*
	ESC 10 Biofungicide Technical	Fungicide		
	Biosave 10LP	Fungicide		
P. syringae strain ESC-11	Biosave 11LP	Fungicide	Post harvest fruits: apples, pears, lemons, sweet potatoes, potatoes	*Penicillium expansum*, *Botrytis cinerea*, *Mucor rot*, *Penicillium italicum*, *P. digitatum*
	ESC 11 Biofungicide Technical	Fungicide		

References: EU pesticide database, EPA pesticide database, biopesticide database, Frohlich *et al.* (2012), Mark *et al.* (2006), Skirvin *et al.* (2000).

approach, the organism of interest is grown under culture conditions favourable to the expression of the biosynthetic genes. A ^{15}N- or ^{15}N-^{13}C-labelled precursor predicted to be present in the metabolic product by bioinformatics tools is added to the culture medium and the label is used to guide fractionation by NMR for purification of the natural product. This genomisotopic approach was applied for the first time to an orphan biosynthetic gene cluster in the *P. protegens* Pf-5 genome and resulted in the isolation of orfamide A, a member of a new subclass of lipopeptides produced by *Pseudomonas* sp. (Gross *et al.*, 2007). The centrality of genomics in natural products discovery has increased over the years. This is clearly shown in the landmark paper by Gross and Loper (2009), which gives an extensive overview of secondary metabolite biosynthesis in the Pseudomonads.

Tools for discovery, prediction and analysis of natural products

An enormous resource of genome and metagenome sequencing projects is offered by several genomic databases including the Genomes Online (GOLD) database (Pagani *et al.*, 2012). The availability of these completed sequences enhances the identification of new NRPS and PKS gene clusters. These new gene clusters are determined by the use of increasingly available bioinformatic tools. Here we give a summary about the online software and databases employed for natural product discovery, prediction and analysis within genomic sequences. Besides the use of NRPS-PKS, for analysis of NRPS/PKS megasynthases (Ansari *et al.*, 2004) and the NRPS Predictor (Rausch *et al.*, 2005) for specificity prediction of adenylation in NRPS, more databases have been developed for scanning genome sequences for either polyketide or non-ribosomal peptides or both. Such databases include ClustScan, also known as the 'Clust Scanner', that quickly and semi-automatically annotates DNA sequences, which encode modular biosynthetic enzymes such as NRPS, PKS and hybrid PKS/NRPS (Starcevic *et al.*, 2008). This package further shows the predicted chemical structures of products and allows the export of these structures to other programs. NP.searcher rapidly scans microbial genomes for biosynthetic gene clusters of secondary metabolites and outputs NRPS and PKS natural products in SMILES (small molecules) format, which enables DNA decoding to give 2D and 3D structures (Li *et al.*, 2009). Bachmann and Ravel (2009) presented an *in silico* method for the prediction of PKS and NRPS from DNA sequence data as well as methods for converting the analyses into structural components. More recently, Prieto *et al.* (2012) launched the NRPSsp website, an NRPS substrate predictor in which an HMM predictor that binds to a specific NRPS was built. Another online program is Antibiotic and Secondary Metabolite Shell (antiSMASH) (Medema *et al.*, 2011). AntiSMASH enables the identification of secondary metabolite biosynthetic gene clusters in bacterial and fungal genomes and also provides information about the structure and function of encoded products. An updated version of antiSMASH has been recently published (Blin *et al.*, 2013). Another database is NORINE, a resource database centred on non-ribosomal peptides and containing up to 1000 peptides that have been discovered so far (Caboche *et al.*, 2008). For each peptide, Norine provides information about its structure, biological activity and the producing organisms, for example. DoBISCUIT is a literature-based, manually curated database of gene clusters for secondary metabolite biosynthesis (Ichikawa *et al.*, 2013), while ClusterMine360, a database of PKS and NRPS gene clusters contains more than 200 gene clusters from more than 185 compound families (Conway and Boddy, 2013). Via crowd-sourcing, ClusterMine 360 allows its members to make inputs in the database while automation is used to ensure data consistency and quality. Some of these bioinformatic softwares/databases have proved vital in genome mining for secondary metabolite biosynthetic gene clusters.

Novel insights in the evolution and functional role of major antibiotics produced by *Pseudomonas* biocontrol agents

Phenazines

Phenazines are nitrogen-containing tricyclic pigments that are produced by *Pseudomonas* spp., a few other Gram-negative proteobacteria including *Burkholderia* spp., *Pectobacterium* spp., *Brevibacterium,* and *Pantoea agglomerans,* but also by Gram-positive Actinobacteria such as *Streptomyces* spp. (Mavrodi *et al.*, 2006). Most phenazine producers are soil inhabitants and/or plant-associated species (Mavrodi *et al.*, 2010). More than 100 different phenazine structures have been described. Phenazine producers associated with biocontrol mainly produce phenazine-1-carboxylic acid (PCA), phenazine-1-carboxamide (PCN), 2-hydroxyphenazine and 2-hydroxyphenazine-1-carboxylic acid. Within the *Pseudomonas* genus, the most prominent producers belong to *P. aeruginosa,* the *P. chlororaphis* subgroup or the *P. fluorescens* subgroup (Table 7.1). Recent research by Mavrodi *et al.* (2012) and Parejko *et al.* (2012, 2013) has revealed an extensive complex of phenazine producers isolated from dryland winter wheat. These *Pseudomonas* spp. mainly produce PCA and could be grouped in five different species, including two new species that have been provisionally named *P. aridus* and *P. cerealis,* while the other isolates belong to the *P. synxantha* and *P. orientalis* clade. All these isolates cluster in the *P. fluorescens* subgroup as defined by Mulet *et al.* (2010) (Table 7.1). Most clinical *P. aeruginosa* isolates produce pyocyanin (PYO), a blue phenazine pigment, but this species can also produce several other phenazine compounds including PCA and PCN (Anjaiah *et al.*, 1998). *P. chlororaphis* isolates have been reported to produce PCA, PCN, 2-hydroxyphenazine and 2-hydroxyphenazine-1-carboxylic acid. *Pseudomonas* CMR12a and CMR5c, biocontrol strains isolated from the roots of tropical tuber crop cocoyam in Cameroon (Perneel *et al.*, 2007; Hofte and Altier, 2010), are taxonomically positioned between *P. chlororaphis* and *P. protegens* in the *P. fluorescens* group. These strains produce PCA and PCN (Perneel *et al.*, 2007) and may represent yet another novel phenazine producing species within the *P. fluorescens* group (Mavrodi *et al.*, 2010).

The core phenazine biosynthesis genes are clustered and highly conserved. A seven-gene locus, named *phzABCDEFG,* is responsible for the synthesis of the first phenazine derivative in pseudomonads, namely PCA. This operon is present in two copies in the genome of *P. aeruginosa.* Phenazine biosynthesis starts from chorismic acid, a product from the shikimate pathway. This aromatic biosynthetic pathway converts primary metabolites to chorismic acid, the last common precursor of the three aromatic amino acids (Phe, Tyr, Trp), via 3-deoxy-D-arabino-heptulosanate-7-phosphate synthase enzymes. The PhzC protein is a variant of DAHP synthase and redirects carbon flow into phenazine biosynthesis. Phenazine diversity results from modification of PCA by dedicated enzymes encoded by genes that are often linked to the phenazine core operon such as PhzO (modifies PCA into 2-OH-PCA), PhzH (modifies PCA into PCN), PhzS (modifies PCA into 1-OH-phenazine) and PhzM (modifies PCA into pyocyanin together with PhzS). In all isolates tested, phenazine production is regulated by quorum sensing. In *P. chlororaphis* and *P. fluorescens,* the phenazine operon is coupled to the regulatory genes *phzI* and *phzR* (Mavrodi *et al.*, 2006; De Maeyer *et al.*, 2011), while in *P. aeruginosa*, phenazine production is regulated via a more complex network utilizing the hierarchical LasI/LasR and RhlI/RhlR quorum sensing systems as well as a third signalling system based on the production of the *Pseudomonas* quinolone signal. Phenazine production also requires a functional GacS/GacA two-component signal transduction system. Many other regulators have been reported to be interconnected with the phenazine quorum sensing regulon. Further details about phenazine biosynthesis can be found in Gross and Loper (2009) and Mavrodi *et al.* (2006).

Phenazines are redox-cycling agents because of their ability to accept or donate electrons. Whether they get reduced or re-oxidized, depends on their redox potential relative to that of other electron transfer molecules. In particular, the oxidation-reduction reactions of PYO, PCA, PCN and 1-OH-PHZ involve a two-electron, two-proton transfer under physiological conditions (Wang and Newman, 2008). Different phenazine structures come along with different physical, chemical and redox properties, which ultimately determine their biological activity (Mavrodi *et al.*, 2006).

Acquisition of the phenazine genes by pseudomonads was an early event, possibly preceding speciation in this group of bacteria, while other bacteria may have obtained the phenazine operon by horizontal gene transfer (Mavrodi *et al.*, 2010). According to Mavrodi *et al.* (2010) this may reflect differences in the roles that phenazines play in these organisms. The multiple roles that phenazines can fulfil for the producing strain have recently been reviewed (Pierson and Pierson, 2010; Mavrodi *et al.*, 2013) and are briefly discussed below. Most of the described effects of phenazines go back to their antagonistic effect as an antibiotic during competition, and as a virulence factor during pathogenesis. This antagonistic effect can be attributed to their ability to generate reactive oxygen species (ROS), which cause oxidative stress in other organisms and tissues (Laursen and Nielsen, 2004; Mavrodi *et al.*, 2006). Because of their toxicity, phenazines can inhibit or kill other microorganisms, hence providing a competitive advantage to the producing strain. Phenazine antibiotics attracted a lot of interest in plant disease control, because of their broad-spectrum antifungal activity. Phenazines have been shown to determine the biological control capacity of a number of soil-borne pseudomonads against a wide range of plant-pathogenic fungi (Chin-A-Woeng *et al.*, 2003; Mavrodi *et al.*, 2006). More specifically, the model strains for phenazine dependent biocontrol, *P. fluorescens* 2-79 (PCA) and *P. chlororaphis* 30-84 (PCA, 2-OH-PCA, 2-OH-PHZ), were shown to suppress the fungal phytopathogen *Gaeumannomyces graminis* var. *tritici* (Pierson and Thomashow, 1992; Thomashow and Weller, 1988; Weller and Cook, 1983) while *P. chlororaphis* PCL1391 (PCN) could suppress *Fusarium oxysporum* f. sp. *radicis-lycopersici* (Chin-A-Woeng *et al.*, 1998). Phenazines are likewise implicated in the biocontrol activity of *Pseudomonas* CMR12a against *Rhizoctonia solani* on bean (D'aes *et al.*, 2011). There have only been limited studies examining the *in vivo* antibiosis effect of pure phenazine compounds. One particular study demonstrates the role of purified PCN in suppression of pre-emergence damping-off by *Pythium splendens* on bean seedlings in inert vermiculite substrate, however, its efficacy was dependent on the presence of surface-active metabolites, such as rhamnolipids (Perneel *et al.*, 2008). Besides the antibiosis effect of phenazines, antagonism against plant pathogens can also be accomplished by induction of a plant defence response, as shown for PYO produced by *P. aeruginosa* 7NSK2 (De Vleesschauwer *et al.*, 2006). Phenazine antibiotics have been shown to contribute to competitiveness and ecological fitness (Mazzola *et al.*, 1992). Phenazine deficient strains of *P. fluorescens* 2-79 and *P. chlororaphis* 30-84 survival was diminished in the wheat rhizosphere, because they had reduced ability to compete with the resident microflora (Mazzola *et al.*, 1992).

Phenazines are thought to control aspects of microbial behaviour, like the structural organization of cellular communities. In *P. chlororaphis* 30-84 phenazines are involved in biofilm formation (Maddula *et al.*, 2006, 2008). Interestingly, different phenazines play different roles in biofilm structure and function. 2–0H-PCA may facilitate cellular adhesion, while PCA may facilitate growth within the biofilm (Maddula *et al.*, 2008; Ramos *et al.*, 2010). Recent work by Mavrodi *et al.* (2012) suggests that PCA producing rhizobacteria are uniquely adapted to the plant rhizosphere under conditions of water stress due to their ability to resist desiccation via formation of robust biofilms.

It has been shown that pseudomonads accumulate NADH when terminal electron acceptors become limiting, resulting in a reduced intracellular redox state (Price-Whelan *et al.*, 2007). Under microaerobic conditions, PYO enables *P. aeruginosa* cells to maintain their steady-state $NADH/NAD^+$ ratio, by acting as an alternative electron acceptor that reoxidizes NADH to NAD^+. Hence, redox homeostasis is maintained, and reduced PYO can be reoxidized after diffusion to more oxygen-rich sites (Price-Whelan *et al.*, 2007). Moreover, under completely anaerobic conditions, phenazines were shown to promote survival, but not growth (Wang *et al.*, 2010). Strikingly, PYO requires oxygen for its production, while PCA can be produced under anaerobic conditions (Dietrich *et al.*, 2006). In the absence of oxygen, alternative electron acceptors such as ferric iron (Fe^{3+}) can also be used. PCN production by *P. chlororaphis* PCL1391 promotes reduction of the insoluble iron mineral $Fe(OH)_3$, hence making iron more bioavailable as ferrous Fe^{2+}. This provides an ecological advantage to the bacterium, since Fe^{2+} is a limited element in soil (Hernandez *et al.*, 2004). Different phenazines react differently with Fe^{3+} minerals and oxygen; PYO is the most reactive with O_2, its electron transfer to Fe^{3+} is even inhibited by oxygen (Cox, 1986), while PCA is more reactive with Fe^{3+} than PYO (Wang and Newman, 2008). Consistently, recent findings showed that PCA promotes biofilm formation by making iron more available, while PYO contributes to biofilm development in an iron-independent way (Wang *et al.*, 2011). PCA can change the redox state of iron (Fe^{3+} to Fe^{2+}) which subsequently allows Fe^{2+} uptake via the cytoplasmic membrane protein FeoB (Wang *et al.*, 2011).

2,4-Diacetylphloroglucinol

DAPG production contributes to biocontrol of plant diseases by various plant-associated *Pseudomonas* isolates. DAPG production has only been observed for a subset of *Pseudomonas* isolates that all belong to the *Pseudomonas fluorescens* group as defined by Mulet *et al.* (2010). Within this group, DAPG producers are typically found in the *P. chlororaphis* and *P. corrugata* subgroup [named subgroup IV and subgroup I by Redondo-Nieto *et al.* (2013)] (Table 7.1). Well-studied producers are *P. protegens* Pf-5 and CHA0, *P. fluorescens* F113 and Q2-87, and *P. brassicacearum* Q8r1-96 (Weller *et al.*, 2007). These last three strains are closely related and group together in subgroup I as defined by Redondo-Nieto *et al.* (2013). Interestingly, *P. fluorescens* NZI7, a mushroom pathogen isolated from a mushroom farm in New Zealand, also produces DAPG and is taxonomically very similar to the *P. protegens* group (Burlinson *et al.*, 2013). DAPG is toxic to a wide range of plant-pathogenic fungi, but also shows anti-bacterial and anti-helminthic activity and is phytotoxic at high concentrations. DAPG producing *Pseudomonas* strains have typically been isolated from soils suppressive to take-all, an important root disease of wheat caused by *G. graminis* var.

tritici (Weller *et al.*, 2007). The 8 kb DAPG gene cluster is highly conserved among DAPG producers and comprises nine genes involved in biosynthesis (*phlACB* and *phlD*), efflux (*phlEI*), degradation (*phlG*) and regulation (*phlH* and *phlF*). Precursors for biosynthesis are three molecules of malonyl CoA. PhlD is a type III polyketide synthase responsible for the biosynthesis of phloroglucinol. The cluster appears to be ancestral in *P. fluorescens* and has not been reported in bacteria other than *Pseudomonas*. The genetics and evolution of DAPG synthesis in fluorescent pseudomonads has recently been reviewed (Troppens, 2013). The *phlACB* operon was acquired from an Archaeon, an event that took place around the origins of the *P. fluorescens* group, while the *phlD* gene appears to have its origins in the Actinobacteria and acquisition was ancient in the genus (Kidarsa *et al.*, 2011).

The target for DAPG activity appears to be the mitochondrion. DAPG may function as a proton ionophore, dissipating the proton motive force and uncoupling respiration and ATP synthesis. Besides its role in antibiosis, DAPG also clearly has a signalling role. DAPG levels rise in late exponential phase and then rapidly decline and do not accumulate as would be expected from an antibiotic. DAPG is an autoinducer and promotes its own activity via modulation of PhlF activity (Troppens *et al.*, 2013). DAPG is also able to signal between bacterial species and enhanced the plant growth-promoting activity of *Azospirillum brasilense* (Combes-Meynet *et al.*, 2011). Moreover, DAPG can promote amino acid exudation from plant roots by blocking amino acid influx by plant roots (Phillips *et al.*, 2004). DAPG has also been reported to induce systemic resistance to *Peronospora parasitica* (Iavicoli *et al.*, 2003) and *Pseudomonas syringae* pv. *tomato* in *Arabidopsis thaliana* (Weller *et al.*, 2012). Brazelton *et al.* (2008) showed that DAPG at concentrations reported to be produced *in situ*, inhibits primary root growth and stimulates root branching in tomato by interacting with an auxin-dependent signalling pathway.

Pyoluteorin

Pyoluteorin is a hybrid non-ribosomal peptide synthase/polyketide synthase natural product and consists of a resorcinol ring attached to a dichlorinated pyrrole moiety. The primary precursors are L-proline and three malonyl-CoA monomers (see Gross and Loper, 2009 for details). Pyoluteorin is toxic against oomycetes, certain bacteria and fungi, and phytotoxic at high concentrations. Pyoluteorin is known to suppress seedling damping off diseases caused by the oomycete pathogen *Pythium ultimum* (Howell and Stipanovic, 1980;mMaurhofer *et al.*, 1994). Pyoluteorin producers can be found in the *P. chlororaphis* subgroup according to Mulet *et al.* (2010) (Table 7.1) and include *P. protegens* Pf-5 and CHA0 and the cocoyam isolate *Pseudomonas* CMR5c. Interestingly, *P. aeruginosa* M18, a well-studied biocontrol strain isolated from the rhizosphere soil of sweet melon in Korea, also produces pyoluteorin (Ge *et al.*, 2004). The pyoluteorin gene cluster contains 17 genes, including nine structural genes, and is 30 kb in length. The gene cluster in *P. aeruginosa* M18 is identical to the cluster in *P. protegens* Pf-5. Pyoluteorin biosynthetic gene clusters have also been detected in some clinical *P. aeruginosa* isolates (Kidarsa *et al.*, 2011). To our knowledge, pyoluteorin production has not been reported in genera other than *Pseudomonas*. The evolutionary origins of pyoluteorin have not been examined. Intriguingly, the pyoluteorin cluster is only present in *Pseudomonas* lineages that also have *phlD*, either as part of the DAPG operon, as is the case in *P. protegens* strains (Ramette *et al.*, 2011) or associated with the pyoluteorin cluster as is the case in certain *P. aeruginosa* strains (Kidarsa *et al.*, 2011). Pyoluteorin and DAPG have been reported to repress each other's production. Kidarsa *et al.* (2011) showed that nanomolar concentrations of phloroglucinol, the gene product of *phlD*, are required for pyoluteorin production, while higher concentrations of phloroglucinol are inhibitory to pyoluteorin production. DAPG had no direct effect on pyoluteorin production. The PhlD homologue of *P. aeruginosa* also functions in phloroglucinol production and regulates pyoluteorin production in *P. aeruginosa* in a similar way. In the biocontrol strain *P. aeruginosa* M18, that produces both pyoluteorin and PCA, PqsR, a LysR type transcriptional regulator, negatively controls pyoluteorin biosynthesis and positively regulates PCA biosynthesis at the transcriptional level (Lu *et al.*, 2009).

Pyrrolnitrin

Pyrrolnitrin was first reported from *Pseudomonas pyrrocinia* (Arima *et al.*, 1964; Imanaka *et al.*, 1965), a bacterium that was later reclassified as *Burkholderia pyrrocinia* (Viallard *et al.*, 1998). Pyrrolnitrin has since been reported from fluorescent pseudomonads, *Myxococcus fulvus*, *Burkholderia* spp., including the opportunistic human pathogen *B. cepacia*, and *Serratia* spp. The pyrrolnitrin producing isolate *Enterobacter agglomerans* IC1270 described in Chernin *et al.* (1996) is actually a *Serratia plymutica* isolate (Ovadis *et al.*, 2004). Pyrrolnitrin is a chlorinated phenylpyrrol which is produced in four steps from the basic precursor L-tryptophan (see Gross and Loper, 2009 for details). Within the genus *Pseudomonas*, pyrrolnitrin is produced by taxonomically related isolates belonging to the *P. fluorescens* group, and more specifically, the *P. chlororaphis* subgroup as defined by Mulet *et al.* (2010) (Table 7.1). Well-studied pyrrolnitrin producers are *P. protegens* Pf-5 and CHA0, several *P. chlororaphis* isolates, and *P. fluorescens* BL915 (Hammer *et al.*, 1999). The cocoyam isolate *Pseudomonas* CMR5c also produces pyrrolnitrin (Perneel *et al.*, 2007). The four pyrrolnitrin biosynthetic genes *prnABCD* are highly conserved among strains of *P. fluorescens* that produce pyrrolnitrin. *Pseudomonas* strains may have obtained the pyrrolnitrin operon by horizontal gene transfer (Paulsen *et al.*, 2005; Costa *et al.*, 2009). Hill *et al.* (1994) showed that pyrrolnitrin produced by *P. fluorescens* BL915 plays a role in the ability of the strain to control damping off of cotton caused by *R. solani*. Pyrrolnitrin has a broad spectrum anti-fungal activity (Chernin *et al.*, 1996) and has been used in the pharmaceutical industry to treat fungal infections. Its mode of action is the inhibition of fungal respiratory chains (Wong and Airall, 1970). Pyrrolnitrin has also been used as a lead structure in the development of a new class of agricultural fungicides, the phenylpyrroles (Nyfeler and Ackermann, 1992). Fenpliconil (4-(2,3-dichlorophenyl)-1*H*-pyrrole-3-carbonitrile) and fludioxonil (4-(2,2-difluoro-1,3-benzodioxol-4-yl)-1*H*-pyrrole-3-carbonitrile) are more stable than pyrrolnitrin and are contact fungicides that are mainly used to control seedborne pathogens including *Fusarium*, *Rhizoctonia* and *Alternaria*. Recently the entire pyrrolnitrin operon of *P. protegens* Pf-5 was expressed in tomato plants. Pyrrolnitrin-expressing plants were resistant to damping-off of tomato seedlings caused by *R. solani* (Mozes-Koch *et al.*, 2012). The ecological role of pyrrolnitrin is not understood (Costa *et al.*, 2009).

Hydrogen cyanide

Various *Pseudomonas* biocontrol strains produce hydrogen cyanide (HCN), a potent poison that inhibits cytochrome c oxidase, the terminal component of the respiratory chain in many organisms. Known HCN producers include *P. aeruginosa*, *P. fluorescens* subgroups I, II and IV as defined by Redondo-Nieto *et al.* (2013) (Table 7.1) and *P. entomophila* (Ryall *et al.*, 2009). A cluster of three genes, *hcnABC*, is responsible for HCN production from the metabolic precursor glycine in these strains (Laville *et al.*, 1998). Cyanogenesis requires microaerophilic conditions and is regulated by the ANR protein (anaerobic regulator of arginine deiminase and nitrate reductase) in both *P. aeruginosa* and *P. protegens*. Iron has a stimulatory effect on HCN production. In addition, cyanogenesis is controlled by the global regulator GacA (Blumer and Haas, 2000). In *P. protegens* CHA0, HCN production accounts for a substantial part of the strain's biocontrol capacity against root rot on tobacco caused by *Thielaviopsis basicola* (Voisard *et al.*, 1989). Devi and Kothamasi (2009) showed that *P. protegens* CHA0 can kill a macroscopic insect pest, *Odontotermes obesus* via cyanide poisoning. HCN, together with DAPG, is also responsible for the biocontrol activity of *Pseudomonas* sp. LBUM300 against bacterial canker of tomato caused by *Clavibacter michiganensis* subsp. *michiganensis* (Lanteigne *et al.*, 2012). Moreover, some *Pseudomonas* strains can scavenge and utilize cyanide as a nitrogen source by excreting keto acids. These compounds combine with cyanide to produce a less toxic cyanohydrin. Cyanohydrins can be degraded intracellularly by cyanide oxidase, a reaction that regenerates the keto acid and produces CO_2 and NH_3 (Kunz *et al.*, 1998; Luque-Almagro *et al.*, 2011). Recent phylogenetic studies have revealed that the HCN biosynthetic genes are ancestral in *Pseudomonas* species associated to roots (Frapolli *et al.*, 2012).

New antibiotics discovered by genome mining

Besides the production of phenazines, DAPG, pyrrolnitrin, pyoluteorin and hydrogen cyanide, the widespread use of genome mining has unravelled new antibiotic products of *Pseudomonas* spp., some of which are beneficial in the control of plant pathogens. Here, we will briefly discuss recently described antibiotics and their bioactive roles.

Rhizoxins

Rhizoxins are polyketide macrolides, originally produced by the endosymbiotic bacterium *Burkholderia rhizoxinica* and isolated from its fungal host, *Rhizopus microspores* (Partida-Martinez and Hertweck, 2007). In *Pseudomonas* spp., this antibiotic is implicated in the anti-fungal activity of *P. chlororaphis* MA342 against seed-borne net blotch of barley caused by the fungal pathogen *Drechslera teres* (Ligon *et al.*, 2000). In *in vitro* assays, rhizoxins also showed toxicity towards the important plant pathogens *Botrytis cinerea* and *Phytophthora ramorum*. About a decade later, genome mining revealed the presence of a rhizoxin gene cluster in *P. protegens* Pf-5 (Loper *et al.*, 2008). Interestingly, the rhizoxin-producer, *P. chlororaphis* strain MA 342, patented by the Swedish company BioAgri AB (Stockholm, Sweden), has been used on a commercial scale in Sweden under the trade name Cedomon since 1998 and has since been registered as a biocontrol agent in several European countries (Tombolini *et al.*, 1999; Mark *et al.*, 2006) (Table 7.2).

Promysalin

Promysalin is an amphipathic antibiotic, composed of salicylic acid and 2,8-dihydroxymyristamide bridged by a 2-pyrroline-5-carboxyl moiety. This antibiotic is produced by the biocontrol strain *P. putida* RW10S1, which was isolated from the rhizosphere of rice, and selectively inhibits several *Pseudomonas* spp. In *in vitro* experiments, promysalin showed activity against *P. aeruginosa* PA14, *P. syringae* pv. *glycinea*, *P. savastanoi* and *P. stutzeri* (Li *et al.*, 2011).

L-Furanomycin

Recently, L-furanomycin, a ninhydrin-reactive non-proteinogenic amino acid with selective antimicrobial properties was isolated from *P. fluorescens* SBW25 (Trippe *et al.*, 2013). L-Furanomycin is a competitive antagonist of L-isoleucine. *In vitro* tests using pure furanomycin showed selective inhibition against plant-pathogenic bacteria including *P. syringae* pv. *tomato, Dickeya dadantii* and *Erwinia amylovora*. Prior to this report, furanomycin was known to be a natural product of *Streptomyces threomyceticus* (Katagiri *et al.*, 1967), but was never reported to be produced among the *Pseudomonas* spp.

Insect toxins

Increased sequencing of microbial genomes has revealed novel insecticidal toxins with potential for crop protection. Additionally, genome comparison of *Pseudomonas* group members showed a number of loci encoding potential virulence factors which possess insecticidal properties (ffrench-Constant *et al.*, 2007; Paulsen *et al.*, 2005). *P. entomophila* was the first bacterium within the *Pseudomonas* group reported to show lethality against insects. When directly injected into the insect's body cavity, *P. entomophila* was observed to initiate a local and systemic immune response in the insect leading to a massive destruction of gut cells. This entomopathogenic bacterium is lethal to both larvae and adult stages of *Drosophila melanogaster* and several other insects. Since *P. entomophila* is non-pathogenic to plants, several studies have emphasized its possible use as biological control agent (Vodovar *et al.*, 2005, 2006; Pechy-Tarr *et al.*, 2008).

Some members of the *P. fluorescens* group also exhibit insecticidal activity. In some cases, the toxicity of certain members of the *P. fluorescens* group to insects has been linked with gene clusters encoding *Mcf* (makes caterpillars floppy) or *Tc* (toxin complexes) toxins, which were first described in entomopathogenic nematode-symbionts/pathogens such as *Photorhabdus* spp. and in insect pathogens like *Serratia entomophilia*. Experiments with *P. protegens* CHA0 and *P. protegens* Pf-5 demonstrated that when injected into the haemocoel, both bacteria can proliferate within the insect host and kill larvae of *Galleria mellonella* and *Manduca sexta*. Subsequently, the *fitD* (fluorescens insect toxin), with close relatedness to *mcf*, was found in the genome of *P. protegens* CHA0 *and P. protegens* Pf-5 and associated with the strain's toxicity against *both insects*. This was confirmed further using *fitD* deletion mutants (Pechy-Tarr *et al.*, 2008). The *fit* locus is also present in the genomes of *P. chlororaphis* O6 and 30-84 (Loper *et al.*, 2012) and was recently also discovered in the genomes of the Cameroonian biocontrol strains *Pseudomonas* CMR12a and *Pseudomonas* CMR5c (Ruffner *et al.*, unpublished). Recent studies further demonstrated that Fit toxin-producing pseudomonads show potent oral activity against larvae of major insect pests of agricultural crops like *Spodoptera littoralis, Heliothis virescens* and *Plutella xylostella*. Seventy to eighty per cent of *Spodoptera* and *Heliothis* larvae were killed when plant leaves were sprayed with suspensions containing only 1000 *Pseudomonas* cells per ml. In this study, *P. fluorescens* CHA0 and *P. chlororaphis* PCL1391 that both harbour the Fit gene cluster, killed insects through oral infection. Oral insecticidal activity of these bacteria was mainly due to their production of Fit toxin since Fit mutants of CHA0 and PCL1391 did not show similar toxicity on insects (Ruffner *et al.*, 2013).

Biosurfactants: new players in the biocontrol field

Pseudomonas biocontrol strains mainly produce two types of biosurfactants, rhamnolipids and cyclic lipopeptides (CLPs). CLPs are comprised of a cyclic oligopeptide lactone ring coupled to a fatty acid tail. Based on their structural characteristics, CLPs have been classified into eight different groups namely, the amphisin, viscosin, putisolvin, orfamide, tolaasin, syringomycin, syringopeptin and entolysin group (Raaijmakers *et al.*, 2006; Gross and Loper, 2009; Roongsawang *et al.*, 2011; Li *et al.*, 2013). Biosurfactants can cause destruction of microbial membranes, leading to death of bacteria, fungi, oomycetes and viruses. Furthermore, biosurfactants perform a diversity of functions in *Pseudomonas* bacteria and can play a role in swarming motility, biofilm formation, environmental adaptation, nutrient availability, and root colonization. These aspects have recently been reviewed by Raaijmakers *et al.* (2010) and D'aes *et al.* (2010) and will not be further discussed. In this review we will provide information about biosurfactants produced by *Pseudomonas* with a reported role in biological control of plant pathogens. An overview is given in Table 7.3.

Rhamnolipids

Rhamnolipids are glycolipids which are typically produced by *P. aeruginosa*. They are composed of one or two β-hydroxy fatty acids linked to one or two rhamnose molecules (Soberon-Chavez *et al.*, 2005). Rhamnolipid production has also been described from *P. chlororaphis, P. putida, P. fluorescens* and *Pseudomonas* sp. GRP(3) (Toribio *et al.*, 2010). Rhamnolipids are intensively studied microbial surfactants and attract a lot of industrial interest (Henkel *et al.*, 2012). These molecules have also been implicated in *Pseudomonas* biocontrol. In a soil assay experiment, *P. aeruginosa* PNA1 demonstrated antibiotic action towards the mycelium of *Pythium myriotylum* on cocoyam. Mutant analysis revealed that the biocontrol activity shown by the bacterium was attributed to the synergistic effect of phenazines and rhamnolipids (Perneel *et al.*, 2008). This synergism was also observed in a *Pythium splendens*–bean pathosystem (Perneel *et al.*, 2008). Here, the effect of rhamnolipids was clearly confirmed by application of the pure compound. Rhamnolipids of *P. aeruginosa* have direct antifungal properties by inhibiting spore germination and mycelium growth of *Botrytis cinerea*. Furthermore, in experiments using a detached leaf assay, rhamnolipids were observed to trigger defence responses against the grey mould disease caused by *B. cinerea* on grapevine (Varnier *et al.*, 2009). Recently, Sanchez *et al.* (2012) showed that pure rhamnolipids trigger immune responses in *Arabidopsis* against biotrophic, hemibiotrophic and necrotrophic pathogens. Rhamnolipids from *Pseudomonas* sp. GRP(3) were attributed to the observed reduction of damping-off disease in field experiments with chilli pepper and tomatoes (Sharma *et al.*, 2007). In this study, methanol extracts of bacterial cultures were screened to obtain the active compound which was confirmed to be rhamnolipids. In another study, a formulation of 25% rhamnolipids in oil was applied in a chicory–*Phytophthora cryptogea* hydroponic system. In both *in vitro* and *in vivo* hydroponic systems, rhamnolipids were effective in controlling brown root rot disease caused by *Phytophthora cryptogea* (De Jonghe *et al.*, 2005).

Viscosin group

The viscosin group contains CLPs having nine amino acids often linked at the N-terminus to 3-hydroxy decanoic acid (3-HDA). Members of this group include viscosin, viscosinamide, massetolide, pseudodesmin, pseudophomin and white line inducing principle (WLIP). The

Table 7.3 Biosurfactants produced by *Pseudomonas* spp. involved in biocontrol

Biocontrol strain	Test plant	Plant pathogen	Lipopeptides involved	Biocontrol mechanism	Experimental setup	Experimental method	Reference
P. putida BW11M1		*Xanthomonas* sp., *Botrytis cinerea*, *Rhizoctonia solani*	Xantholysin	Antibiotic	*In vitro*	Mutant analysis	Li *et al*. (2013)
P. putida RW10S2		*Xanthomonas* sp.	WLIP	Antibiotic	*In vitro*	Mutant analysis	Rokni-Zadeh *et al*. (2012)
P. putida 267		*Phytophthora capsici*	Putisolvins	Lysis of zoospores	*In vitro*	Mutant analysis	Kruijt *et al*. (2009)
P. aeruginosa PNA1	Cocoyam	*Pythium myriotylum*	Rhamnolipids and phenazines	Antibiotic	Soil assay	Mutant analysis	Perneel *et al*. (2008)
	Bean	*Pythium splendens*	Rhamnolipids and phenazines	Antibiotic	Soil assay	Mutant analysis, pure compound	Perneel *et al*. (2008)
	Witloof chicory	*Phytophthora cryptogea*	Rhamnolipids	Antibiotic	*In vitro*, *in vivo* hydroponic system	Pure compound	De Jonghe *et al*. (2005)
P. aeruginosa BP35	Black pepper	*Phytophthora capsici*, *P. palmivora*, *P. megakarya*	Rhamnolipids	Zoosporicidal	*In vitro*	MeOH extract of cell-free supernatant	Kumar *et al*. (2013)
P. aeruginosa	Grapevine	*Botrytis cinerea*	Rhamnolipids	Induced Systemic Resistance (ISR)	Detached leaf assay	Pure compound	Varnier *et al*. (2009)
Pseudomonas sp. GRP(3)	Chili, tomato	*Pythium aphanidermatum*, *Phytophthora nicotianae*	Rhamnolipids		Field trials	Methanolic extracts	Sharma *et al*.(2007)
P. fluorescens DR54	Sugar beet	*Pythium ultimum*	Viscosinamide	Antibiotic, reduced mycelial density, radial growth, oospore formation	Soil assay, *in vitro*	Pure compound	Thrane *et al*. (1999, 2000), Nielsen *et al*. (1999)

Table 7.3 Continued

Biocontrol strain	Test plant	Plant pathogen	Lipopeptides involved	Biocontrol mechanism	Experimental setup	Experimental method	Reference
	Sugar beet	*R. solani*	Viscosinamide	Antibiotic, reduces sclerotia formation, mycelium biomass development	Soil assay, pot experiments, *in vitro*		Thrane *et al.* (1999, 2001), Nielsen *et al.* (1998, 1999)
P. fluorescens SS101	Tomato	*Phytophthora infestans*	Massetolide A	ISR	Soil assay, leaf application of bacteria	Mutant analysis, pure compound	Tran *et al.* (2007)
	Hyacinth bulbs	*Pythium intermedium*	Massetolide A		Soil assay, plant bulbs	Mutant analysis	de Souza *et al.* (2003)
		Pythium spp., *P. infestans*, *Albugo candida*	Massetolide A	Zoospore lysis	*In vitro*	Mutant analysis	de Souza *et al.* (2003)
P. fluorescens SBW25		*P. infestans*	Viscosin	Antibiotic, Zoospore lysis	*In vitro*	Mutant analysis	De Bruijin *et al.* (2007)
P. fluorescens BRG100		*Leptosphaeria maculans*, *Sclerotinia sclerotiorum*	Pseudophomin A	Antibiotic	*In vitro*	Pure compound application	Pedras *et al.* (2003)
P. fluorescens		*R. solani*, *P. ultimum*	Tensin		*In vitro*	Pure compound	Nielsen *et al.* (2002)
P. fluorescens		*R. solani*, *P. ultimum*	Lokisin		*In vitro*	Pure compound	Nielsen *et al.* (2002)
P. fluorescens sp. 96.578		*R. solani*	Tensin	Reduced radial mycelium	*In vitro*	Pure compound application	Nielsen *et al.* (2000)
P. 'reactans' NCPPB1311		*Erwinia carotovora* subsp. *carotovora*, *Agaricus bisporus*	WLIP		*In vitro*	Pure compound	Lo Cantore *et al.* (2006)

Biocontrol strain	Test plant	Plant pathogen	Lipopeptides involved	Biocontrol mechanism	Experimental setup	Experimental method	Reference
P. 'reactans'		*Pseudomonas tolaasii*	WLIP		*In vitro*, application on mushroom cap	Pure compound	Soler-Rivas *et al*. (1999)
P. tolaasii NCPPB2192		*Escherichia coli*, *Erwinia*, *Agrobacterium*, *Pseudomonas*, *Xanthomonas*, *Pleurotus* spp., *Agaricus bisporus*	Tolaasin 1		*In vitro*	Pure compound	Lo Cantore *et al*. (2003, 2006)
Pseudomonas sp. CMR12a	Bean	*R. solani*	Sessilin and orfamide	Antibiotic	Soil assay	Mutant analysis	D'aes *et al*. (2011)
Pseudomonas sp. SH-C52	Groundnut	*Sclerotium rolfsii*	Thanamycin	Antibiotic	Nethouse and field	Mutant analysis	Le *et al*. (2012)
	Sugar beet	*R. solani*	Thanamycin	Antibiotic	Soil assay	Mutant analysis	Mendes *et al*. (2011)
Pseudomonas sp. DSS73		*R. solani*, *P. ultimum*	Amphisin		*In vitro*	Pure compound, mutant analysis	Nielsen *et al*. (2002); Sorensen *et al*. (2001); Andersen *et al*. (2003)
Pseudomonas sp. DF41	Canola	*Sclerotinia sclerotiorum*	Sclerosin	Disruption of electric potential of plasma membrane	Soil assay, plant sprayed with bacteria	Mutant analysis	Berry *et al*. (2010, 2012)

discovery of these CLPs has been reported in various groups of antagonistic *Pseudomonas* spp. that were isolated from diverse origins. Although viscosins were originally isolated from pectolytic stains of *P. fluorescens* causing head rot of broccoli (Hildebrand *et al.*, 1998) and from another *Pseudomonas* strain isolated from a marine environment (Gerard *et al.*, 1997), viscosin was also described in *P. fluorescens* SBW25, a biocontrol strain isolated from the sugar beet plant phyllosphere (de Bruijn *et al.*, 2007; Deleij *et al.*, 1995). Massetolide A was originally described from a *Pseudomonas* sp. isolated from the surface of leafy red algae in the sea (Gerard *et al.*, 1997) and later reported for *P. fluorescens* SS101, an isolate from the rhizosphere of wheat (de Souza *et al.*, 2003). The WLIP terminology is based on the capacity of the producing bacteria to form a white-line-in-agar when grown in association with the mushroom pathogen *P. tolaasii*. This CLP was first reported in *Pseudomonas 'reactans'* (Mortishire-Smith *et al.*, 1991) and has since been described for several *Pseudomonas* strains including *Pseudomonas putida* RW10S2, a rice rhizosphere isolate (Rokni-Zadeh *et al.*, 2012), the biocontrol strain *Pseudomonas aurantiaca* PB-St2, an isolate from sugarcane stems (Mehnaz *et al.*, 2013) and *Pseudomonas 'reactans'* NCPPB1311 (Lo Cantore *et al.*, 2006). Viscosinamide was first reported in *P. fluorescens* DR54 originally isolated from the rhizosphere of sugar beet (Nielsen *et al.*, 1998; Nielsen *et al.*, 1999). Pseudophomins A and B were produced by *P. fluorescens* strain BRG100, which was isolated from the rhizosphere of green foxtail (Caldwell *et al.*, 2012; Quail *et al.*, 2002). The most recently reported member of this group, Pseudodesmins A and B, were isolated from *Pseudomonas* bacteria obtained from the mucus layer in the skin of the black belly salamander (Sinnaeve *et al.*, 2009).

Of all the CLP structural groups, members of the viscosin group appear to be most prominently reported to exhibit biocontrol capacity against diverse plant pathogens including bacteria, fungi and oomycetes. *In vitro* assays were conducted using cell cultures and cell-free culture supernatants of *P. fluorescens* SBW25 against zoospores of *Phytophthora infestans* (de Bruijn *et al.*, 2007). Immobilization and eventual lysis of the whole zoospore population was observed. This result was similar to what was obtained when the purified CLP of *P. fluorescens* SBW25, viscosin, was tested. Since viscosin-deficient mutants did not show such positive results, viscosin is clearly important in the lysis of the zoospores. Massetolide A production by *P. fluorescens* SS101 was shown to be the vital component in the biocontrol activity of this bacterium against late blight of tomato caused by *Phytopthora infestans* (Tran *et al.*, 2007). This was confirmed by using massetolide-deficient mutants and the pure compound in biocontrol experiments. Local and systemic control of the pathogen via induced resistance was shown by *P. fluorescens* SS101. However, the production of massetolide A by *P. fluorescens* SS101 was not required for the capacity of this bacterium to suppress root rot mediated by *Pythium* spp. on apple and wheat (Mazzola *et al.*, 2007). Viscosinamide, produced by *Pseudomonas* sp. strain DR54, also indicates that CLPs are important in the biocontrol of plant pathogens. *In vitro* tests with viscosinamide against *P. ultimum* and *R. solani* showed reduction in biomass and radial growth of mycelium. Application of viscosinamide *in situ* also led to decreased oospore formation and sclerotia formation in *P. ultimum* and *R. solani* respectively (Nielsen *et al.*, 1998; Nielsen *et al.*, 1999; Thrane *et al.*, 1999, 2000, 2001). A cell-free crude extract containing the WLIP produced by *P. 'reactans'* SPC 8907 showed inhibition against browning of mushrooms caused by *P. tolaasii* (Soler-Rivas *et al.*, 1999). *P. 'reactans'* NCPPB1311 showed antagonism against *Erwinia carotovora* subsp. *carotovora* and *Agaricus bisporus* (Lo Cantore *et al.*, 2006). This bioactivity was attributed to WLIP produced by this strain. More recently, WLIP-producing *Pseudomonas putida* RW10S2 demonstrated antagonism against plant-pathogenic *Xanthomonas* species (Rokni-Zadeh *et al.*, 2012). Pure WLIP compound inhibited the growth of several *Xanthomonas* strains and some Gram-positive bacteria. Although these experiments were only conducted *in vitro*, it further lends weight to the supposed biocontrol capacity of this CLP.

Another biocontrol strain is the *P. fluorescens* BRG100 from which two lipopeptides, Pseudophomins A and B, were isolated (Pedras *et al.*, 2003). Pseudophomin A appeared to be the main inhibitory component in the anti-fungal activity of this strain towards *Leptosphaeria maculans* and *Sclerotinia sclerotiorum*. Although pseudodesmins A and B reportedly show bioactivity against several human-pathogenic gram positive bacteria, this compound was not tested for antimicrobial activity with plant pathogens (Sinnaeve *et al.*, 2009). It is apparent that almost all members of the viscosin group except the Pseudodesmins have been implicated in the biocontrol of economically important plant pathogens.

Putisolvin group

A unique feature of the putisolvins is the peptide moiety of 12 amino acids and a hexanoic lipid tail. So far, Putisolvin I and II have been described only from *P. putida* PCL 1445, isolated from a site polluted with polycyclic aromatic hydrocarbons (Kuiper *et al.*, 2004) and *P. putida* 267, isolated from black pepper rhizosphere (Kruijt *et al.*, 2009). These data suggest that putisolvins are mainly produced by fluorescent bacteria within the *Pseudomonas putida* group. The role of putisolvins in the biocontrol of plant pathogens is yet to be clearly elucidated. When biocontrol strains *P. putida* strain 267 was tested for antagonism against pre- and post-emergence damping-off of cucumber caused by *Phytophthora capsici*, effective biocontrol capacity by this strain was demonstrated. However, growth chamber experiments showed that CLP-deficient mutants were as effective in biocontrol as the parental strains. Therefore, it appears that the biocontrol capacity of these bacteria is independent of their CLP production. In contrast, *in vitro* tests with partially purified putisolvins obtained from *P. putida* 267 resulted in the lysis of zoospores of *Phytophthora capsici* within 90 seconds (Kruijt *et al.*, 2009).

Amphisin group

The amphisin group consists of CLPs containing an 11 amino acid cyclic peptide that is linked at the N-terminal end to β-hydroxydecanoyl (Sorensen *et al.*, 2001). Members of this group include amphisin, tensin, lokisin, arthrofactin and pholipeptin A. Pertaining to biocontrol activity, amphisin, lokisin and tensin appear to be more interesting members of the group. Nielsen *et al.* (2000), using the CLP tensin, isolated from *Pseudomonas fluorescens* strain 96.578, demonstrated antagonistic efficiency against *R. solani* via reduction of the pathogen's radial mycelium extension. An extensive survey was conducted to investigate lipopeptide production among approximately 600 fluorescent *Pseudomonas* spp. isolated from the sugar beet rhizosphere. Based on the different molecular weights and compositions of CLPs identified among the *Pseudomonas* spp. isolates, a total of eight subgroups of strains producing CLPs were made all of which were affiliated to *P. fluorescens* biovars. In this study, amphisin-, lokisin- and tensin-producing *Pseudomonas* strains demonstrated a high level of antagonism against *P. ultimum* and *R. solani*. Pure compounds of amphisin, lokisin and tensin were also highly antagonistic when tested with these pathogens (Nielsen *et al.*, 2002). Andersen *et al.* (2003) also observed that the ability of *Pseudomonas* sp. DSS73 to inhibit the root pathogenic fungus *R. solani* is due in part to its production of amphisin.

Tolaasin group

The composition and length of the peptide chain of members of this group ranges between 19 to 25 amino acids with the lipid tail comprising of 3-HDA or 3-hydroxyoctanoic acid (3-HOA). Members include tolaasins, fuscopeptins, corpeptins, syringopeptins, sclerosin and sessilin (Raaijmakers *et al.*, 2006; Roongsawang *et al.*, 2011; Berry *et al.*, 2012; D'aes *et al.*, 2011). Of these CLPs, tolaasin, sessilin and sclerosin have demonstrated biocontrol capacity against plant pathogens. Tolaasin I, produced by *P. tolaasii* NCPPB2192, exhibited growth inhibition against the edible fungi *Agaricus bisporus, Lentinus edodes* and *Pleurotus* spp. Interestingly, purified tolaasin I reportedly inhibited the growth of Gram-negative bacteria belonging to the genera *Erwinia, Agrobacterium, Xanthomonas, Escherichia* and *Pseudomonas* (Lo Cantore *et al.*, 2006). The capacity of *Pseudomonas* sp. DF41 to suppress *Sclerotinia* stem rot of canola was shown to be dependent on the production of the linear lipopeptide sclerosin, which is closely related to corpeptin, a lipopeptide produced by the tomato pathogen *P. corrugata* (Berry *et al.*, 2010, 2012). When ascopores and sclerotia of *S. sclerotinia* were incubated with purified sclerosin, the germination of both cell types was inhibited. Sclerosin also showed antimicrobial activity against *Bacillus* species. The CLP sessilin, produced by *Pseudomonas* CMR12a, a biocontrol strain isolated from the roots of tropical tuber crop cocoyam, shows structural relatedness to the tolaasin group and differs from tolaasin I by only one amino acid. Sessilin is implicated in the biocontrol of *R. solani* on bean. Using mutant analysis and soil assay experiments, the biocontrol mechanism could be attributed to the combined effect of phenazines and sessilin (D'aes *et al.*, 2011).

Orfamides

Orfamides, isolated from cultures of *P. protegens* Pf-5, were the first compounds mined from *Pseudomonas* genomes using the genomisotopic approach discussed above. They consist of ten amino acids which bear close resemblance to the CLPs of the viscosin group. CLPs in this group include orfamide A-C characterized by a 3-hydroxy-dodecanoic or tetradecanoic acid connected to the *N*-terminus of the amino acids. Of these 3 CLPs, orfamide A appears to be dominant (Gross *et al.*, 2007). Orfamide A can lyse zoospores produced by an oomycete plant pathogen (Gross *et al.*, 2007). Also, *Pseudomonas* sp. CMR12a produces CLPS which are structurally similar to orfamides. Recent data suggest that both sessilins and orfamides are needed for effective biocontrol against *R. solani* on cabbage (Khuong and Höfte, unpublished).

Entolysin and xantholysins

Entolysin, described from *P. entomophila* consists of a peptide moiety of 14 amino acids and a typical cyclization in which the lactone ring is formed between the C-terminal carboxyl group and the 10th amino acid rather than one of the first amino acids. Although entolysin was shown to be important for the swarming motility and haemolytic activity of *P. entomophila,* it is not involved in the biocontrol activity of this *Pseudomonas* strain. An entolysin mutant, *etlC,* showed similar biocontrol when compared with that shown by the wild type strain in a cucumber-*Pythium ultimum* pathosystem (Vallet-Gely *et al.*, 2010). The recently discovered xantholysins consist of four lipopeptide compounds designated xantholysin A-D. The banana rhizosphere isolate, *P. putida* BW11M1, produces xantholysins consisting of 14 amino acids, compounds that appeared most similar to the entolysins (Li *et al.*, 2013). Apart from a role in swarming, biofilm formation, anti-fungal activity and toxicity to Gram-positive bacteria, a distinct feature of this lipopeptide is antagonism against some Gram-negative bacteria including several *Xanthomonads*. In addition, broad anti-fungal activity towards several Ascomycetes including *R. solani* and *B. cinerea* was demonstrated by the capacity of xantholysin to delay or stop radial mycelia outgrowth. These attributes of xantholysin were further established in *in vitro* experiments using xantholysin-deficient mutants.

Thanamycin

Pseudomonas strain SH-C52 was isolated from a *R. solani* suppressive soil in the Netherlands (Mendes *et al.*, 2011). *In silico* analysis of the genome sequence of *Pseudomonas* sp. SH-C52 led to the description of a 9-amino acid chlorinated lipopeptide designated thanamycin. Mutation of the biosynthesis genes and experiments with the pure compound revealed that thanamycin was implicated in the biocontrol capacity shown by this strain towards *Sclerotium rolfsii* on groundnut and *R. solani* on sugarbeet (Le *et al.*, 2012; Watrous *et al.*, 2012).

Pathogens or biocontrol agents?

The examples shown in this review reveal that it is not easy to draw a clear distinction between pathogenic and plant beneficial *Pseudomonas* strains. Pathogenic and beneficial *Pseudomonas* strains are interspersed in the same taxonomic groups (Table 7.1) and some of their secondary metabolites can be implicated in both biocontrol and disease. The mushroom pathogen *P. fluorescens* NZI7 appears to be very similar to biocontrol strains such as *P. protegens* Pf-5 and CHA0 and also produces DAPG. The biocontrol strain *Pseudomonas* CMR12a produces sessilin, a CLP implicated in biocontrol, that is very similar to tolaasin, a major virulence factor produced by the mushroom pathogens *P. tolaasii and P. fluorescens* NZI7. The biocontrol strain *Pseudomonas* sp. DF41 produces sclerosin, a compound related to corpeptin, a virulence toxin produced by the tomato pathogen *P. corrugata*. The CLP viscosin has been implicated in causing head rot on broccoli, but is also involved in biocontrol against major fungal pathogens. *P. syringae* pathovars are major pathogens on various plants, but *P. syringae* isolates can also be potent biocontrol agents, some of which have been commercialized. Various *P. aeruginosa* strains with potent biocontrol activity have been isolated from the rhizosphere of plants. These strains produce similar phenazine antibiotics as biocontrol strains belonging to the *P. chlororaphis* group. However, Kumar *et al.* (2013) revealed that the plant-associated *P. aeruginosa* strain PaBP35 is as virulent in animals as clinical *P. aeruginosa* type strains. Taxonomically, *P. aeruginosa* strains of environmental sources cluster with clinical strains of *P. aeruginosa* (Kumar *et al.*, 2013). Alonso *et al.* (1999) likewise showed that clinical and non-clinical *P. aeruginosa* strains are functionally equivalent in traits relevant for their virulence or environmental properties. Kumar *et al.* (2013) conclude that these organisms should not be further developed as biocontrol agents.

Conclusions

Pseudomonas biocontrol strains can be found in the *P. fluorescens*, *P. syringae*, *P. putida* and *P. aeruginosa* group. The *P. fluorescens* group, however, appears to be the most versatile and contains many antibiotic producing biocontrol strains. Taxonomically, the *P. fluorescens* group is diverse and its classification is in need of revision. Some antibiotics found in *Pseudomonas* biocontrol strains, such as phenazines, DAPG and HCN are produced by specific taxonomic groups and appear to be ancestral. Their production is usually tightly regulated and they may have important physiological role besides their antibiotic activity. Phenazines are involved in biofilm formation and can act as alternative electron acceptors and iron-scavenging agents in microaerophilic conditions. DAPG has a signalling role in the interaction of the producing strain with other micro-organisms and the plant root, while HCN can be used as a nitrogen source by certain bacteria. Other antibiotics such as pyoluteorin and pyrrolnitrin, may have been obtained by horizontal gene transfer. Genome sequencing has revealed many orphan biosynthetic gene clusters in the genomes of *Pseudomonas* biocontrol strain. New bioinformatics tools and the genomoisotopic approach have led to the discovery of various new secondary metabolites in these strains. Production of biosurfactants, mainly cyclic lipopeptides, appears to be widespread in the *P. fluorescens* lineage and for many of these compounds biocontrol activity has been demonstrated. Moreover, certain *Pseudomonas* biocontrol strains produce potent insect toxins. Similar secondary metabolites are produced by beneficial and pathogenic *Pseudomonas* strains and the distinction between these strains is not so clear cut which may have implications for the practical use of these organisms. Fluorescent *Pseudomonads* are an amazing group of bacteria and will continue to inspire researchers for many years to come. *Pseudomonas* isolates from tropical environments have been underexplored and may form an untapped reservoir of interesting secondary metabolites.

Acknowledgements

Pseudomonas work in our laboratory is supported by a grant from the Fund for Scientific Research-Flanders (FWO project 3G.002.10N) and by the INTERREG IV programme France-Wallonie-Vlaanderen (Phytobio project). F.E.O. is funded by a Schlumberger Faculty for the Future Fellowship.

Web resources

antiSMASH: www.secondarymetabolites.org/
Validly published names of prokaryotes: www.dsmz.de.
ClusterMine360: http://www.clustermine360.ca/
ClustScan database: http://bioserv.pbf.hr/cms/
DoBISCUIT: www.bio.nite.go.jp/pks/
Genomes online database: www.genomesonline.org.
NORINE database of non-ribosomal peptides: http://bioinfo.lifl.fr/norine/
NP.searcher: http://dna.sherman.lsi.umich.edu/
NRPS predictor: http://www-ab.informatik.uni-tuebingen.de/software/NRPSpredictor.
NRPSsp: www.nrpssp.com/
PseudoMLSA database: www.uib.es/microbiologiaBD/Welcome.html.

References

Andersen, J.B., Koch, B., Nielsen, T.H., Sorensen, D., Hansen, M., Nybroe, O., Christophersen, C., Sorensen, J., Molin, S., and Givskov, M. (2003). Surface motility in *Pseudomonas* sp. DSS73 is required for efficient biological containment of the root-pathogenic microfungi *Rhizoctonia solani* and *Pythium ultimum*. Microbiology *149*, 37–46.

Anjaiah, V., Koedam, N., Nowak-Thompson, B., Loper, J.E., Hofte, M., Tambong, J.T., and Cornelis, P. (1998). Involvement of phenazines and anthranilate in the antagonism of *Pseudomonas aeruginosa* PNA1 and Tn5 derivatives toward *Fusarium* spp. and *Pythium* spp. Mol. Plant Microbe Interact. *11*, 847–854.

Ansari, M.Z., Yadav, G., Gokhale, R.S., and Mohanty, D. (2004). NRPS-PKS: a knowledge-based resource for analysis of NRPS/PKS megasynthases. Nucleic Acids Res. *32,* W405–W413.

Arima, K., Fukuta, A., Imanaka, H., Kousaka, M., and Tamura, G. (1964). Pyrrolnitrin new antibiotic substance produced by *Pseudomonas*. Agric. Biol. Chem. Tokyo *28,* 575–576.

Audenaert, K., Pattery, T., Cornelis, P., and Hofte, M. (2002). Induction of systemic resistance to *Botrytis cinerea* in tomato by *Pseudomonas aeruginosa* 7NSK2: Role of salicylic acid, pyochelin, and pyocyanin. Mol. Plant Microbe Interact. *15,* 1147–1156.

Bachmann, B.O., and Ravel, J. (2009). Methods for *in silico* prediction of microbial polyketide and nonribosomal peptide biosynthetic pathways from DNA sequence data. Methods Enzymol. *458,* 181–217.

Bennasar, A., Mulet, M., Lalucat, J., and Garcia-Valdes, E. (2010). PseudoMLSA: a database for multigenic sequence analysis of *Pseudomonas* species. BMC Microbiol. *10,* 118.

Berry, C., Fernando, W.G.D., Loewen, P.C., and de Kievit, T.R. (2010). Lipopeptides are essential for Pseudomonas sp DF41 biocontrol of *Sclerotinia sclerotiorum*. Biol. Control 55, 211–218.

Berry, C.L., Brassinga, A.K.C., Donald, L.J., Fernando, W.G.D., Loewen, P.C., and de Kievit, T.R. (2012). Chemical and biological characterization of sclerosin, an antifungal lipopeptide. Can. J. Microbiol. *58,* 1027–1034.

Blin, K., Medema, M.H., Kazempour, D., Fischbach, M.A., Breitling, R., Takano, E., and Weber, T. (2013). antiSMASH 2.0-a versatile platform for genome mining of secondary metabolite producers. Nucleic Acids Res. *41,* W204–W212.

Blumer, C., and Haas, D. (2000). Mechanism, regulation, and ecological role of bacterial cyanide biosynthesis. Arch. Microbiol. *173,* 170–177.

Brazelton, J.N., Pfeufer, E.E., Sweat, T.A., Gardener, B.B.M., and Coenen, C. (2008). 2,4-diacetylphloroglucinol alters plant root development. Mol. Plant Microbe Interact. *21,* 1349–1358.

de Bruijn, I., de Kock, M.J.D., Yang, M., de Waard, P., van Beek, T.A., and Raaijmakers, J.M. (2007). Genome-based discovery, structure prediction and functional analysis of cyclic lipopeptide antibiotics in Pseudomonas species. Mol. Microbiol. *63,* 417–428.

Bull, C.T., Stack, J.P., and Smilanick, J.L. (1997). *Pseudomonas syringae* strains ESC-10 and ESC-11 survive in wounds on citrus and control green and blue molds of citrus. Biol. Control 8, 81–88.

Burlinson, P., Studholme, D., Cambray-Young, J., Heavens, D., Rathjen, J., Hodgkin, J., and Preston, G.M. (2013). *Pseudomonas fluorescens* NZI7 repels grazing by C. elegans, a natural predator. ISME J. *7,* 1126–1138.

Caboche, S., Pupin, M., Leclere, V., Fontaine, A., Jacques, P., and Kucherov, G. (2008). NORINE: a database of nonribosomal peptides. Nucleic Acids Res. *36,* D326–D331.

Caldwell, C.J., Hynes, R.K., Boyetchko, S.M., and Korber, D.R. (2012). Colonization and bioherbicidal activity on green foxtail by *Pseudomonas fluorescens* BRG100 in a pesta formulation. Can. J. Microbiol. *58,* 1–9.

Chernin, L., Brandis, A., Ismailov, Z., and Chet, I. (1996). Pyrrolnitrin production by an *Enterobacter agglomerans* strain with a broad spectrum of antagonistic activity towards fungal and bacterial phytopathogens. Curr. Microbiol. 32, 208–212.

Chin-A.-Woeng, T.F.C., Bloemberg, G.V., van der Bij, A.J., van der Drift, K.M.G.F., Schripsema, J., Kroon, B., Scheffer, R.J., Keel, C., Bakker, P.A.H.M., Tichy, H.V., *et al.* (1998). Biocontrol by phenazine-1-carboxamide-producing *Pseudomonas chlororaphis* PCL1391 of tomato root rot caused by Fusarium oxysporum f. sp. radicis-lycopersici. Mol. Plant Microbe Interact. *11,* 1069–1077.

Chin-A.-Woeng, T.F.C., Bloemberg, G.V., and Lugtenberg, B.J.J. (2003). Phenazines and their role in biocontrol by *Pseudomonas* bacteria. New Phytol. *157,* 503–523.

Combes-Meynet, E., Pothier, J.F., Moenne-Loccoz, Y., and Prigent-Combaret, C. (2011). The Pseudomonas secondary metabolite 2,4-diacetylphloroglucinol is a signal inducing rhizoplane expression of *Azospirillum* genes involved in plant-growth promotion. Mol. Plant Microbe Interact. *24,* 271–284.

Conway, K.R., and Boddy, C.N. (2013). ClusterMine360: a database of microbial PKS/NRPS biosynthesis. Nucleic Acids Res. *41,* D402–D407.

Costa, R., van Aarle, I.M., Mendes, R., and van Elsas, J.D. (2009). Genomics of pyrrolnitrin biosynthetic loci: evidence for conservation and whole-operon mobility within Gram-negative bacteria. Environ. Microbiol. *11,* 159–175.

Cox, C.D. (1986). Role of Pyocyanin in the acquisition of iron from transferrin. Infect. Immun. *52,* 263–270.

D'aes, J., De Maeyer, K., Pauwelyn, E., and Hofte, M. (2010). Biosurfactants in plant–Pseudomonas interactions and their importance to biocontrol. Env Microbiol Rep 2, 359–372.

D'aes, J., Gia, K.H.H., De Maeyer, K., Pannecoucque, J., Forrez, I., Ongena, M., Dietrich, L.E.P., Thomashow, L.S., Mavrodi, D.V., and Hofte, M. (2011). Biological control of *Rhizoctonia* root rot on bean by phenazine- and cyclic lipopeptide-producing *Pseudomonas* CMR12a. Phytopathology *101*, 996–1004.

De Jonghe, K., De Dobbelaere, I., Sarrazyn, R., and Hofte, M. (2005). Control of *Phytophthora cryptogea* in the hydroponic forcing of witloof chicory with the rhamnolipid-based biosurfactant formulation PRO1. Plant Pathol. *54*, 219–226.

De Maeyer, K., D'aes, J., Hua, G.K.H., Perneel, M., Vanhaecke, L., Noppe, H., and Höfte, M. (2011). N-Acylhomoserine lactone quorum-sensing signalling in antagonistic phenazine-producing *Pseudomonas* isolates from the red cocoyam rhizosphere. Microbiology *157*, 459–472.

De Vleesschauwer, D., and Hofte, M. (2009). Rhizobacteria-induced systemic resistance. Adv. Bot. Res. *51*, 223–281.

De Vleesschauwer, D., Cornelis, P., and Hofte, M. (2006). Redox-active pyocyanin secreted by *Pseudomonas aeruginosa* 7NSK2 triggers systemic resistance to *Magnaporthe grisea* but enhances *Rhizoctonia solani* susceptibility in rice. Mol. Plant Microbe Interact. *19*, 1406–1419.

Deleij, F.A.A.M., Sutton, E.J., Whipps, J.M., Fenlon, J.S., and Lynch, J.M. (1995). Impact of field release of genetically-modified *Pseudomonas fluorescens* on indigenous microbial populations of wheat. Appl. Environ. Microbiol. *61*, 3443–3453.

Devi, K.K., and Kothamasi, D. (2009). *Pseudomonas fluorescens* CHA0 can kill subterranean termite *Odontotermes obesus* by inhibiting cytochrome c oxidase of the termite respiratory chain. FEMS Microbiol. Lett. *300*, 195–200.

Dietrich, L.E.P., Price-Whelan, A., Petersen, A., Whiteley, M., and Newman, D.K. (2006). The phenazine pyocyanin is a terminal signalling factor in the quorum sensing network of *Pseudomonas aeruginosa*. Mol. Microbiol. *61*, 1308–1321.

ffrench-Constant, R.H., Dowling, A., and Waterfield, N.R. (2007). Insecticidal toxins from *Photorhabdus* bacteria and their potential use in agriculture. Toxicon. *49*, 436–451.

Frapolli, M., Pothier, J.F., Defago, G., and Moenne-Loccoz, Y. (2012). Evolutionary history of synthesis pathway genes for phloroglucinol and cyanide antimicrobials in plant-associated fluorescent pseudomonads. Mol. Phylogenet. Evol. *63*, 877–890.

Frohlich, A., Buddrus-Schiemann, K., Durner, J., Hartmann, A., and von Rad, U. (2012). Response of barley to root colonization by *Pseudomonas* sp. DSMZ 13134 under laboratory, greenhouse, and field conditions. J. Plant. Interact. 7, 1–9.

Ge, Y.H., Huang, X.Q., Wang, S.L., Zhang, X.H., and Xu, Y.Q. (2004). Phenazine-l-carboxylic acid is negatively regulated and pyoluteorin positively regulated by gacA in *Pseudomonas* sp. M18. FEMS Microbiol. Lett. 237, 41–47.

Gross, H., and Loper, J.E. (2009). Genomics of secondary metabolite production by *Pseudomonas* spp. Nat. Prod. Rep. *26*, 1408–1446.

Gross, H., Stockwell, V.O., Henkels, M.D., Nowak-Thompson, B., Loper, J.E., and Gerwick, W.H. (2007). The genomisotopic approach: a systematic method to isolate products of orphan biosynthetic gene clusters. Chem. Biol. *14*, 53–63.

Haas, D., and Defago, G. (2005). Biological control of soil-borne pathogens by fluorescent pseudomonads. Nat. Rev. Microbiol. 3, 307–319.

Henkel, M., Muller, M.M., Kugler, J.H., Lovaglio, R.B., Contiero, J., Syldatk, C., and Hausmann, R. (2012). Rhamnolipids as biosurfactants from renewable resources: concepts for next-generation rhamnolipid production. Process. Biochem. *47*, 1207–1219.

Hernandez, M.E., Kappler, A., and Newman, D.K. (2004). Phenazines and other redox-active antibiotics promote microbial mineral reduction. Appl. Environ. Microbiol. *70*, 921–928.

Hildebrand, P.D., Braun, P.G., McRae, K.B., and Lu, X. (1998). Role of the biosurfactant viscosin in broccoli head rot caused by a pectolytic strain of *Pseudomonas fluorescens*. Can. J. Plant Pathol. *20*, 296–303.

Hill, D.S., Stein, J.I., Torkewitz, N.R., Morse, A.M., Howell, C.R., Pachlatko, J.P., Becker, J.O., and Ligon, J.M. (1994). Cloning of genes involved in the synthesis of pyrrolnitrin from *Pseudomonas fluorescens* and role of Pyrrolnitrin synthesis in biological control of plant disease. Appl. Environ. Microb. *60*, 78–85.

Hofte, M., and Altier, N. (2010). Fluorescent pseudomonads as biocontrol agents for sustainable agricultural systems. Res. Microbiol. *161*, 464–471.

Howell, C.R., and Stipanovic, R.D. (1980). Suppression of *Pythium ultimum*-induced damping-off of cotton seedlings by *Pseudomonas fluorescens* and its antibiotic, pyoluteorin. Phytopathology *70*, 712–715.

Iavicoli, A., Boutet, E., Buchala, A., and Metraux, J.P. (2003). Induced systemic resistance in *Arabidopsis thaliana* in response to root inoculation with *Pseudomonas fluorescens* CHA0. Mol. Plant Microbe Interact. *16*, 851–858.

Ichikawa, N., Sasagawa, M., Yamamoto, M., Komaki, H., Yoshida, Y., Yamazaki, S., and Fujita, N. (2013). DoBISCUIT: a database of secondary metabolite biosynthetic gene clusters. Nucleic Acids Res. *41*, D408–D414.

Imanaka, H., Kousaka, M., Tamura, G., and Arima, K. (1965). Studies on pyrrolnitrin a new antibiotic. 2. Taxonomic studies on pyrrolnitrin-producing strain. J. Antibiot. *18*, 205.

Janisiewicz, W.J., and Korsten, L. (2002). Biological control of postharvest diseases of fruits. Annu. Rev. Phytopathol. *40*, 411–441.

Kamilova, F., Validov, S., Azarova, T., Mulders, I., and Lugtenberg, B. (2005). Enrichment for enhanced competitive plant root tip colonizers selects for a new class of biocontrol bacteria. Environ. Microbiol. *7*, 1809–1817.

Katagiri, K., Tori, K., Kimura, Y., Yoshida, T., Nagasaki, T., and Minato, H. (1967). A new antibiotic. Furanomycin an isoleucine antagonist. J. Med. Chem. *10*, 1149.

Kidarsa, T.A., Goebel, N.C., Zabriskie, T.M., and Loper, J.E. (2011). Phloroglucinol mediates cross-talk between the pyoluteorin and 2,4-diacetylphloroglucinol biosynthetic pathways in *Pseudomonas fluorescens* Pf-5. Mol. Microbiol. *81*, 395–414.

Kruijt, M., Tran, H., and Raaijmakers, J.M. (2009). Functional, genetic and chemical characterization of biosurfactants produced by plant growth-promoting *Pseudomonas putida* 267. J. Appl. Microbiol. *107*, 546–556.

Kuiper, I., Lagendijk, E.L., Pickford, R., Derrick, J.P., Lamers, G.E.M., Thomas-Oates, J.E., Lugtenberg, B.J.J., and Bloemberg, G.V. (2004). Characterization of two *Pseudomonas putida* lipopeptide biosurfactants, putisolvin I and II, which inhibit biofilm formation and break down existing biofilms. Mol. Microbiol. *51*, 97–113.

Kumar, A., Munder, A., Aravind, R., Eapen, S.J., Tummler, B., and Raaijmakers, J.M. (2013). Friend or foe: genetic and functional characterization of plant endophytic *Pseudomonas aeruginosa*. Environ. Microbiol. *15*, 764–779.

Kunz, D.A., Chen, J.L., and Pan, G.L. (1998). Accumulation of alpha-keto acids as essential components in cyanide assimilation by *Pseudomonas fluorescens* NCIMB 11764. Appl. Environ. Microbiol. *64*, 4452–4459.

Lanteigne, C., Gadkar, V.J., Wallon, T., Novinscak, A., and Filion, M. (2012). Production of DAPG and HCN by *Pseudomonas* sp LBUM300 contributes to the biological control of bacterial canker of tomato. Phytopathology *102*, 967–973.

Laursen, J.B., and Nielsen, J. (2004). Phenazine natural products: biosynthesis, synthetic analogues, and biological activity. Chem. Rev. *104*, 1663–1685.

Laville, J., Blumer, C., Von Schroetter, C., Gaia, V., Defago, G., Keel, C., and Haas, D. (1998). Characterization of the hcnABC gene cluster encoding hydrogen cyanide synthase and anaerobic regulation by ANR in the strictly aerobic biocontrol agent *Pseudomonas fluorescens* CHAO. J. Bacteriol. *180*, 3187–3196.

Le, C.N., Kruijt, M., and Raaijmakers, J.M. (2012). Involvement of phenazines and lipopeptides in interactions between *Pseudomonas* species and *Sclerotium rolfsii*, causal agent of stem rot disease on groundnut. J. Appl. Microbiol. *112*, 390–403.

Leong, J. (1986). Siderophores – Their biochemistry and possible role in the biocontrol of plant pathogens. Annu. Rev. Phytopathol. *24*, 187–209.

Li, M.H.T., Ung, P.M.U., Zajkowski, J., Garneau-Tsodikova, S., and Sherman, D.H. (2009). Automated genome mining for natural products. BMC Bioinformatics *10*.

Li, W., Estrada-de los Santos, P., Matthijs, S., Xie, G.L., Busson, R., Cornelis, P., Rozenski, J., and De Mot, R. (2011). Promysalin, a salicylate-containing *Pseudomonas putida* antibiotic, promotes surface colonization and selectively targets other *Pseudomonas*. Chem. Biol. *18*, 1320–1330.

Li, W., Rokni-Zadeh, H., De Vleeschouwer, M., Ghequire, M.G.K., Sinnaeve, D., Xie, G.L., Rozenski, J., Madder, A., Martins, J.C., and De Mot, R. (2013). The antimicrobial compound xantholysin defines a new group of *Pseudomonas* cyclic lipopeptides. PLoS One *8*, e62946.

Ligon, J.M., Hill, D.S., Hammer, P.E., Torkewitz, N.R., Hofmann, D., Kempf, H.J., and van Pee, K.H. (2000). Natural products with antifungal activity from *Pseudomonas* biocontrol bacteria. Pest Manag. Sci. *56*, 688–695.

Liu, H.M., He, Y.J., Jiang, H.X., Peng, H.S., Huang, X.Q., Zhang, X.H., Thomashow, L.S., and Xu, Y.Q. (2007). Characterization of a phenazine-producing strain *Pseudomonas chlororaphis* GP72 with broad-spectrum antifungal activity from green pepper rhizosphere. Curr. Microbiol. *54*, 302–306.

Lo Cantore, P., Lazzaroni, S., Evidente, A., Cafarchia, C., and Iacobellis, N.S. (2003). Production of tolaasin I and WLIP by *Pseudomonas tolaasii* and *P. 'reactans'*, their antimicrobial activity and possible role in the virulence of the pathogens. In *Pseudomonas syringae* Pathovars and Related Pathogens: Biology and Genetics, Iacobellis, N.S., Collmer, A., Hutcheson, S.W., Mansfield, J.W., Morris, C.E., Murillo, J., Schaad, N.W., Stead, D.E., Surico, G., and Ullrich, M.S., eds. (Kluwer Academic Publishers, Dordrecht, The Netherlands), pp. 255–262.

Lo Cantore, P., Lazzaroni, S., Coraiola, M., Dalla Serra, M., Cafarchia, C., Evidente, A., and Iacobellis, N.S. (2006). Biological characterization of white line-inducing principle (WLIP) produced by *Pseudomonas reactans* NCPPB1311. Mol. Plant Microbe Interact. *19*, 1113–1120.

Loper, J.E., Henkels, M.D., Shaffer, B.T., Valeriote, F.A., and Gross, H. (2008). Isolation and identification of rhizoxin analogs from *Pseudomonas fluorescens* Pf-5 by using a genomic mining strategy. Appl. Environ. Microbiol. *74*, 3085–3093.

Loper, J.E., Hassan, K.A., Mavrodi, D.V., Davis, E.W., Lim, C.K., Shaffer, B.T., Elbourne, L.D.H., Stockwell, V.O., Hartney, S.L., Breakwell, K., *et al.* (2012). Comparative genomics of plant-associated *Pseudomonas* spp.: insights into diversity and inheritance of traits involved in multitrophic interactions. PLoS Genet. *8*, e1002784.

Lu, J.S., Huang, X.Q., Li, K., Li, S.N., Zhang, M.Y., Wang, Y., Jiang, H.X., and Xu, Y.Q. (2009). LysR family transcriptional regulator PqsR as repressor of pyoluteorin biosynthesis and activator of phenazine-1-carboxylic acid biosynthesis in *Pseudomonas* sp M18. J. Biotechnol. *143*, 1–9.

Lugtenberg, B., and Kamilova, F. (2009). Plant-growth-promoting Rhizobacteria. Annu. Rev. Microbiol. *63*, 541–556.

Lugtenberg, B.J.J., Malfanova, N., Kamilova, F., and Berg, G. (2013). Microbial control of plant root diseases. In Molecular Microbial Ecology of the Rhizosphere, F.J. De Bruijn, ed. (Wiley Blackwell, Singapore), pp. 575–586.

Luque-Almagro, V.M., Merchan, F., Blasco, R., Igeno, M.I., Martinez-Luque, M., Moreno-Vivian, C., Castillo, F., and Roldan, M.D. (2011). Cyanide degradation by *Pseudomonas pseudoalcaligenes* CECT5344 involves a malate: quinone oxidoreductase and an associated cyanide-insensitive electron transfer chain. Microbiology *157*, 739–746.

Maddula, V.S.R.K., Zhang, Z., Pierson, E.A., and Pierson, L.S. (2006). Quorum sensing and phenazines are involved in biofilm formation by *Pseudomonas chlororaphis (aureofaciens)* strain 30-84. Microb. Ecol. *52*, 289–301.

Maddula, V.S.R.K., Pierson, E.A., and Pierson, L.S. (2008). Altering the ratio of phenazines in *Pseudomonas chlororaphis (aureofaciens)* strain 30–84: Effects on biofilm formation and pathogen inhibition. J. Bacteriol. *190*, 2759–2766.

Mark, G.L., Morrissey, J.P., Higgins, P., and O'Gara, F. (2006). Molecular-based strategies to exploit *Pseudomonas* biocontrol strains for environmental biotechnology applications. FEMS Microbiol. Ecol. *56*, 167–177.

Maurhofer, M., Keel, C., Haas, D., and Defago, G. (1994). Pyoluteorin production by *Pseudomonas fluorescens* strain CHA0 is involved in the suppression of Pythium damping-off of cress but not of cucumber. Eur. J. Plant Pathol. *100*, 221–232.

Mavrodi, D.V., Ksenzenko, V.N., Bonsall, R.F., Cook, R.J., Boronin, A.M., and Thomashow, L.S. (1998). A seven-gene locus for synthesis is of phenazine-1-carboxylic acid by *Pseudomonas fluorescens* 2-79. J. Bacteriol. 180, 2541–2548.

Mavrodi, D.V., Blankenfeldt, W., and Thomashow, L.S. (2006). Phenazine compounds in fluorescent *Pseudomonas* spp. biosynthesis and regulation. Annu. Rev. Phytopathol. *44*, 417–445.

Mavrodi, D.V., Peever, T.L., Mavrodi, O.V., Parejko, J.A., Raaijmakers, J.M., Lemanceau, P., Mazurier, S., Heide, L., Blankenfeldt, W., Weller, D.M., *et al.* (2010). Diversity and evolution of the phenazine biosynthesis pathway. Appl. Environ. Microbiol. *76*, 866–879.

Mavrodi, D.V., Parejko, J.A., Mavrodi, O.V., Kwak, Y.S., Weller, D.M., Blankenfeldt, W., and Thomashow, L.S. (2013). Recent insights into the diversity, frequency and ecological roles of phenazines in fluorescent *Pseudomonas* spp. Environ. Microbiol. *15*, 675–686.

Mavrodi, O.V., Mavrodi, D.V., Parejko, J.A., Thomashow, L.S., and Weller, D.M. (2012). Irrigation differentially impacts populations of indigenous antibiotic-producing *Pseudomonas* spp. in the rhizosphere of wheat. Appl. Environ. Microbiol. *78*, 3214–3220.

Mazzola, M., Cook, R.J., Thomashow, L.S., Weller, D.M., and Pierson, L.S. (1992). Contribution of phenazine antibiotic biosynthesis to the ecological competence of fluorescent Pseudomonads in soil habitats. Appl. Environ. Microbiol. *58*, 2616–2624.

Mazzola, M., Zhao, X., Cohen, M.F., and Raaijmakers, J.M. (2007). Cyclic lipopeptide surfactant production by *Pseudomonas fluorescens* SS101 is not required for suppression of complex Pythium spp. populations. Phytopathology 97, 1348–1355.

Medema, M.H., Blin, K., Cimermancic, P., de Jager, V., Zakrzewski, P., Fischbach, M.A., Weber, T., Takano, E., and Breitling, R. (2011). antiSMASH: rapid identification, annotation and analysis of secondary metabolite biosynthesis gene clusters in bacterial and fungal genome sequences. Nucleic Acids Res. *39*, W339–W346.

Mehnaz, S., Saleem, R.S.Z., Yameen, B., Pianet, I., Schnakenburg, G., Pietraszkiewicz, H., Valeriote, F., Josten, M., Sahl, H.G., Franzblau, S.G., *et al.* (2013). Lahorenoic Acids A-C, ortho-dialkyl-substituted aromatic acids from the biocontrol strain *Pseudomonas aurantiaca* PB-St2. J. Nat. Prod. *76*, 135–141.

Mendes, R., Kruijt, M., de Bruijn, I., Dekkers, E., van der Voort, M., Schneider, J.H.M., Piceno, Y.M., DeSantis, T.Z., Andersen, G.L., Bakker, P.A.H.M., *et al.* (2011). Deciphering the rhizosphere microbiome for disease-suppressive bacteria. Science *332*, 1097–1100.

Meyer, J.M. (2000). Pyoverdines: pigments, siderophores and potential taxonomic markers of fluorescent *Pseudomonas* species. Arch Microbiol *174*, 135–142.

Mortishire-Smith, R.J., Nutkins, J.C., Packman, L.C., Brodey, C.L., Rainey, P.B., Johnstone, K., and Williams, D.H. (1991). Determination of the structure of an extracellular peptide produced by the mushroom saprotroph *Pseudomonas reactans*. Tetrahedron *47*, 3645–3654.

Mozes-Koch, R., Gover, O., Tanne, E., Peretz, Y., Maori, E., Chernin, L., and Sela, I. (2012). Expression of an entire bacterial operon in plants. Plant Physiol. *158*, 1883–1892.

Mulet, M., Lalucat, J., and Garcia-Valdes, E. (2010). DNA sequence-based analysis of the *Pseudomonas* species. Environ. Microbiol. *12*, 1513–1530.

Nielsen, M.N., Sorensen, J., Fels, J., and Pedersen, H.C. (1998). Secondary metabolite- and endochitinase-dependent antagonism toward plant-pathogenic microfungi of *Pseudomonas fluorescens* isolates from sugar beet rhizosphere. Appl. Environ. Microbiol. *64*, 3563–3569.

Nielsen, T.H., Christophersen, C., Anthoni, U., and Sorensen, J. (1999). Viscosinamide, a new cyclic depsipeptide with surfactant and antifungal properties produced by *Pseudomonas fluorescens* DR54. J. Appl. Microbiol. *87*, 80–90.

Nielsen, T.H., Thrane, C., Christophersen, C., Anthoni, U., and Sorensen, J. (2000). Structure, production characteristics and fungal antagonism of tensin – a new antifungal cyclic lipopeptide from *Pseudomonas fluorescens* strain 96.578. J. Appl. Microbiol. *89*, 992–1001.

Nielsen, T.H., Sorensen, D., Tobiasen, C., Andersen, J.B., Christophersen, C., Givskov, M., and Sorensen, J. (2002). Antibiotic and biosurfactant properties of cyclic lipopeptides produced by fluorescent *Pseudomonas* spp. from the sugar beet rhizosphere. Appl. Environ. Microbiol. *68*, 3416–3423.

Nyfeler, R., and Ackermann, P. (1992). Phenylpyrroles, a new class of agricultural fungicides related to the natural antibiotic pyrrolnitrin. ACS Sym. Ser. *504*, 395–404.

Ovadis, M., Liu, X.G., Gavriel, S., Ismailov, Z., Chet, I., and Chernin, L. (2004). The global regulator genes from biocontrol strain *Serratia plymuthica* IC1270: cloning, sequencing, and functional studies. J. Bacteriol. *186*, 4986–4993.

Pagani, I., Liolios, K., Jansson, J., Chen, I.M.A., Smirnova, T., Nosrat, B., Markowitz, V.M., and Kyrpides, N.C. (2012). The Genomes OnLine Database (GOLD) v.4: status of genomic and metagenomic projects and their associated metadata. Nucleic Acids Res. *40*, D571–D579.

Palleroni, N., Kunisawa, R., Contopou.R., and Doudorof. M. (1973). Nucleic acid homologies in genus *Pseudomonas*. Int. J. Syst. Bacteriol. 23, 333–339.

Parejko, J.A., Mavrodi, D.V., Mavrodi, O.V., Weller, D.M., and Thomashow, L.S. (2012). Population structure and diversity of phenazine-1-carboxylic acid producing fluorescent *Pseudomonas* spp. from dryland cereal fields of central Washington state (USA). Microb. Ecol. *64*, 226–241.

Parejko, J.A., Mavrodi, D.V., Mavrodi, O.V., Weller, D.M., and Thomashow, L.S. (2013). Taxonomy and Distribution of phenazine-producing Pseudomonas spp. in the dryland agroecosystem of the inland Pacific Northwest, United States. Appl. Environ. Microbiol. *79*, 3887–3891.

Park, J.Y., Oh, S.A., Anderson, A.J., Neiswender, J., Kim, J.C., and Kim, Y.C. (2011). Production of the antifungal compounds phenazine and pyrrolnitrin from *Pseudomonas chlororaphis* O6 is differentially regulated by glucose. Lett. Appl. Microbiol. *52*, 532–537.

Partida-Martinez, L.P., and Hertweck, C. (2007). A gene cluster encoding rhizoxin biosynthesis in "*Burkholderia rhizoxina*", the bacterial endosymbiont of the fungus *Rhizopus microsporus*. Chembiochem. *8*, 41–45.

Paulsen, I.T., Press, C.M., Ravel, J., Kobayashi, D.Y., Myers, G.S.A., Mavrodi, D.V., DeBoy, R.T., Seshadri, R., Ren, Q.H., Madupu, R., *et al.* (2005). Complete genome sequence of the plant commensal *Pseudomonas fluorescens* Pf-5. Nat. Biotechnol. *23*, 873–878.

Pechy-Tarr, M., Bruck, D.J., Maurhofer, M., Fischer, E., Vogne, C., Henkels, M.D., Donahue, K.M., Grunder, J., Loper, J.E., and Keel, C. (2008). Molecular analysis of a novel gene cluster encoding an insect toxin in plant-associated strains of *Pseudomonas fluorescens*. Environ. Microbiol. *10*, 2368–2386.

Pedras, M.S.C., Ismail, N., Quail, J.W., and Boyetchko, S.M. (2003). Structure, chemistry, and biological activity of pseudophomins A and B, new cyclic lipodepsipeptides isolated from the biocontrol bacterium *Pseudomonas fluorescens*. Phytochemistry *62*, 1105–1114.

Peix, A., Ramirez-Bahena, M.H., and Velazquez, E. (2009). Historical evolution and current status of the taxonomy of genus *Pseudomonas*. Infect. Genet. Evol. *9*, 1132–1147.

Perneel, M., Heyrman, J., Adiobo, A., De Maeyer, K., Raaijmakers, J.M., De Vos, P., and Hofte, M. (2007). Characterization of CMR5c and CMR12a, novel fluorescent *Pseudomonas* strains from the cocoyam rhizosphere with biocontrol activity. J. Appl. Microbiol. *103*, 1007–1020.

Perneel, M., D'hondt, L., De Maeyer, K., Adiobo, A., Rabaey, K., and Hofte, M. (2008). Phenazines and biosurfactants interact in the biological control of soil-borne diseases caused by *Pythium* spp. Environ. Microbiol. *10*, 778–788.

Phillips, D.A., Fox, T.C., King, M.D., Bhuvaneswari, T.V., and Teuber, L.R. (2004). Microbial products trigger amino acid exudation from plant roots. Plant Physiol. *136*, 2887–2894.

Pierson, L.S., and Pierson, E.A. (2010). Metabolism and function of phenazines in bacteria: impacts on the behavior of bacteria in the environment and biotechnological processes. Appl. Microbiol. Biotechnol. *86*, 1659–1670.

Pierson, L.S., and Thomashow, L.S. (1992). Cloning and heterologous expression of the phenazine biosynthetic locus from *Pseudomonas aureofaciens* 30–84. Mol. Plant Microbe Interact. *5*, 330–339.

Price-Whelan, A., Dietrich, L.E.P., and Newman, D.K. (2007). Pyocyanin alters redox Homeostasis and carbon flux through central metabolic pathways in *Pseudomonas aeruginosa* PA14. J. Bacteriol. *189*, 6372–6381.

Prieto, C., Garcia-Estrada, C., Lorenzana, D., and Martin, J.F. (2012). NRPSsp: non-ribosomal peptide synthase substrate predictor. Bioinformatics *28*, 426–427.

Quail, J.W., Ismail, N., Pedras, M.S.C., and Boyetchko, S.M. (2002). Pseudophomins A and B, a class of cyclic lipodepsipeptides isolated from a *Pseudomonas* species. Acta Crystallogr. C *58*, o268–o271.

Raaijmakers, J.M., de Bruijn, I., and de Kock, M.J.D. (2006). Cyclic lipopeptide production by plant-associated *Pseudomonas* spp.: diversity, activity, biosynthesis, and regulation. Mol. Plant Microbe Interact. *19*, 699–710.

Raaijmakers, J.M., de Bruijn, I., Nybroe, O., and Ongena, M. (2010). Natural functions of lipopeptides from *Bacillus* and *Pseudomonas*: more than surfactants and antibiotics. FEMS Microbiol. Rev. *34*, 1037–1062.

Ramette, A., Frapolli, M., Fischer-Le Saux, M., Gruffaz, C., Meyer, J.M., Defago, G., Sutra, L., and Moenne-Loccoz, Y. (2011). *Pseudomonas protegens* sp nov., widespread plant-protecting bacteria producing the biocontrol compounds 2,4-diacetylphloroglucinol and pyoluteorin. Syst. Appl. Microbiol. *34*, 180–188.

Ramos, I., Dietrich, L.E.P., Price-Whelan, A., and Newman, D.K. (2010). Phenazines affect biofilm formation by *Pseudomonas aeruginosa* in similar ways at various scales. Res. Microbiol. *161*, 187–191.

Rausch, C., Weber, T., Kohlbacher, O., Wohlleben, W., and Huson, D.H. (2005). Specificity prediction of adenylation domains in nonribosomal peptide synthetases (NRPS) using transductive support vector machines (TSVMs). Nucleic Acids Res. *33*, 5799–5808.

Redondo-Nieto, M., Barret, M., Morrissey, J., Germaine, K., Martinez-Granero, F., Barahona, E., Navazo, A., Sanchez-Contreras, M., Moynihan, J.A., Muriel, C., *et al.* (2013). Genome sequence reveals that *Pseudomonas fluorescens* F113 possesses a large and diverse array of systems for rhizosphere function and host interaction. BMC Genomics *14*, 54.

Rokni-Zadeh, H., Li, W., Sanchez-Rodriguez, A., Sinnaeve, D., Rozenski, J., Martins, J.C., and De Mot, R. (2012). Genetic and functional characterization of cyclic lipopeptide white-line-inducing principle (WLIP) production by rice rhizosphere isolate *Pseudomonas putida* RW10S2. Appl. Environ. Microbiol. *78*, 4826–4834.

Roongsawang, N., Washio, K., and Morikawa, M. (2011). Diversity of nonribosomal peptide synthetases involved in the biosynthesis of lipopeptide biosurfactants. Int. J. Mol. Sci. *12*, 141–172.

Ruffner, B., Pechy-Tarr, M., Ryffel, F., Hoegger, P., Obrist, C., Rindlisbacher, A., Keel, C., and Maurhofer, M. (2013). Oral insecticidal activity of plant-associated pseudomonads. Environ. Microbiol. *15*, 751–763.

Ryall, B., Mitchell, H., Mossialos, D., and Williams, H.D. (2009). Cyanogenesis by the entomopathogenic bacterium *Pseudomonas entomophila*. Lett. Appl. Microbiol. *49*, 131–135.

Sanchez, L., Courteaux, B., Hubert, J., Kauffmann, S., Renault, J.H., Clement, C., Baillieul, F., and Dorey, S. (2012). Rhamnolipids elicit defense responses and induce disease resistance against biotrophic, hemibiotrophic, and necrotrophic pathogens that require different signaling pathways in *Arabidopsis* and highlight a central role for salicylic acid. Plant Physiol. *160*, 1630–1641.

Seaton, S.C., Silby, M.W., and Levy, S.B. (2013). Pleiotropic effects of GacA on *Pseudomonas fluorescens* Pf0-1 *in vitro* and in soil. Appl. Environ. Microbiol. *79*, 5405–5410.

Selin, C., Fernando, W.G.D., and de Kievit, T. (2012). The PhzI/PhzR quorum-sensing system is required for pyrrolnitrin and phenazine production, and exhibits cross-regulation with RpoS in *Pseudomonas chlororaphis* PA23. Microbiology *158*, 896–907.

Sharma, A., Jansen, R., Nimtz, M., Johri, B.N., and Wray, V. (2007). Rhamnolipids from the rhizosphere bacterium *Pseudomonas* sp GRP(3) that reduces damping-off disease in chilli and tomato nurseries. J. Nat. Prod. *70*, 941–947.

Silby, M.W., Cerdeno-Tarraga, A.M., Vernikos, G.S., Giddens, S.R., Jackson, R.W., Preston, G.M., Zhang, X.X., Moon, C.D., Gehrig, S.M., Godfrey, S.A.C., *et al.* (2009). Genomic and genetic analyses of diversity and plant interactions of *Pseudomonas fluorescens*. Genome Biol. *10*, R51.

Sinnaeve, D., Michaux, C., Van Hemel, J., Vandenkerckhove, J., Peys, E., Borremans, F.A.M., Sas, B., Wouters, J., and Martins, J.C. (2009). Structure and X-ray conformation of pseudodesmins A and B, two new cyclic lipodepsipeptides from *Pseudomonas* bacteria. Tetrahedron *65*, 4173–4181.

Skirvin, R.M., Kohler, E., Steiner, H., Ayers, D., Laughnan, A., Norton, M.A., and Warmund, M. (2000). The use of genetically engineered bacteria to control frost on strawberries and potatoes. Whatever happened to all of that research? Sci. Hortic. *84*, 179–189.

Soberon-Chavez, G., Lepine, F., and Deziel, E. (2005). Production of rhamnolipids by *Pseudomonas aeruginosa*. Appl. Microbiol. Biotechnol. *68*, 718–725.

Soler-Rivas, C., Arpin, N., Olivier, J.M., and Wichers, H.J. (1999). WLIP, a lipodepsipeptide of *Pseudomonas 'reactans'*, as inhibitor of the symptoms of the brown blotch disease of *Agaricus bisporus*. J. Appl. Microbiol. *86*, 635–641.

Sorensen, D., Nielsen, T.H., Christophersen, C., Sorensen, J., and Gajhede, M. (2001). Cyclic lipoundecapeptide amphisin from *Pseudomonas* sp strain DSS73. Acta Crystallogr. C *57*, 1123–1124.

de Souza, J.T., de Boer, M., de Waard, P., van Beek, T.A., and Raaijmakers, J.M. (2003). Biochemical, genetic, and zoosporicidal properties of cyclic lipopeptide surfactants produced by *Pseudomonas fluorescens*. Appl. Environ. Microbiol. *69*, 7161–7172.

Starcevic, A., Zucko, J., Simunkovic, J., Long, P.F., Cullum, J., and Hranueli, D. (2008). ClustScan: an integrated program package for the semi-automatic annotation of modular biosynthetic gene clusters and in silico prediction of novel chemical structures. Nucleic Acids Res. *36*, 6882–6892.

Thomashow, L.S., and Weller, D.M. (1988). Role of a phenazine antibiotic from *Pseudomonas fluorescens* in biological control of *Gaeumannomyces graminis* var. *tritici*. J. Bacteriol. *170*, 3499–3508.

Thrane, C., Olsson, S., Nielsen, T.H., and Sorensen, J. (1999). Vital fluorescent stains for detection of stress in *Pythium ultimum* and *Rhizoctonia solani* challenged with viscosinamide from *Pseudomonas fluorescens* DR54. FEMS Microbiol. Ecol. *30*, 11–23.

Thrane, C., Nielsen, T.H., Nielsen, M.N., Sorensen, J., and Olsson, S. (2000). Viscosinamide-producing *Pseudomonas fluorescens* DR54 exerts a biocontrol effect on *Pythium ultimum* in sugar beet rhizosphere. FEMS Microbiol. Ecol. *33*, 139–146.

Thrane, C., Nielsen, M.N., Sorensen, J., and Olsson, S. (2001). *Pseudomonas fluorescens* DR54 reduces sclerotia formation, biomass development, and disease incidence of *Rhizoctonia solani* causing damping-off in sugar beet. Microb. Ecol. *42*, 438–445.

Tombolini, R., van der Gaag, D.J., Gerhardson, B., and Jansson, J.K. (1999). Colonization pattern of the biocontrol strain *Pseudomonas chlororaphis* MA 342 on barley seeds visualized by using green fluorescent protein. Appl. Environ. Microbiol. *65*, 3674–3680.

Toribio, J., Escalante, A.E., and Soberon-Chavez, G. (2010). Rhamnolipids: production in bacteria other than *Pseudomonas aeruginosa*. Eur. J. Lipid Sci. Technol. *112*, 1082–1087.

Tran, H., Ficke, A., Asiimwe, T., Hofte, M., and Raaijmakers, J.M. (2007). Role of the cyclic lipopeptide massetolide A in biological control of *Phytophthora infestans* and in colonization of tomato plants by *Pseudomonas fluorescens*. New Phytol. *175*, 731–742.

Trippe, K., McPhail, K., Armstrong, D., Azevedo, M., and Banowetz, G. (2013). *Pseudomonas fluorescens* SBW25 produces furanomycin, a non-proteinogenic amino acid with selective antimicrobial properties. BMC Microbiol. *13*, 111.

Troppens, D.M., Moynihan, J.A., Barret, M., O'Gara, F., and Morrissey, J. (2013). Genetics and evolution of 2,4-diacetylphloroglucinon synthesis in *Pseudomonas fluorescens*. In Molecular Microbial Ecology of the Rhizosphere, F.J. De Bruijn, ed. (Wiley Blackwell, Singapore), pp. 593–605.

Vallet-Gely, I., Novikov, A., Augusto, L., Liehl, P., Bolbach, G., Pechy-Tarr, M., Cosson, P., Keel, C., Caroff, M., and Lemaitre, B. (2010). Association of hemolytic activity of *Pseudomonas entomophila*, a versatile soil bacterium, with cyclic lipopeptide production. Appl. Environ. Microbiol. *76*, 910–921.

Varnier, A.L., Sanchez, L., Vatsa, P., Boudesocque, L., Garcia-Brugger, A., Rabenoelina, F., Sorokin, A., Renault, J.H., Kauffmann, S., Pugin, A., *et al.* (2009). Bacterial rhamnolipids are novel MAMPs conferring resistance to *Botrytis cinerea* in grapevine. Plant Cell Environ. *32*, 178–193.

Viallard, V., Poirier, I., Cournoyer, B., Haurat, J., Wiebkin, S., Ophel-Keller, K., and Balandreau, J. (1998). *Burkholderia graminis* sp, nov., a rhizospheric *Burkholderia* species, and reassessment of [*Pseudomonas*] *phenazinium*, [*Pseudomonas*] *pyrrocinia* and [*Pseudomonas*] *glathei* as *Burkholderia*. Int. J. Syst. Bacteriol. *48*, 549–563.

Vodovar, N., Vinals, M., Liehl, P., Basset, A., Degrouard, J., Spellman, P., Boccard, F., and Lemaitre, B. (2005). *Drosophila* host defense after oral infection by an entomopathogenic *Pseudomonas* species. Proc. Natl. Acad. Sci. U.S.A. *102*, 11414–11419.

Vodovar, N., Vallenet, D., Cruveiller, S., Rouy, Z., Barbe, V., Acosta, C., Cattolico, L., Jubin, C., Lajus, A., Segurens, B., *et al.* (2006). Complete genome sequence of the entomopathogenic and metabolically versatile soil bacterium *Pseudomonas entomophila*. Nat. Biotechnol. *24*, 673–679.

Voisard, C., Keel, C., Haas, D., and Defago, G. (1989). Cyanide production by *Pseudomonas fluorescens* helps suppress black root rot of Tobacco under gnotobiotic conditions. EMBO J. *8*, 351–358.

Volksch, B., and May, R. (2001). Biological control of *Pseudomonas syringae* pv. *glycinea* by epiphytic bacteria under field conditions. Microb. Ecol. *41*, 132–139.

Wang, Y., and Newman, D.K. (2008). Redox reactions of phenazine antibiotics with ferric (hydr)oxides and molecular oxygen. Environ. Sci. Technol. *42*, 2380–2386.

Wang, Y., Kern, S.E., and Newman, D.K. (2010). Endogenous phenazine antibiotics promote anaerobic survival of *Pseudomonas aeruginosa* via extracellular electron transfer. J. Bacteriol. *192*, 365–369.

Wang, Y., Wilks, J.C., Danhorn, T., Ramos, I., Croal, L., and Newman, D.K. (2011). Phenazine-1-carboxylic acid promotes bacterial biofilm development via ferrous iron acquisition. J. Bacteriol. *193*, 3606–3617.

Watrous, J., Roach, P., Alexandrov, T., Heath, B.S., Yang, J.Y., Kersten, R.D., van der Voort, M., Pogliano, K., Gross, H., Raaijmakers, J.M., *et al.* (2012). Mass spectral molecular networking of living microbial colonies. Proc. Natl. Acad. Sci. U.S.A. *109*, E1743–E1752.

Weller, D.M., and Cook, R.J. (1983). Suppression of take-all of wheat by seed treatments with fluorescent Pseudomonads. Phytopathology *73*, 463–469.

Weller, D.M., Landa, B.B., Mavrodi, O.V., Schroeder, K.L., De La Fuente, L., Bankhead, S.B., Molar, R.A., Bonsall, R.F., Mavrodi, D.V., and Thomashow, L.S. (2007). Role of 2,4-diacetylphloroglucinol-producing fluorescent *Pseudomonas* spp. in the defense of plant roots. Plant Biol. *9*, 4–20.

Weller, D.M., Mavrodi, D.V., van Pelt, J.A., Pieterse, C.M.J., van Loon, L.C., and Bakker, P.A.H.M. (2012). Induced systemic resistance in *Arabidopsis thaliana* against *Pseudomonas syringae* pv. *tomato* by 2,4-diacetylphloroglucinol-producing *Pseudomonas fluorescens*. Phytopathology *102*, 403–412.

Wensing, A., Braun, S.D., Buttner, P., Expert, D., Volksch, B., Ullrich, M.S., and Weingart, H. (2010). Impact of siderophore production by *Pseudomonas syringae* pv. *syringae* 22d/93 on epiphytic fitness and biocontrol activity against *Pseudomonas syringae* pv. *glycinea* 1a/96. Appl. Environ. Microbiol. *76*, 2704–2711.

Wong, D.T., and Airall, J.M. (1970). Mode of action of antifungal agents – Effect of pyrrolnitrin on mitochondrial electron transport. J. Antibiot. 23, 55.

The Potential Role of Bacteriophages in Shaping Plant–Bacteria Interactions

8

Britt Koskella and Tiffany B. Taylor

Abstract

Bacteriophages are ubiquitous and abundant across ecosystems. For bacteria living in or on plant hosts, phages are also likely to have significant impact on the plant–bacteria interaction. Potential mechanisms shaping this interaction include lysing of bacterial cells, horizontal gene transfer among bacterial genomes, and the alteration of bacterial phenotype. In this chapter we first review the current understanding of bacteria–phage ecology and evolution gained both from experiments performed under controlled, laboratory conditions and from observations in natural settings. We then discuss the extent to which this understanding can help inform predictions about the effects of phages on plant-associated bacterial populations and communities, and highlight key studies setting out to test these predictions. Finally, we explore the potential use of phages as biocontrol agents for plant pathogens and highlight the need for more data regarding the evolutionary implications of these treatments and the ecological impact of phages on bacterial community-level processes.

Introduction

The success and spread of a bacterium colonizing a plant depends on the specific adaptations of the bacterium, the response of the host, local ecological interactions (including other microbial species), and the abiotic environment. Attempting to explain interactions among plants and plant-associated bacteria without taking into account each of these potentially interrelating factors is likely to fall short, but attempts to incorporate all factors simultaneously can lead to an unruly and relatively useless model. There is therefore a trade-off when studying plant–microbe interactions, and a first step should be to assess the role that each factor is likely to play in shaping the interaction in an attempt to identify those factors that will contribute more substantially than others. We argue that for plant-associated bacteria, including plant pathogens, interactions with bacteriophage viruses (phages) are likely to play an important role in determining the fitness, competitive success, and interaction with the plant host for colonizing bacteria. This is both because phages can act as obligate killers of their host cells, thereby reducing densities and selecting for resistant bacterial types with potentially altered phenotype, and also because phages are capable of integrating into the genome of their bacterial host and of moving bacterial genes among unrelated bacterial strains and species.

Phages have been shown to change the competitive hierarchy of bacterial strains/species (Bohannan and Lenski, 2000; Koskella *et al.*, 2012), mediate horizontal gene transfer among bacteria (Kidambi *et al.*, 1994; Canchaya *et al.*, 2003), and alter bacterial virulence on hosts (i.e. harm caused to the host; Levin and Bull, 1996; Boyd and Brüssow, 2002; Munro *et al.*, 2003; Rice *et al.*, 2008). Given the paucity of research on phages in the phyllosphere and rhizosphere to date, there is great need for targeted studies both on natural interactions and using experimental approaches. The overarching goals should enhance our understanding of how phages alter the evolution of their bacterial hosts, affect the assembly and stability of microbial communities within eukaryotic hosts, and ultimately influence disease – from the level of individual susceptibility to spread across populations.

Fortunately, we have a head start in this demanding task; there is a large body of literature focusing on bacteria–phage interactions in the laboratory and in non-plant environments. By identifying which previously established patterns and processes are also likely to hold true for bacteria–phage interactions within the plant host, we can make and clear predictions about the role phages might play in shaping plant-associated bacterial populations and communities, and subsequently plant fitness. For example, phages might target and select against colonizing pathogenic bacteria or promote the growth of commensals either directly, by shaping the evolution of the microbe, or indirectly, by altering apparent competition in the bacterial community (Bohannan and Lenski, 2000). Better characterization of these processes will advance our understanding of natural plant–microbe interactions, but will also allow for a critical evaluation of the application of phages as control agents of bacterial plant pathogens (i.e. phage therapy). In this chapter, we first introduce the life history, evolution and ecology of phages, then review the current evidence regarding the role phages play in shaping the phenotype of their bacterial hosts, and finally discuss the potential direct and indirect impact of phages on plant-associated bacteria, with a particular focus on their utility as biocontrol agents.

Overview of bacteria–phage interactions

Phages have an obligate association with their bacterial hosts, as they are unable to reproduce outside of host cells. Initial infection of bacterial cells occurs when a phage attaches to the cell (so-called 'adsorption') and injects its own genome into that of the host cell. Adsorption requires phage recognition of bacterial receptors that are most often located on the cell wall, although some phages recognize and attach to flagella or pili (Samuel *et al.*, 1999; Mattick, 2002). This recognition is made by the phage tail fibres and resembles a 'lock-and-key' association, although the exact underlying mechanisms remain poorly understood (Labrie *et al.*, 2010). Once inside the host cell, phages either hijack the bacterial cell's metabolic machinery to synthesize phage-encoded enzymes and structural components (if they are 'lytic' phages) or else integrate their own genomes into that of their host (if they are 'temperate' phages). Lytic phages then replicate and mature within their host cell and eventually burst the cell open releasing the next generation of viral particles into the environment, facilitating transmission to new host cells. Temperate phages, on the other hand, remain in the host genome as so-called 'prophages' and are replicated along with the host cell during division; they retain the ability to enter the lytic cycle, however, and often do so under stressful conditions (Choi *et al.*, 2010). It has been suggested that temperate phages will only be tolerated in the genome if they provide some kind of selective benefit to the bacterium – such as virulence genes that will enhance the fitness of a pathogen (Rankin *et al.*, 2011). A third type of phage, 'filamentous' phages, infect and replicate within the cell in a similar way to lytic phages but do not require cell lysis in order to release into the environment (Rakonjac *et al.*, 2011). For phages in the lytic cycle, the time to cell lysis after initial phage infection will depend on host density, adsorption rate, the genetic details of the interaction, and phage genome size (Bull *et al.*, 2004; Shao and Wang, 2008). Both theoretical and empirical examination of the optimal time to lysis suggest a trade-off between lysis time and burst size, i.e. the number of phage particles released, such that phages with an intermediate burst size have the highest fitness (Wang, 2006).

Although good estimates of phage abundance remain scarce, there is evidence that they can outnumber their bacterial hosts 10:1 in many ecosystems (Suttle, 2005; Srinivasiah *et al.*, 2008; Rodriguez-Valera *et al.*, 2009). This alone suggests that phages are likely important players in microbial communities, as the presence of phages necessarily implies the presence of susceptible bacteria (in the case of lytic phages) or the presence of lysogens (in the case of temperate phages). Although phages often have negligible direct effects on bacterial density due to the rapid emergence of resistant bacterial strains (Lenski and Levin, 1985; Waterbury and Valois, 1993; Middelboe *et al.*, 2001; but see Harcombe and Bull, 2005), they can have large effects on bacterial population dynamics (Pantastico-Caldas *et al.*, 1992), diversity (Rodriguez-Valera *et al.*, 2009), sensitivity to antibiotics (Hagens *et al.*, 2006; Jalasvuori *et al.*, 2011), and bacterial fitness (Bohannan *et al.*, 2002). The life history of a phage is a key determinant of the ecological and evolutionary effects it will have on the host population. For example, lytic phages have been shown to select against common bacterial types [the so-called 'kill the winner hypothesis' (Weinbauer and Rassoulzadegan, 2004)] and to select for bacterial resistance (Lenski and Levin, 1985; Buckling and Rainey, 2002). Temperate phages, on the other hand, can in fact drive the spread of bacterial cells harboring them if they confer an advantage to the hosts that they infect (Bossi *et al.*, 2003). The effect of a given phage type on the population of its bacterial host is therefore a function of the underlying genetics of the interaction, the life history of the phage, and the ecological setting in which the interaction occurs.

Phage specificity

Any given phage is only capable of infecting a subset of the bacteria found in the same environment, and the subset it is able to infect is termed the phage's 'host range.' A phage's host range is a function of its ability to recognize and bind to receptors on the surface of each bacterial cell (adsorption), as well as its ability to escape detection and silencing or degradation once inside the cell (Labrie *et al.*, 2010). Over coevolutionary time, therefore, the host range will be shaped both by phage adaptation to specific receptors and by evolved host defence. For lytic bacteriophages, fitness will depend on the rate of encounter with susceptible host cells in the environment, and will be impacted as a function of population and community diversity, dispersal, and exposure to the harsh conditions outside of a cell. It seems clear, therefore, that those phages with a larger host range should be at an inherent advantage. Indeed, experimental evolution studies suggest that phages evolve towards host specialization (i.e. a more narrow host range) when host populations are held constant over time, but evolve towards a more broad host range when host populations are allowed to coevolve (Hall *et al.*, 2011; Poullain *et al.*, 2007).

In natural populations, phage host range falls along a continuum from highly specific phages, capable of infecting only a narrow subset of bacterial strains of a given species, to very general phages, capable of infecting bacteria spanning genera (Holmfeldt *et al.*, 2007;

Hyman and Abedon, 2010; Flores et al., 2013; Koskella and Meaden, 2013). A recent meta-analysis of the statistical structure of bacteria–phage interactions, primarily from lab-based studies, shows that the infection networks are non-random and are typically nested (Flores *et al.*, 2011), where highly resistant bacteria are primarily infected by more 'generalist' phages. There exist very few datasets examining the specificity of natural phages on bacterial hosts from the same environment in which they were found, but the studies that have characterized host range in this way seem more likely to find a signature of modularity, where one subset of phages are capable of infecting an independent subset of bacterial hosts than another. For example, analysis of 215 phages on 286 potential hosts sampled from geographically separated sites in the Atlantic Ocean found evidence for strong modularity that is driven, at least in part, by geographic spread (Flores *et al.*, 2013). Similarly, an interaction network of phages and bacteria isolated from the horse chestnut phyllosphere shows clear evidence of modularity with very little nestedness (Koskella and Meaden, 2013). Overall, it is likely that phage host range will be shaped both by the relatedness among potential bacterial hosts and the previous coevolutionary history of the bacteria and phage, which is likely to be driven in part by spatial proximity.

The evolution of bacterial resistance

Bacteria have evolved a number of defence mechanisms against invading phages, including those that prevent phage adsorption, degrade phage DNA inside the cell, and initiate cell death upon infection (Fig. 8.1). To prevent adsorption, bacteria can lose or alter the target receptor, produce an extracellular matrix of polysaccharides, or produce competitive inhibitors that bind to the phage attachment site (reviewed by Labrie *et al.*, 2010). Phages can counter-adapt by altering their tail fibres used for attachment, by producing enzymes to degrade the matrix, or altering their attachment sites. Once phages have successfully injected their genome into the host cell, bacteria can still respond either using a restriction-modification system, whereby recognized phage DNA is degraded by restriction enzymes, or through the CRISPR-Cas system. Clustered regularly interspaced short palindromic repeats (CRISPR) regions and the CRISPR-associated (*cas*) genes are present in nearly half of all bacterial species (Jore *et al.*, 2012), and allow the bacteria to integrate short segments of phage DNA into spacers within its own genome which allows the bacterium to subsequently cleave newly invading phages at these sites. Phages can avoid having their DNA degraded via these mechanisms through loss of the site recognized by the bacteria's endonuclease, by acquiring a bacterial methylase gene, or via the recently discovered anti-CRISPR genes (Bondy-Denomy *et al.*, 2012). This 'arms race' of defence and counter-defence has led to a great diversity of mechanisms, as outlined above, and to polymorphisms within and among bacterial and phage populations (Bohannan and Lenski, 2000). The evolutionary arms race is likely to follow different trajectories among different interacting bacteria and phage populations. This can lead both to phages that are specialized on only a subset of bacterial hosts (Koskella and Meaden, 2013) and to phage 'local adaptation', whereby phages are better able to infect bacterial hosts from their own local environment (Vos *et al.*, 2009; Koskella *et al.*, 2011a). We explore both of these ideas in further detail below.

Resistance of bacteria to bacteriophages has been associated with substantial fitness costs (Lenski, 1988a; Bohannan *et al.*, 1999), including an increased cost of deleterious mutations (Buckling *et al.*, 2006) and decreased competitive ability (Brockhurst *et al.*, 2005; Lennon *et al.*, 2007; Quance and Travisano, 2009). Given the ubiquity of bacteria–phage interactions,

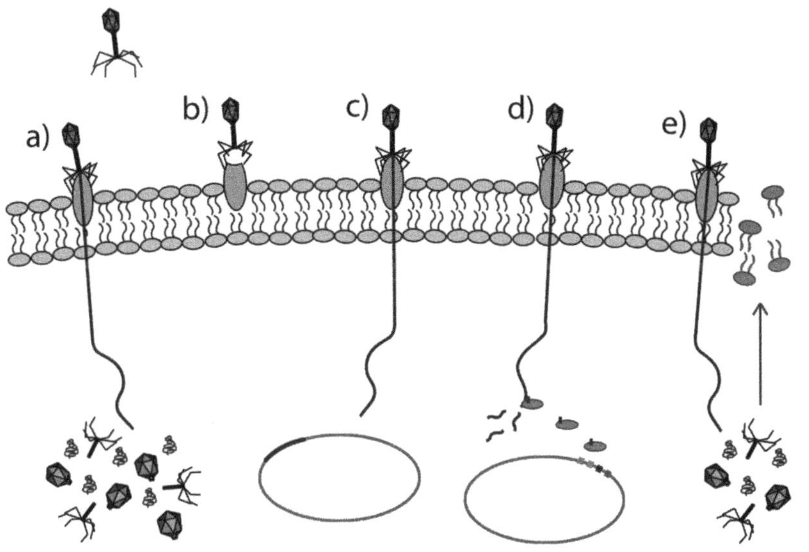

Figure 8.1 Bacteria–phage interactions are successful if (a) the phage is successfully able to bind to the bacterial cell via receptor-recognition, inject its RNA or DNA into the host cell and replicate using host cell machinery. (b) Bacteria can evolve resistance to a given phage by loss or alteration of the target receptor. (c) Upon successful infection, some phage genomes are integrated into the genome of their host, in which case they can replicate concurrently with their host and transmit vertically through a bacterial lineage (these are referred to as temperate phages). However, bacteria have evolved a number of systems for degrading phage DNA upon successful binding. (d) For example, in some cases small segments of the phage genome are integrated into the host genome (inserted within spacers of the CRISPR region of the genome) and these segments are used to block future replication of that phage lineage. (e) Finally, many bacterial species have mechanisms that lead to cell death, for example by degrading the cell wall, upon successful phage infection, thereby protecting neighbouring cells from further infection. As discussed in the text, phages are capable of overcoming each of these resistance mechanisms via a diverse set of counter-adaptation. A colour version of this figure is located in the plate section at the back of the book.

and their key role in all ecosystems, these costs are likely to have important ecological consequences (Bohannan and Lenski, 2000; Fuhrman and Schwalbach, 2003). In particular, because they are obligate killers lytic bacteriophages have the potential to impose strong selection on host populations. The receptors to which phages bind are often important for bacterial function (Whitchurch and Mattick, 1994; Brockhurst *et al.*, 2005). Bacteriophages commonly exploit bacterial surface motility appendages (flagella and pili). For example, flagellatropic phages are known to reversibly bind to helical grooves on the bacterial flagellum and use the rotation of the flagellum to spiral towards the cell surface (Samuel *et al.*, 1999), and similarly, pilus-specific phages will attach and fuse their membrane with that of the bacterial cell during pili retraction (Romantschuk and Bamford, 1985; Mattick, 2002). Resistance can therefore be gained by the loss or alteration of these structures, which will undoubtedly have consequences for bacterial motility (Koskella *et al.*, 2011a). Phage-resistant mutants often show defective flagella that are unable to rotate (Icho and Iino, 1978), and abnormal unpiliated or hyperpiliated bacteria may arise to prevent phage attachment (Bradley, 1980; Mattick, 2002; Brockhurst *et al.*, 2005).

Reduced motility function is likely to have important implications for bacterial fitness in both pathogenic and non-pathogenic bacteria (Korber *et al.*, 1994). For example, immotile mutants of the opportunistic animal pathogen *Pseudomonas aeruginosa* show reduced infectivity on human hosts and impaired biofilm formation (Drake and Montie, 1988; O'Toole and Kolter, 1998). Similarly, epiphytic, non-motile strains of the plant pathogen *Pseudomonas syringae* were found to have reduced fitness and competitive ability compared with more motile strains, especially under conditions of environmental stress (Haefele and Lindow, 1987). Indeed, motility is a key component of fitness for bacteria in the plant phyllosphere and is necessary for successful pathogenicity, as bacteria colonizing leaf surfaces are better able to invade the leaf interior through the stomata if they are motile (Panopoulos and Schroth, 1974; Beattie and Lindow, 1999; Melotto *et al.*, 2006). Bacteria that have evolved resistance to phages using lipopolysaccharides (LPS) as receptors have also been seen to exhibit reduced motility (across twitching, swarming and swimming). Furthermore, a microbial population that evolved resistance to a heterogeneous phage population consisting of both LPS and pili-specific phages, similarly showed decreased motility across all motility types (Hosseinidoust *et al.*, 2013).

Bacteria–phage coevolution

As bacteria evolve resistance to their phages, they will impose strong selection on phage populations to overcome this resistance given the obligate association phages have with their hosts. This bacterial adaptation and phage reciprocal adaptation can lead to long bouts of coevolution and drive divergence among populations (Buckling and Rainey, 2002). The relatively shorter generation times and larger populations sizes of phages compared to their bacterial host leads to the prediction that phages should be more 'locally adapted' to their host population than the hosts are to the phages (Gandon and Michalakis, 2002). In other words, phages are predicted to be more infective to their local bacterial hosts compared to hosts from other populations. Indeed, Vos *et al.* (2009) found evidence to support this prediction across natural phage-bacteria populations separated by mere centimetres in the soil. Koskella *et al.* (2011b) also found a strong signature of phage local adaptation, but this time at a much larger scale. They compared phages collected from multiple leaves of eight horse chestnut trees and found that phages were no more or less infective to bacteria from other leaves within the same tree, but were much more infective to bacterial hosts from the same tree relative to those from neighbouring trees. Understanding these patterns of local adaptation can reveal how coevolutionary interactions drive adaptive differences among populations (Thompson, 1999), and studies of natural populations need to be conducted across a range of environments to determine the influence of other abiotic and biotic factors on the pattern observed.

It is also important to consider the different coevolutionary dynamics that are predicted when phages infect their hosts in isolation, compared with when multiple phage clones infect host cells simultaneously (known as 'coinfection'). The ubiquity and abundance of phages in many habitats, including soil (Vos *et al.*, 2010), water (Suttle, 2005; Maranger and Bird, 1995) and the phyllosphere (Koskella *et al.*, 2011b) mean coinfection is likely to be a common phenomenon. Experimental evidence suggests that phages are able to recognize whether they are the sole invader of a host cell, or whether they are sharing the host with another phage type (Leggett *et al.*, 2012). This suggests that phages have the potential to interact with each other, and adjust their behaviour accordingly. The evolutionary outcome, in terms of how quickly host cells are lysed, will depend on how the phages interact and how selection is acting on that interaction (Brown *et al.*, 2002). For example, if the interaction is antagonistic the phages might interfere with one another by increasing their own relative frequency compared to other phages within the host. The consequence of this is that the reproducing phages should burst the host cells more quickly in order to thwart the competitor's opportunity to maximize replication (Chao *et al.*, 2000). Alternatively, phages might work synergistically and act prudently, i.e. kill the host more slowly in order to maximize overall yield (Chao *et al.*, 2000). Experimental evidence obtained by Leggett *et al.* (2012) suggests that when lytic phages are allowed to coevolve by forcing coinfection of a host, they will kill the host more rapidly than they would under conditions of single infection. Moreover, this response is 'plastic', such that a phage infecting a host cell singly will maximize viral particle load by delaying burst time compared to one that is coinfecting a host. This strategy supports both predictions made by Chao *et al.* (2000) and allows the phage to maximize its fitness based on the infection conditions it is under.

Coinfection among temperate phages is also common (e.g. Ramirez *et al.*, 1999), and bacteria can accumulate multiple prophages that will lie dormant in the host genome until lysis is induced (Chen *et al.*, 2006). This creates opportunity for conditions of coinfection by multiple and diverse phage strains, and moreover, the potential for conflict between the phage strains, because entry of a new phage into the lytic cycle will destroy any resident prophages along with the host (Berngruber *et al.*, 2010). Competition between coinfecting temperate phages was studied experimentally by Refardt (2011), who found that the most competitive phages normally 'win' by producing more viral particles (i.e. the phage that replicates fastest and bursts the cell open before the competitor phage has had time to excise from the genome and replicate is at a clear advantage). Interestingly, Refardt (2011) also found the order of arrival of the temperate phage was very important in determining competitive outcome, and some combinations of phages would cooperate more readily and behave more prudently than others. These novel and highly informative experiments suggest that the social life of phages is surprisingly complex given their relatively small genome. Thus, to truly understand the coevolutionary dynamics of phages and bacteria we not only have to understand how phages interact and coevolve with their host, but also how they interact and coevolve with one other.

The role phages play in maintaining bacterial diversity

Many microorganisms on the planet cannot be successfully cultured in the laboratory. However, the recent rise of metagenomics allows us to explore the vast microbial diversity that exists in natural communities (Riesenfeld *et al.*, 2004), the underlying viral diversity within and across these communities, and the dynamics of each over time. Phages have the potential to directly influence microbial populations [e.g. by regulating bacterial densities (Levin *et al.*, 2013; Wei *et al.*, 2011)] and microbial communities [e.g. by altering the competitive hierarchy among species (Bohannan and Lenski, 2000)], but we have very little evidence quantifying the magnitude and importance of any such effects. The 'kill the winner' hypothesis posits that phages will target those bacterial species with the highest growth rate in a given environment and therefore offer a competitive advantage to more rare species (a form of negative frequency-dependence), thereby increasing diversity by enabling coexistence of microbial species (Weinbauer and Rassoulzadegan, 2004). Moreover, if dominant bacterial species utilize antagonistic mechanisms to control competitor densities (e.g. via production of bacteriocins), the reduction/elimination of the dominant species will have an indirect impact on the population dynamics of competitor species. In the case where two or more bacteria are susceptible to the same phage, the more dominant bacterial species can out compete the other indirectly by increasing the overall density of the phage environment [this process is termed 'apparent competition' (Holt, 1977)]. Finally, phenotypic changes associated with resistance can influence key life-history traits, such as bacterial colonization success or quorum sensing ability, which can make an otherwise dominant bacterial species less competitive, thereby increasing the probability of coexistance among species.

Phage-mediated horizontal gene transfer

Whole-genome sequencing has revealed the enormous extent of horizontal gene transfer (HGT) in natural communities, indicating that gene transfer among organisms is not limited to laboratories but a constitutive factor in diversification (Weinbauer and Rassoulzadegan, 2003; Smillie *et al.*, 2011). Phages play a key role in the evolution of bacterial genomes as they provide a mechanism for HGT among bacterial hosts. The interchange of genes among bacterial strains and species holds the potential to improve bacterial fitness, including changing virulence and pathogenicity (Jackson *et al.*, 2011). The process of transduction, i.e. phage-mediated transfer of bacterial DNA from one bacterial cell to another, can occur either when prophages excise from the host genome and take along adjoining regions of the genome when infecting new cells (so-called specialized transduction), or when lytic phages repackage random segments of bacterial DNA, along with their own, into their capsid before lysing the host cell (so-called general transduction). Certain regions of the bacterial genome are adapted to aid the integration of prophages (Julio *et al.*, 2000) resulting in consistency of integration sites, thus strengthening coevolutionary relationships between bacteria and prophage genome structures (Bobay *et al.*, 2012). The potential for phage-mediated transfer of virulence, toxins and antibiotic resistant genes are of particular concern to human and agricultural health, and this has led to intensive research and alarming evidence of phage-mediated transfer of pathogenicity islands (Fineran *et al.*, 2009), as well as transfer of antibiotic resistance genes (Fard *et al.*, 2011). Moreover, HGT can occur across distantly related species (Chen and Novick, 2009), and even among highly conserved genes (Beumer and Robinson, 2005). Thus it seems, in order to understand the potential consequences of bacteriophage-mediate transduction, we must better understand the extent of bacteriophage specificity between strains and between species.

Interactions among plants, bacteria, and their phages

As discussed throughout the chapter so far, phages are known to affect the ecology and evolution of bacterial host populations, shape bacterial community dynamics, alter bacterial phenotypes, and play an important role in shaping the evolution of bacterial genomes. For the remainder of this chapter we will examine the potential importance of these processes in shaping interactions among bacteria and their plant hosts. Plant-associated bacteria face strong and often conflicting selection pressures to adapt to their plant host environments, compete against other microbes, and evolve resistance against lytic phages. For plant pathogens, many adaptations that confer an advantage for infecting the plant host also carry a potential cost for resisting phages, and vice versa (Table 8.1). Thus, there are a number of *a priori* predictions regarding how phage-mediated selection might influence the bacteria's ability to infect the host plant and/or alter the infectivity of the bacterial pathogen across plant host genotypes. In addition, many bacterial traits which are likely to be affected by phage-mediated selection may be similarly important for fitness of more mutualistic bacterial species. For example, the attachment of soil bacteria to plant roots is a function of receptors on the bacterial surface, including lipopolysaccharides (LPS; Rodríguez-Navarro *et al.*, 2007), that are known attachment sites for phages. These potential trade-offs would likely generate selection mosaics across space, as the relative strength of the plant immune response and the prevalence of phages will differ among plant host genotypes and species, as well as across space (Thompson, 1999). These contrasting selection pressures could lead to differential patterns of plant susceptibility to pathogens, shifts in microbial diversity or community composition among neighbouring plants, or differing levels of pathogenicity among populations of the same bacterial species across plant hosts.

Influence of plant environment on bacteria–phage interactions

The plant host environment is a complex and relatively harsh habitat for bacteria and phages, both of which are sensitive to UV exposure and fluctuations in temperature and humidity (Lindow and Brandl, 2003). This is especially true of the phyllosphere relative to the rhizosphere, and indeed there is some evidence for higher phage abundance below ground than in aerial parts of plants; phages infecting the causal agent of fire blight, *Erwinia amylovora*, were more readily isolated from soil surrounding fruit trees than from above-ground cuttings

Table 8.1 Key bacterial traits and their known/proposed roles in bacterial virulence on plant hosts and bacterial resistance against phage. Arrows indicate the direction in which a given bacterial trait is expected to influence bacterial virulence to the plant or resistance against phage (i.e. increased bacterial growth rate is expected to increase the virulence of the bacterial pathogen on its plant host but has also been found to be negatively correlated with resistance to phages given the commonly observed fitness costs of resistance)

Bacterial trait	Effect on virulence	Effect on resistance	Reference
Increased growth rate	↑	↓	Hirano and Upper (2000), Quiñones *et al.* (2005), Poisot *et al.* (2013)
Decreased cell size	?	↑	O'Flynn *et al.* (2004)
Exopolysaccharide production	↑	↓	Romanenko *et al.* (2003), Hammad (1998), Scanlan and Buckling (2011)
Quorum sensing ability	↑	↑	Von Bodman *et al.* (2003), Quiñones *et al.* (2005), Høyland-Kroghsbo *et al.* (2013)
Type III secretion system	↑	?	Collmer *et al.* (2000), Chisholm *et al.* (2006)
Type IV pili expression	↑	↑	Roine *et al.* (1998), Burdman *et al.* (2011), Taguchi and Ichinose (2011), Kim *et al.* (2012)
Density/function of LPS receptors	↑↓	↑	Evans *et al.* (2010), Chowdhury and Jagannadham (2013), Garcia-Doval and van Raaij (2012)
Increased motility	↑	↑↓	Hrabak and Willis (1992), Rich *et al.* (1992), Melotto et al. (2006), Koskella *et al.* (2011a)
Pyocyanin production	↑	↑	Rahme *et al.* (1997), Hosseinidoust *et al.* (2013)

(Gill *et al.*, 2003). Many bacteria respond to the plant environment by altering expression of surface receptors, changing growth patterns, or altering cell size. This 'phase variation' is likely to influence the interactions among bacterial cells and phages in the plant environment, for example by increasing or decreasing density of receptors to which phages attach or altering cell motility and therefore the rate of encounter between bacteria and phages (Labrie *et al.*, 2010). Thus we have good reason to predict both that phages will influence the interaction among bacteria and plants, and also that the plant environment will alter the interaction between bacteria and phages. If this holds true, results from bacteria–phage interactions in the laboratory might not be informative for predicting the role of phages in the phyllosphere or rhizosphere.

Theoretical and experimental studies in the laboratory have shown that lytic phages can play a role in driving bacterial population diversity due to the strong selective pressures that phages impose on their hosts (Buckling and Rainey, 2002; Weinbauer and Rassoulzadegan, 2004; Rodriguez-Valera *et al.*, 2009; Williams, 2013). Whether these results translate to natural populations will depend on both the prevalence and host range of phages living in/on plants. There is building evidence that phages are highly prevalent in the plant phyllosphere and rhizosphere (Koskella *et al.*, 2011b; Romero-Suarez *et al.*, 2012). For example, estimates of viral abundance from the soil of rice paddies found approximately 3×10^8 viruses from 1 g of soil (Kim *et al.*, 2008), and estimates from the rhizosphere of mature sugar beet plants were found to be 1.5×10^7 per gram of soil (Ashelford *et al.*, 2003). The phages found in these environments span the continuum of broad to narrow host ranges, some infecting multiple genera and others being restricted to only a few isolates of a single species (Koskella and Meaden, 2013). Since plants interact with an eclectic community of bacteria, from mutualists to antagonists, the effect of shared phages on competitive interactions among bacterial species within these communities will depend on how specific phages are to their bacterial hosts. A study of 26 phages isolated from walnut trees and capable of infecting *Xanthomonas arboricola* pv. *juglandis*, the causal agent of walnut blight disease, found clear genotype specificity of the phages for a subset of bacterial isolates all belonging to the same pathovar. However, although some of the phages had very broad host range, spanning multiple bacterial species, no single phage was capable of infecting all genotypes of *X. arboricola* pv. *juglandis* (Romero-Suarez *et al.*, 2012). Similarly, a study of 14 phages isolated from the leaves of horse chestnut trees found that nearly ⅓ were capable of infecting both *Pseudomonas* and *Erwinia* species and that most were capable of infecting multiple pathovars within *P. syringae* (Koskella and Meaden, 2013). The observed specificity of those phages tested to date (i.e. the fact that no phages infect all bacterial hosts), suggests the clear potential of phages for shaping the diversity and composition of microbial communities. Although no clear evidence for negative frequency-dependent, or Kill-the-Winner type dynamics of bacteria and their phages yet exist for plant-associated bacterial communities, there is evidence of temporal succession of phages in the rhizosphere of sugar beets over the course of nine months (Ashelford *et al.*, 1999) and evidence of phage-mediated evolution of bacterial communities within horse chestnut tree leaves over the course of a single growing season (Koskella, 2013).

Potential role of phages in shaping plant–microbe interactions

Selection on populations of plant-associated microbes by lytic phages will likely lead to the evolution of resistance, but as outlined earlier in the chapter, this resistance often comes at a cost to bacterial fitness. There are a number of studies that find phenotypic changes of key bacterial fitness traits in response to phage-mediated selection, such as bacterial motility (Brockhurst *et al.*, 2005; Hosseinidoust *et al.*, 2013), biofilm production (Rice *et al.*, 2008), production of extracellular compounds (such as pyocyanin, pyoverdin, and rhamnolipid; Hosseinidoust *et al.*, 2013), and resistance to antibiotics (Hosseinidoust *et al.*, 2013; Zhang and Buckling, 2012). These changes can cascade up to affect plant fitness if the function being altered is one directly influencing the interaction between the plant and bacterium. For example, phage-resistant mutants of the nitrogen-fixing bacteria, *Rhizobium trifolii*, tended to be less efficient at fixing nitrogen than the ancestral strains (Kleczkowska, 1950). Similarly, phage-resistant mutants of *Bradyrhizobium japonicum* were found to be defective in nodulation (Stacey *et al.*, 1982). In both cases the decreased bacterial function could lead to decreased plant fitness, especially under more stressful environmental conditions. On the other hand, if phage resistance carries a fitness cost for plant-pathogenic bacteria this could lead to positive effects on plant fitness. As outlined in Table 8.1, there are a number of possible mechanisms for phage-mediated decreased virulence of bacteria. However the current evidence for this effect is equivocal at best; phage-resistant isolates of the broad host range phytopathogen, *Erwinia carotovora,* were no more or less virulent to tobacco plants despite decreased LPS expression (Pirhonen *et al.*, 1988) and phage-resistant strains of the causal agent of crown gall, *Agrobacterium tumefaciens,* showed no change in tumour-inducing virulence relative to sensitive strains (Bradley *et al.*, 1984). It is therefore important to systematically check, rather than assume, the true cost of resistance in a natural system, as the evolutionary outcome of phage–bacteria interactions will depend on the biotic environment in which they occur – something that laboratory experiments cannot simulate. For example, Koskella *et al.* (2011a) surveyed 80 natural bacterial isolates from horse chestnut leaves against a phage cocktail made from locally associated phages in order to test for costs of evolved resistance in terms of reduced bacterial motility. Surprisingly, we found no clear association, and even some suggestion of a positive association, between phage resistance and bacterial motility for pseudomonads from the phyllosphere (Fig. 8.2). While this does not rule out the importance of phage-imposed selection on the evolution of bacterial motility, it does suggest that its effect is unpredictable in a natural setting, and is likely to depend on the natural phage community and bacterial environment.

The majority of studies that aim to determine population dynamics of interacting phage and bacterial populations conduct experiments in artificial chemostats. While this has provided vital and detailed information regarding the ecological and microevolutionary processes that shape this interaction (e.g. Horne, 1970; Chao *et al.*, 1977; Levin *et al.*, 1977; Levin and Lenski, 1983, 1985; Lenski, 1984, 1988b; Stewart and Levin, 1984; Lenski and Levin, 1985; Buckling and Rainey, 2002), this optimized, artificial environment does not capture the spatial and temporal heterogeneity of natural environments. For example, the physical and biochemical properties of the soil have been found to be important in determining the types of phages that can exist and the encounter rate with host populations (Williamson *et al.*, 2003). These effects can be both beneficial and deleterious, for example acidic soils have been found to steadily and permanently inactivate free phage particles (Pantastico-Caldas *et al.*, 1992); however, clay particles provide a highly active surface for bacteria–phage interactions. Clay particles have positively charged molecules (cations) covering their surface, which bind to negatively charged molecules on the surface of the phages and bacteria, as well as to proteins and other nutrient molecules that provide nutrients for

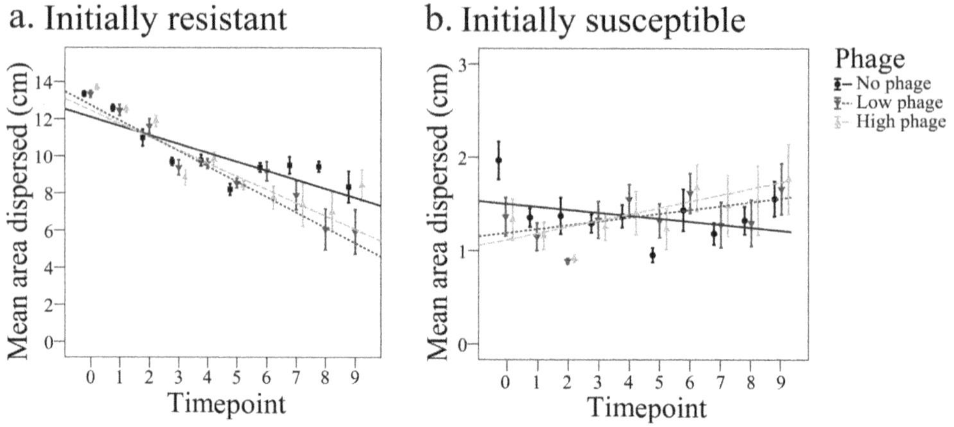

Figure 8.2 Unpublished results from Koskella *et al.* (2011a) where a subset of natural *Pseudomonas syringae* isolates from the horse chestnut phyllosphere were used to search for a causal link between bacterial motility and resistance to phages. Replicate bacterial lines were selected for dispersal (over ten serial transfers) in the presence or absence of naturally co-occurring phages. Change in dispersal area over the course of the experiment for (a) the initially susceptible and (b) the initially resistant isolates are shown (note different scales for dispersal on parts (a) and (b)). Bacterial lines were evolved in the absence (closed circles and a solid regression line) or presence of phage either at a low titre (dark grey inverted triangles and dotted regression line) or high titre (light grey triangles and dashed regression line). Error bars represent ± 1 SEM. We observed that initially resistant lines showed an overall decrease in dispersal ($F=297.30$, $P<0.001$), whereas initially susceptible isolates, showed less net change over time ($F=4.684$, $P=0.031$). In addition, resistant lines evolved in the presence of phages showed a moderately greater decrease in dispersal overtime ($F_{2288}=6.641$, $P=0.002$), whereas initially susceptible lines evolved in the presence of phages showed a moderate increase in dispersal overtime ($F_{2290}=6.260$, $P=0.002$). This mirrored our observations from a natural population survey that showed a positive relationship between bacterial motility and resistance to phages.

host growth. This brings host and phage populations together, prolonging infectivity and increasing the adsorption rate of lytic phages (Marsh and Wellington, 1994). Within a soil environment host and phage populations appear to be spatially structured (Vos *et al.*, 2009). The low, and fluctuating frequencies of resistant bacterial hosts that have been observed in soil populations suggests that the habitat structure may impede encounter rate of phages with viable hosts (Chibani-Chennoufi *et al.*, 2004). In contrast, within a marine environment it is thought that the circulation of nutrients and hosts keeps both bacterial and phage populations high (Yong Yu *et al.*, 2011).

Comparisons between marine and soil phage communities have highlighted some other important differences between the environments. For example, it seems that despite vast differences in compositions and geographical location, phage abundance in the soil is fairly stable, at around $0.8–3.9\times10^9$ viral particles, and this value seems strangely independent of overall bacterial abundance (Williamson *et al.*, 2003, 2007). In comparison, phage abundance in marine habitats (from deep sea to fresh water marshes) can vary up to 2000-fold (Srinivasiah *et al.*, 2008). Furthermore, turnover of phage populations seems slower within the soil compared to marine environments, with a greater tendency towards temperate phage lifecycles. This effect is likely due to the increased heterogeneity of the soil environment, resulting in reduced phage–bacteria interactions, and periods of stationary phase for starved bacterial populations. When bacterial populations are not in the active growth phase the associated phage population must also remain dormant, either as free phage particles in the environment or else as prophages integrated within the host genome. The reduced encounter rate and intermittent growth phases of the host population means a temperate phage lifecycle is a more effective strategy, because phages can still be maintained in stationary host populations and phages are protected from degradation in the environment.

In comparison, when encounter rate is higher due to either high level of population mixing (such as in the marine environment and in a chemostat) or in host populations in active growth phase, a lytic lifecycle might be favoured as this will maximize phage reproduction. Surveys of natural populations are in line with these predictions and suggest that virulent, lytic phages are more common within aquatic ecosystems, whereas temperate phages are more common among soil populations (Marsh and Wellington, 1994). Lysogeny can have a large impact on the host population in the form of lysogenic conversion. This is a process whereby expression of genes on the phage genomes alter host phenotype, often resulting in superinfection immunity. Here, the expression of the phage-encoded repressor protein(s) inhibits further infections by the same phage type or different phages with specific binding sites for the repressor protein. This will have consequences for phage specificity and can cause prophage-induced changes to cell surface receptors (Marsh and Wellington, 1994). These prophages are also likely to have consequences for the frequency of genetic exchange between bacterial populations, which can influence fitness associated traits in plant-associated bacteria. Finally, prophages within the bacterial genome can directly impact the bacterial phenotype in a way that alters interactions with the plant host. A recent analysis of 37 plant pathogen genomes found that 5169 genes were potentially of phage origin, and many of these genes are involved in strain differentiation or the induction of disease symptoms (Varani *et al.*, 2013).

Transduction among plant-associated bacteria

Phages have been shown to play an important role in mediating the transfer of genes among plant-associated bacteria via both specific and general transduction. Phage-mediated gene transfer can occur in the soil (Herron, 2004), and experimental studies have shown that plant leaves can act as a viable surface for which transduction can take place among microorganisms (Kidambi *et al.*, 1994). All these factors indicate that phage-mediated gene transfer has the potential to be an important factor in natural microbial communities. Evidence for generalized transduction by a phage capable of infecting over 25% of tested strains of *Erwinia carotovora* subsp. *atroseptica*, the causative agent of soft rot and blackleg potato diseases, suggests a frequency of transduction between 2.0×10^{-6} to 1.8×10^{-8} (Toth *et al.*, 1997). The on-going challenge is to find direct evidence for its impact on important plant-associated microbes. The consequences of phage-mediated gene transfer will greatly depend on whether the specificity of phages is limited to within species, or if it also occurs between species (Torsvik *et al.*, 2002). The growing body of molecular data will shed light on such questions and on common phage-mediated elements that are important for the fitness of both the bacterium, and as such, its plant host. Phages are just one form of genetic mobile element that can mediate HGT, but given their ubiquitous nature are likely to act as potent force of gene flow among bacterial genomes. This area of research is particularly important when considering the use of genetically modified microorganisms for agricultural practices, where genetically modified genes could generate very different outcomes if they were able to jump between hosts.

Phages as biocontrol agents

Given that lytic phages can lyse their host cells, reach high abundance in the plant environment, and are non-toxic, they hold the potential to act as biocontrol agents against plant-pathogenic bacterial populations. Phages have shown promise for use in the treatment of bacterial spot of peach (Civerolo and Kiel, 1969), soft rot of calla lily (Ravensdale *et al.*, 2007), palea browning of rice (Azegami, 2013), bacterial spot disease on tomato (Balogh *et al.*, 2003), citrus canker and citrus bacterial spot of grapefruit (Balogh *et al.*, 2008), rotting and blackleg in potato (Adriaenssens, 2012) and fire blight of apple (Schnabel *et al.*, 1998). Early uses of phage therapy to control plant-pathogenic bacterial populations generated a great deal of interest but fell out of widespread favour given the relatively minor effects observed, inconsistent success rate, and ability of bacteria to evolve resistance. However, modern applications of phage therapy take a more sophisticated approach by considering the timing of application (Iriarte et al., 2007), the use of engineered phages to decrease the possibility of bacterial resistance (Lu and Collins, 2009), the use of carrier bacterial species that maintain phage densities in the plant (Azegami, 2013), the combination of multiple phages to increase the host range of the treatment (Schnabel et al., 1998), the combination of phages and chemical control (Borah *et al.*, 2000), and the costs associated with bacterial resistance in treatment design (Hall *et al.*, 2012; Koskella *et al.*, 2012). In addition, the use of phage-encoded enzymes, such as lysins, to break down bacterial cell walls without necessitating the introduction of self-replicating phages into the environment is increasingly discussed as a control strategy (O'Flaherty *et al.*, 2009).

The timing of phage application is proving to be a particularly critical factor for success of treatment, both in terms of application relative to timing of infection and in terms of phage survival in the environment. Phage activity and ability to control bacterial spot has been shown to negatively correlate with UV exposure (Iriarte *et al.*, 2007), and in line with this, phage treatment was found to be more effective when applied in the evening than in the morning (Balogh *et al.*, 2003). The success of phage in controlling bacterial densities can also depend on whether phage is applied prior to, during, or post-infection. For example, control of *Ralstonia solanacearum* by pre-treatment of tomato plants with phages led to a complete reduction of bacterial wilt symptoms relative to control plants (Fujiwara *et al.*, 2011). Similarly, pre-treatment of peach foliage with phages reduced bacterial spot symptoms, while application post-infection had little effect on disease development (Civerolo and Kiel, 1969). Finally, treatment of diseased potato tubers with phages showed evidence of decreased disease symptoms, both in the plants treated and the offspring of these plants (McKenna *et al.*, 2001). Another factor that affects the success of phage treatment as biocontrol is the frequency of application. However, as with the timing of application, there does not seem to be a simple relationship between frequency of application and reduction of disease, and this is a factor that will likely have to be optimized for every potential phage treatment (Frampton *et al.*, 2012). One factor that does seem to correlate with success, as might be intuited, is the concentration of phage. Although the success rate typically does not increase linearly with phage concentration, there is almost certainly a lower threshold below which phages are unable to control the bacterial population. Understanding this threshold is key for at least two reasons: both because the fewer phages that need to be produced the less expensive production of treatment will be, and also because the weaker the selection pressure imposed on the bacterial population the less likely the bacteria will be to evolve resistance.

Two approaches which have shown particularly high success rates in the next generation of phage treatments are the use of non-virulent carrier bacterial species and the combination of phages with chemical control strategies. Given that phage population densities are inherently limited by host availability in the local environment, and that high concentrations of phages are more likely to control the target bacterial hosts, any mechanism that increases phage survival and reproduction should increase the success of treatment. Indeed, there is good evidence that the use of protective formulas to reduce the effect of UV can aid in the efficacy of treatment (Iriarte *et al.*, 2007). Alternatively, phage concentrations can be bolstered by the co-application of a 'carrier' bacterium that is both harmless to the plant host and susceptible to the bacteriophage(s) being used. For example, co-application of phage and the epiphytic bacterium, *Pantoea agglomerans*, reduced infection of fire blight on apple flowers by 56% (Boulé *et al.*, 2011) and pre-inoculation of tobacco plants with an avirulent strain of *Ralstonia solanacearum* and a bacteriophage reduced the severity of wilt relative to either treatment alone (Tanaka *et al.*, 1990). In these cases, the use of a carrier bacterial strain meant that phage densities could remain high even after the target host bacterial populations decreased in density. The combination of phage and chemical treatment has similarly shown increased efficacy relative to phage treatment alone. For example, combined application of the bacteriophage XMP-1 and streptomycin was successful in controlling bacterial leaf spot of mung bean (Borah *et al.*, 2000). Combined treatment of onion leaf blight, caused by *Xanthomonas axonopodis* pv. *allii*, with the plant defence activator, acibenzolar-S-methyl was also more effective in controlling disease progression than either treatment alone and matched that of traditional copper bactericides (Lang *et al.*, 2007).

Finally, a good understanding of the costs associated with bacterial resistance to phages can also play a clear role in designing treatment. For example, strains of *Pseudomonas syringae* pv. *tomato* that were experimentally evolved in the presence of multiple bacteriophages were found to have lower fitness (measured as density after 24 hours of growth) in the absence of phage than those evolved in the presence of a single phage (Koskella *et al.*, 2012). Thus even if bacteria are able to readily evolve resistance to all phages in a given cocktail, the treatment may still be effective in that it reduces the growth rate of resistant bacteria.

Conclusions and future direction

In this chapter we have outlined key findings on how bacteria and phages interact in the laboratory, and worked to extend these results to predicting how this interaction might differ in the plant phyllosphere and rhizosphere. The strong spatial structure and heterogeneous resources associated with the plant environment, as well as the actions of the plant immune response are all likely to affect the trajectory of bacteria–phage coevolution. In addition, we have described the many ways in which phages might alter the fitness of plants, both directly by controlling the density of bacterial pathogens (thereby increasing plant health) and altering the community of commensals (potentially decreasing plant health), and also indirectly by selecting for bacterial traits that make a given bacterium either more pathogenic or less efficient as a mutualist. Much of our current understanding, however, comes from combined knowledge about pairwise interactions among bacteria and plants, and bacteria and phages. It is therefore unclear which aspects of the predicted three-way interaction will be supported or how important phages will prove to be in shaping plant–microbe interactions. There is no doubt that laboratory experiments have been vital in identifying key features of bacteria–phage interactions and important coevolutionary processes, but in order to successfully apply this knowledge we must understand such processes in the exponentially noisier environment that is the field.

Recently, we have seen an increase in soil based phage research, primarily due to the potential economic applications in agronomical and epidemiological industries. These studies focus on the autecology of specific viruses in the soil which relate to useful or undesirable plant-associated bacteria (Kimura *et al.*, 2008). In soils, there are 'hot spots' of intensive bacterial proliferation, such as at the rhizosphere and surrounding plant residues (where increased nutrients allow faster bacteria turnover; Buée *et al.*, 2009). Therefore, it seems that phage ecology in the soil will be exceptionally important for plant-associated microbes, and research should be most extensive in these areas. Phages have the potential to be utilized as biological control agents that can act to eradicate plant pathogens and promote plant commensals. In addition, they have the power to coevolve with their host – thus maintaining their effectiveness as a control agent without necessary intervention. Bacteriophages can shape every aspect of the microbial community and this will have far reaching consequences for the multitude of plants with which they interact. By gaining a greater understanding of this microecosystem, we can understand how phages influence important plant-associated bacteria, and utilize their potential as biocontrol agents.

Acknowledgments

We would like to thank Jesús Murillo, Rob Jackson, Dawn Arnold, and Boris Vinatzer for the invitation to participate in this volume, and Sean Meaden for comments on a previous version of this chapter. BK would like to acknowledge funding by a NERC research fellowship (R16150).

Web resources

http://textbookofbacteriology.net/phage.html

References.

Adriaenssens, E.M., Van Vaerenbergh, J., Vandenheuvel, D., Dunon, V., Ceyssens, P.J., De Proft, M., Kropinski, A.M., Noben, J.-P., Maes, M., and Lavigne, R. (2012). T4-related bacteriophage LIMEstone isolates for the control of soft rot on potato caused by '*Dickeya solani*'. PLoS One *7*, e33227.

Ashelford, K.E., Day, M.J., Bailey, M.J., Lilley, A.K., and Fry, J.C. (1999). *In situ* population dynamics of bacterial viruses in a terrestrial environment. Appl. Environ. Microbiol. *65*, 169–174.

Ashelford, K.E., Day, M.J., and Fry, J.C. (2003). Elevated abundance of bacteriophage infecting bacteria in soil. Appl. Environ. Microbiol. *69*, 285–289.

Azegami, K. (2013). Suppressive effect of bacteriophage on bacterial palea browning of rice caused by *Pantoea ananatis*. J. Gen. Plant Pathol. *79*, 145–154.

Balogh, B., Jones, J.B., Momol, M.T., Olson, S.M., Obradovic, A., King, P., and Jackson, L.E. (2003). Improved efficacy of newly formulated bacteriophages for management of bacterial spot on tomato. Plant Dis. *87*, 949–954.

Balogh, B., Canteros, B.I., Stall, R.E., and Jones, J.B. (2008). Control of citrus canker and citrus bacterial spot with bacteriophages. Plant Dis. *92*, 1048–1052.

Beattie, G.A., and Lindow, S.E. (1999). Bacterial colonization of leaves: a spectrum of strategies. Phytopathology *89*, 353–359.

Berngruber, T.W., Weissing, F.J., and Gandon, S. (2010). Inhibition of superinfection and the evolution of viral latency. J. Virol. *84*, 10200–10208.

Beumer, A., and Robinson, J.B. (2005). A broad-host-range, generalized transducing phage (SN-T) acquires 16S rRNA genes from different genera of bacteria. Appl. Environ. Microbiol. *71*, 8301–8304.

Bobay, L.-M., Rocha, E.P.C., and Touchon, M. (2012). The adaptation of temperate bacteriophages to their host genomes. Mol. Biol. Evol. *30*, 737–751.

Bohannan, B.J.M., and Lenski, R.E. (2000). Linking genetic change to community evolution: insights from studies of bacteria and bacteriophage. Ecol. Lett. *3*, 362–377.

Bohannan, B.J.M., Travisano, M., and Lenski, R.E. (1999). Epistatic interactions can lower the cost of resistance to multiple consumers. Evolution *53*, 292–295.

Bohannan, B.M., Kerr, B., Jessup, C., Hughes, J., and Sandvik, G. (2002). Trade-offs and coexistence in microbial microcosms. Antonie Van Leeuwenhoek. *81*, 107–115.

Bondy-Denomy, J., Pawluk, A., Maxwell, K.L., and Davidson, A.R. (2012). Bacteriophage genes that inactivate the CRISPR/Cas bacterial immune system. Nature *493*, 429–432.

Borah, P.K., Jindal, J.K., and Verma, J.P. (2000). Integrated management of bacterial leaf spot of mungbean with bacteriophages of Xav and chemicals. J. Mycol. Plant Pathol. *30*, 19–21.

Bossi, L., Fuentes, J.A., Mora, G., and Figueroa-Bossi, N. (2003). Prophage contribution to bacterial population dynamics. J. Bacteriol. *185*, 6467–6471.

Boulé, J., Sholberg, P.L., Lehman, S.M., O'gorman, D.T., and Svircev, A.M. (2011). Isolation and characterization of eight bacteriophages infecting *Erwinia amylovora* and their potential as biological control agents in British Columbia, Canada. Can. J. Plant Pathol. *33*, 308–317.

Boyd, E.F., and Brüssow, H. (2002). Common themes among bacteriophage-encoded virulence factors and diversity among the bacteriophages involved. Trends Microbiol. *10*, 521–529.

Bradley, D.E. (1980). A function of *Pseudomonas aeruginosa* PAO polar pili: twitching motility. Can. J. Microbiol. *26*, 146–154.

Bradley, D.E., Douglas, C.J., and Peschon, J. (1984). Flagella-specific bacteriophages of *Agrobacterium tumefaciens*: demonstration of virulence of nonmotile mutants. Can. J. Microbiol. *30*, 676–681.

Brockhurst, M.A., Buckling, A., and Rainey, P.B. (2005). The effect of a bacteriophage on diversification of the opportunistic bacterial pathogen, *Pseudomonas aeruginosa*. Proc. Biol. Sci. *272*, 1385–1391.

Brown, S.P., Hochberg, M.E., and Grenfell, B.T. (2002). Does multiple infection select for raised virulence? Trends Microbiol. *10*, 401–405.

Buckling, A., and Rainey, P.B. (2002). Antagonistic coevolution between a bacterium and a bacteriophage. Proc. Biol. Sci. *269*, 931–936.

Buckling, A., Wei, Y., Massey, R.C., Brockhurst, M.A., and Hochberg, M.E. (2006). Antagonistic coevolution with parasites increases the cost of host deleterious mutations. Proc. Biol. Sci. *273*, 45–49.

Buée, M., De Boer, W., Martin, F., van Overbeek, L., and Jurkevitch, E. (2009). The rhizosphere zoo: an overview of plant-associated communities of microorganisms, including phages, bacteria, archaea, and fungi, and of some of their structuring factors. Plant Soil *321*, 189–212.

Bull, J., Pfennig, D.W., and Wang, I.-N. (2004). Genetic details, optimization and phage life histories. Trends Ecol. Evol. *19*, 76–82.

Burdman, S., Bahar, O., Parker, J.K., and De La Fuente, L. (2011). Involvement of type IV pili in pathogenicity of plant-pathogenic bacteria. Genes *2*, 706–735.

Canchaya, C., Fournous, G., Chibani-Chennoufi, S., Dillmann, M.-L., and Brüssow, H. (2003). Phage as agents of lateral gene transfer. Curr. Opin. Microbiol. *6*, 417–424.

Chao, L., Levin, B.R., and Stewart, F.M. (1977). A complex community in a simple habitat: an experimental study with bacteria and phage. Ecology *58*, 369–378.

Chao, L., Hanley, K.A., Burch, C.L., Dahlberg, C., and Turner, P.E. (2000). Kin selection and parasite evolution: higher and lower virulence with hard and soft selection. Q. Rev. Biol. *75*, 261–275.

Chen, F., Wang, K., Stewart, J., and Belas, R. (2006). Induction of multiple prophages from a marine bacterium: a genomic approach. Appl. Environ. Microbiol. *72*, 4995–5001.

Chen, J., and Novick, R.P. (2009). Phage-mediated intergeneric transfer of toxin genes. Science *323*, 139–141.

Chibani-Chennoufi, S., Bruttin, A., Dillmann, M., and Brüssow, H. (2004). Phage–host interaction: an ecological perspective. J. Bacteriol. *186*, 3677–3686.

Chisholm, S.T., Coaker, G., Day, B., and Staskawicz, B.J. (2006). Host–microbe interactions: shaping the evolution of the plant immune response. Cell *124*, 803–814.

Choi, J., Kotay, S.M., and Goel, R. (2010). Various physico-chemical stress factors cause prophage induction in *Nitrosospira multiformis* 25196 – an ammonia oxidizing bacteria. Water Res. *44*, 4550–4558.

Chowdhury, C., and Jagannadham, M.V. (2013). Virulence factors are released in association with outer membrane vesicles of *Pseudomonas syringae* pv. tomato T1 during normal growth. Biochim. Biophys. Acta *1834*, 231–239.
Civerolo, E.L., and Kiel, H.L. (1969). Inhibition of bacterial spot of peach foliage by *Xanthomonas pruni* bacteriophage. Phytopathology *59*, 67.
Collmer, A., Badel, J.L., Charkowski, A.O., Deng, W.L., Fouts, D.E., Ramos, A.R., Rehn, A.H., Anderson, D.M., Schneewind, O., van Dijk, K., and Alfano, J.R. (2000). *Pseudomonas syringae* Hrp type III secretion system and effector proteins. Proc. Natl. Acad. Sci. U.S.A. *97*, 8770–8777.
Drake, D., and Montie, T.C. (1988). Flagella, motility and invasive virulence of *Pseudomonas aeruginosa*. J. Gen. Microbiol. *134*, 43–52.
Evans, T.J., Ind, A., Komitopoulou, E., and Salmond, G.P.C. (2010). Phage-selected lipopolysaccharide mutants of *Pectobacterium atrosepticum* exhibit different impacts on virulence. J. Appl. Microbiol. *109*, 505–514.
Fard, M.N.R., Barton, M.D., and Heuzenroeder, M.W. (2011). Bacteriophage-mediated transduction of antibiotic resistance in enterococci. Lett. Appl. Microbiol. *52*, 559–564.
Fineran, P.C., Petty, N.K., and Salmond, G.P.C. (2009). Transduction: host DNA transfer by bacteriophages. In The Encyclopedia of Microbiology, M. Schaechter, ed. (Elsevier, Oxford, UK), pp. 666–679.
Flores, C.O., Meyer, J.R., Valverde, S., Farr, L., and Weitz, J.S. (2011). Statistical structure of host–phage interactions. Proc. Natl. Acad. Sci. U.S.A. *108*, E288–E297.
Flores, C.O., Valverde, S., and Weitz, J.S. (2013). Multi-scale structure and geographic drivers of cross-infection within marine bacteria and phages. ISME J. *7*, 520–532.
Frampton, R.A., Pitman, A.R., and Fineran, P.C. (2012). Advances in bacteriophage-mediated control of plant pathogens. Int. J. Microbiol. *2012*, 1–11.
Fuhrman, J.A., and Schwalbach, M. (2003). Viral influence on aquatic bacterial communities. Biol. Bull. *204*, 192–195.
Fujiwara, A., Fujisawa, M., Hamasaki, R., Kawasaki, T., Fujie, M., and Yamada, T. (2011). Biocontrol of *Ralstonia solanacearum* by treatment with lytic bacteriophages. Appl. Environ. Microbiol. *77*, 4155–4162.
Gandon, S., and Michalakis, Y. (2002). Local adaptation, evolutionary potential and host–parasite coevolution: interactions between migration, mutation, population size and generation time. J. Evol. Biol. *15*, 451–462.
Garcia-Doval, C., and van Raaij, M.J. (2012). Structure of the receptor-binding carboxy-terminal domain of bacteriophage T7 tail fibers. Proc. Natl. Acad. Sci. U.S.A. *109*, 9390–9395.
Gill, J.J., Svircev, A.M., Smith, R., and Castle, A.J. (2003). Bacteriophages of *Erwinia amylovora*. Appl. Environ. Microbiol. *69*, 2133–2138.
Haefele, D.M., and Lindow, S.E. (1987). Flagellar motility confers epiphytic fitness advantages upon *Pseudomonas syringae*. Appl. Environ. Microbiol. *53*, 2528–2533.
Hagens, S., Habel, A., and Bläsi, U. (2006). Augmentation of the antimicrobial efficacy of antibiotics by filamentous phage. Microb. Drug Resist. *12*, 164–168.
Hall, A.R., Scanlan, P.D., and Buckling, A. (2011). Bacteria–phage coevolution and the emergence of generalist pathogens. Am. Nat. *177*, 44–53.
Hall, A.R., De Vos, D., Friman, V., Pirnay, J., and Buckling, A. (2012). Effects of sequential and simultaneous application of bacteriophages on populations of *Pseudomonas aeruginosa in vitro* and in waxmoth larvae. Appl. Environ. Microbiol. *78*, 5646–5652.
Hammad, A.M.M. (1998). Evaluation of alginate-encapsulated *Azotobacter chroococcum* as a phage-resistant and an effective inoculum. J. Basic Microbiol. *38*, 9–16.
Harcombe, W.R., and Bull, J.J. (2005). Impact of phages on two-species bacterial communities. Appl. Environ. Microbiol. *71*, 5254–5259.
Herron, P.R. (2004). Phage ecology and genetic exchange in soil. In Molecular Microbial Ecology Manual, G.A. Kowalchuk, F.J. de Bruijn, I.M. Head, A.D.L. Akkermans and J.D. van Elsas, eds. (Springer, Dordrecht, The Netherlands), pp. 427–438.
Hirano, S.S., and Upper, C.D. (2000). Bacteria in the leaf ecosystem with emphasis on *Pseudomonas syringae*—a pathogen, ice nucleus, and epiphyte. Microbiol. Mol. Biol. Rev. *64*, 624–653.
Holmfeldt, K., Middelboe, M., Nybroe, O., and Riemann, L. (2007). Large variabilities in host strain susceptibility and phage host range govern interactions between lytic marine phages and their flavobacterium hosts. Appl. Environ. Microbiol. *73*, 6730–6739.
Holt, R.D. (1977). Predation, apparent competition, and the structure of prey communities. Theor. Pop. Biol. *12*, 197–229.

Horne, M.T. (1970). Coevolution of *Escherichia coli* and bacteriophages in chemostat culture. Science *168*, 992–993.

Hosseinidoust, Z., Tufenkji, N., and Van De Ven, T.G. (2013). Predation by homogeneous and heterogeneous phage environments affects virulence determinants of *Pseudomonas aeruginosa*. Appl. Environ. Microbiol. *79*, 2862–2871.

Høyland-Kroghsbo, N.M., Mærkedahl, R.B., and Svenningsen, S.L. (2013). A quorum-sensing-induced bacteriophage defense mechanism. Mbio. *4*, e00362-12.

Hrabak, E.M., and Willis, D.K. (1992). The *lemA* gene required for pathogenicity of *Pseudomonas syringae* pv *syringae* on bean is a member of a family of 2-component regulators. J. Bacteriol. *174*, 3011–3020.

Hyman, P., and Abedon, S.T. (2010). Bacteriophage host range and bacterial resistance. Adv. Appl. Microbiol. *70*, 217–248.

Icho, T., and Iino, T. (1978). Isolation and characterization of motile *Escherichia coli* mutants resistant to bacteriophage chi. J. Bacteriol. *134*, 854–860.

Iriarte, F.B., Balogh, B., Momol, M.T., Smith, L.M., Wilson, M., and Jones, J.B. (2007). Factors affecting survival of bacteriophage on tomato leaf surfaces. Appl. Environ. Microbiol. *73*, 1704–1711.

Jackson, R.W., Vinatzer, B., Arnold, D.L., Dorus, S., and Murillo, J. (2011). The influence of the accessory genome on bacterial pathogen evolution. Mob. Genet. Elements. *1*, 55–65.

Jalasvuori, M., Friman, V.-P., Nieminen, A., Bamford, J.K.H., and Buckling, A. (2011). Bacteriophage selection against a plasmid-encoded sex apparatus leads to the loss of antibiotic-resistance plasmids. Biol. Lett. *7*, 902–905.

Jore, M.M., Brouns, S.J., and van der Oost, J. (2012). RNA in defense: CRISPRs protect prokaryotes against mobile genetic elements. Cold Spring Harb. Perspect. Biol. *4*, a003657.

Julio, S.M., Heithoff, D.M., and Mahan, M.J. (2000). ssrA (tmRNA) Plays a role in *Salmonella enterica* serovar typhimurium pathogenesis. J. Bacteriol. *182*, 1558–1563.

Kidambi, S.P., Ripp, S., and Miller, R.V. (1994). Evidence for phage-mediated gene transfer among *Pseudomonas aeruginosa* strains on the phylloplane. Appl. Environ. Microbiol. *60*, 496–500.

Kim, K.H., Chang, H.W., Nam, Y.D., Roh, S.W., Kim, M.S., Sung, Y., Oh, H.M., and Bae, J.W. (2008). Amplification of uncultured single-stranded DNA viruses from rice paddy soil. Appl. Environ. Microbiol. *74*, 5975–5985.

Kim, S., Rahman, M., Seol, S.Y., Yoon, S.S., and Kim, J. (2012). *Pseudomonas aeruginosa* bacteriophage PA1Ø requires type IV pili for infection and shows broad bactericidal and biofilm removal activities. Appl. Environ. Microbiol. *78*, 6380–6385.

Kimura, M., Jia, Z.-J., Nakayama, N., and Asakawa, S. (2008). Ecology of viruses in soils: past, present and future perspectives. Soil Sci. Plant Nutr. *54*, 1–32.

Kleczkowska, J. (1950). A study of phage-resistant mutants of *Rhizobium trifolii*. J. Gen. Microbiol. *4*, 298–310.

Korber, D.R., Lawrence, J.R., and Caldwell, D.E. (1994). Effect of motility on surface colonization and reproductive success of *Pseudomonas fluorescens* in dual-dilution continuous culture and batch culture systems. Appl. Environ. Microbiol. *60*, 1421–1429.

Koskella, B. (2013). Phage-mediated selection on microbiota of a long-lived host. Curr. Biol. *23*, 1256–1260.

Koskella, B., and Meaden, S. (2013). Understanding bacteriophage specificity in natural microbial communities. Viruses *5*, 806–823.

Koskella, B., Taylor, T.B., Bates, J., and Buckling, A. (2011a). Using experimental evolution to explore natural patterns between bacterial motility and resistance to bacteriophages. ISME J. *5*, 1809–1817.

Koskella, B., Thompson, J.N., Preston, G.M., and Buckling, A. (2011b). Local biotic environment shapes the spatial scale of bacteriophage adaptation to bacteria. Am. Nat. *177*, 440–451.

Koskella, B., Lin, D.M., Buckling, A., and Thompson, J.N. (2012). The costs of evolving resistance in heterogeneous parasite environments. Proc. Biol. Sci. *279*, 1896–1903.

Labrie, S.J., Samson, J.E., and Moineau, S. (2010). Bacteriophage resistance mechanisms. Nature Rev. Microbiol. *8*, 317–327.

Lang, J.M., Gent, D.H., and Schwartz, H.F. (2007). Management of *Xanthomonas* leaf blight of onion with bacteriophages and a plant activator. Plant Dis. *91*, 871–878.

Leggett, H.C., Benmayor, R., Hodgson, D.J., and Buckling, A. (2012). Experimental evolution of adaptive phenotypic plasticity in a parasite. Curr. Biol. *23*, 139–142.

Lennon, J.T., Khatana, S.A.M., Marston, M.F., and Martiny, J.B.H. (2007). Is there a cost of virus resistance in marine cyanobacteria? ISME J. *1*, 300–312.

Lenski, R.E. (1984). Coevolution of bacteria and phage: are there endless cycles of bacterial defenses and phage counterdefenses? J. Theor. Biol. *108*, 319–325.
Lenski, R.E. (1988a). Experimental studies of pleiotropy and epistasis in *Escherichia coli*. I. Variation in competitive fitness among mutants resistant to virus T4. Evolution *42*, 425–432.
Lenski, R.E. (1988b). Dynamics of interactions between bacteria and virulent bacteriophage. Adv. Microb. Ecol. *10*, 1–44.
Lenski, R.E., and Levin, B.R. (1985). Constraints on the coevolution of bacteria and virulent phage: a model, some experiments, and predictions for natural communities. Am. Nat. *125*, 585–602.
Levin, B.R., and Bull, J.J. (1996). Phage therapy revisited: the population biology of a bacterial infection and its treatment with bacteriophage and antibiotics. Am. Nat. *147*, 881–898.
Levin, B.R., and Lenski, R.E. (1983). Coevolution in bacteria and their viruses and plasmids. In Coevolution, D.J. Futuyma and M. Slatkin, eds. (Sinauer, Sunderland, Massachusetts, USA), pp. 99–127.
Levin, B.R., and Lenski, R.E. (1985). Bacteria and phage: a model system for the study of the ecology and co-evolution of hosts and parasites. In Ecology and Evolution of Host–Parasite Interactions, D. Rollinson and R.M., Anderson, eds. (Academic Press, London, England), pp. 227–242.
Levin, B.R., Stewart, F.M., and Chao, L. (1977). Resource limited growth, competition, and predation: a model and experimental studies with bacteria and bacteriophage. Am. Nat. *111*, 3–24.
Levin, B.R., Moineau, S., Bushman, M., and Barrangou, R. (2013). The population and evolutionary dynamics of phage and bacteria with CRISPR-mediated immunity. PLoS Genet. *9*, e1003312.
Lindow, S.E., and Brandl, M.T. (2003). Microbiology of the phyllosphere. Appl. Environ. Microbiol. *69*, 1875–1883.
Lu, T.K., and Collins, J.J. (2009). Engineered bacteriophage targeting gene networks as adjuvants for antibiotic therapy. Proc. Natl. Acad. Sci. U.S.A. *106*, 4629–4634.
McKenna, F., El-Tarabily, K.A., Hardy, G.S., and Dell, B. (2001). Novel *in vivo* use of a polyvalent *Streptomyces* phage to disinfest *Streptomyces scabies*-infected seed potatoes. Plant Pathol. *50*, 666–675.
Maranger, R., and Bird, D.F. (1995). Viral abundance in aquatic systems: a comparison between marine and fresh waters. Mar. Ecol. Prog. Ser. *121*, 217–226.
Marsh, P., and Wellington, E.M.H. (1994). Phage–host interactions in soil. FEMS Microbiol. Ecol. *15*, 99–108.
Mattick, J.S. (2002). Type IV pili twitching motility. Annu. Rev. Microbiol. *56*, 289–314.
Melotto, M., Underwood, W., Koczan, J., Nomura, K., and He, S.Y. (2006). Plant stomata function in innate immunity against bacterial invasion. Cell *126*, 969–980.
Middelboe, M., Hagström, A., Blackburn, N., Sinn, B., Fischer, U., Borch, N.H., Pinhassi, J., Simu, K., and Lorenz, M.G. (2001). Effects of bacteriophages on the population dynamics of four strains of pelagic marine bacteria. Microb. Ecol. *42*, 395–406.
Munro, J., Oakey, J., Bromage, E., and Owens, L. (2003). Experimental bacteriophage-mediated virulence in strains of *Vibrio harveyi*. Dis. Aquat. Org. *54*, 187–194.
O'Flaherty, S., Ross, R.P., and Coffey, A. (2009). Bacteriophage and their lysins for elimination of infectious bacteria. FEMS Microbiol. Rev. *33*, 801–819.
O'Flynn, G., Ross, R.P., Fitzgerald, G.F., and Coffey, A. (2004). Evaluation of a cocktail of three bacteriophages for biocontrol of *Escherichia coli* O157: H7. Appl. Environ. Microbiol. *70*, 3417–3424.
O'Toole, G.A., and Kolter, R. (1998). Flagellar and twitching motility are necessary for *Pseudomonas aeruginosa* biofilm development. Mol. Microbiol. *30*, 295–304.
Panopoulos, N.J., and Schroth, M.N. (1974). Role of flagellar motility in the invasion of bean leaves by *Pseudomonas phaseolicola*. Phytopathology *64*, 1389–1397.
Pantastico-Caldas, M., Duncan, K.E., Istock, C.A., and Bell, J.A. (1992). Population dynamics of bacteriophage and *Bacillus subtilis* in soil. Ecology 73, 1888–1902.
Pirhonen, M., Heino, P., Helander, I., Harju, P., and Palva, E.T. (1988). Bacteriophage T4 resistant mutants of the plant pathogen, *Erwinia carotovora*. Micro. Pathogen. *4*, 359–367.
Poisot, T., Bell, T., Martinez, E., Gougat-Barbera, C., and Hochberg, M.E. (2013). Terminal investment induced by a bacteriophage in a rhizosphere bacterium [v2; ref status: indexed, http://f1000r.es/zh]. F1000Research *1*, 21.
Poullain, V., Gandon, S., Brockhurst, M.A., Buckling, A., and Hochberg, M.E. (2007). The evolution of specificity in evolving and coevolving antagonistic interactions between a bacteria and its phage. Evolution *62*, 1–11.
Quance, M.A., and Travisano, M. (2009). Effects of temperature on the fitness cost of resistance to bacteriophage T4 in *Escherichia coli*. Evolution *63*, 1406–1416.

Quiñones, B., Dulla, G., and Lindow, S.E. (2005). Quorum sensing regulates exopolysaccharide production, motility, and virulence in *Pseudomonas syringae*. Mol. Plant Microbe Interact. *18*, 682–693.

Rahme, L.G., Tan, M.W., Le, L., Wong, S.M., Tompkins, R.G., Calderwood, S.B., and Ausubel, F.M. (1997). Use of model plant hosts to identify *Pseudomonas aeruginosa* virulence factors. Proc. Natl. Acad. Sci. U.S.A. *94*, 13245–13250.

Rakonjac, J., Bennett, N.J., Spagnuolo, J., Gagic, D., and Russel, M. (2011). Filamentous bacteriophage: biology, phage display and nanotechnology applications. Curr. Issues Mol. Biol. *13*, 51.

Ramirez, M., Severina, E., and Tomasz, A. (1999). A high incidence of prophage carriage among natural isolates of *Streptococcus pneumoniae*. J. Bacteriol. *181*, 3618–3625.

Rankin, D.J., Rocha, E.P.C., and Brown, S.P. (2011). What traits are carried on mobile genetic elements, and why? Heredity *106*, 1–10.

Ravensdale, M., Blom, T.J., Gracia-Garza, J.A., Svircev, A.M., and Smith, R.J. (2007). Bacteriophages and the control of *Erwinia carotovora* subsp. carotovora. Can. J. Plant Pathol. *29*, 121–130.

Refardt, D. (2011). Within-host competition determines reproductive success of temperate bacteriophages. ISME J. *5*, 1451–1460.

Rice, S.A., Tan, C.H., Mikkelsen, P.J., Kung, V., Woo, J., Tay, M., Hauser, A., McDougald, D., Webb, J.S., and Kjelleberg, S. (2008). The biofilm life cycle and virulence of *Pseudomonas aeruginosa* are dependent on a filamentous prophage. ISME J. *3*, 271–282.

Rich, J.J., Hirano, S.S., and Willis, D.K. (1992). Pathovar-specific requirement for the *Pseudomonas syringae lemA* gene in disease lesion formation. Appl. Environ. Microbiol. *58*, 1440–1446.

Riesenfeld, C., Schloss, P., and Handelsman, J. (2004). Metagenomics: genomic analysis of microbial communities. Annu. Rev. Genetics *38*, 525–552.

Rodríguez-Navarro, D.N., Dardanelli, M.S., and Ruíz-Saínz, J.E. (2007). Attachment of bacteria to the roots of higher plants. FEMS Microbiol. Lett. 272, 127–136.

Rodriguez-Valera, F., Martin-Cuadrado, A.-B., Rodriguez-Brito, B., Pasic, L., Thingstad, T.F., Rohwer, F., and Mira, A. (2009). Explaining microbial population genomics through phage predation. Nat. Rev. Micro. 7, 828–836.

Roine, E., Raineri, D.M., Romantschuk, M., Wilson, M., and Nunn, D.N. (1998). Characterization of type IV pilus genes in *Pseudomonas syringae* pv. tomato DC3000. Mol. Plant Microbe Interact. *11*, 1048–1056.

Romanenko, A.S., Lomovatskaya, L.A., Shafikova, T.N., Borovskii, G.B., and Krivolapova, N.V. (2003). Potato cell plasma membrane receptors to ring rot pathogen extracellular polysaccharides. J. Phytopathology *151*, 1–6.

Romantschuk, M., and Bamford, D.H. (1985). Function of pili in bacteriophage ϕ6 penetration. J. Gen. Virol. *66*, 2461–2469.

Romero-Suarez, S., Jordan, B., and Heinemann, J.A. (2012). Isolation and characterization of bacteriophages infecting *Xanthomonas arboricola* pv. *juglandis*, the causal agent of walnut blight disease. World J. Microbiol. Biotechnol. *28*, 1917–1927.

Samuel, A.D.T., Pitta, T.P., Ryu, W.S., Danese, P.N., Leung, E.C.W., and Berg, H.C. (1999). Flagellar determinants of bacterial sensitivity to chi-phage. Proc. Natl. Acad. Sci. U.S.A. *96*, 9863–9866.

Scanlan, P.D., and Buckling, A. (2011). Co-evolution with lytic phage selects for the mucoid phenotype of *Pseudomonas fluorescens* SBW25. ISME J. *6*, 1148–1158.

Schnabel, E.L., Fernando, W.G.D., Meyer, M.P., Jones, A.L., and Jackson, L.E. (1998). Bacteriophage of *Erwinia amylovora* and their potential for biocontrol. Acta Hort. *489*, 649–654.

Shao, Y., and Wang, N. (2008). Bacteriophage adsorption rate and optimal lysis time. Genetics *180*, 471–482.

Smillie, C.S., Smith, M.B., Friedman, J., Cordero, O.X., David, L.A., and Alm, E.J. (2011). Ecology drives a global network of gene exchange connecting the human microbiome. Nature *480*, 241–244.

Stacey, G., Paau, A.S., Noel, K.D., Maier, R.J., Silver, L.E., and Brill, W.J. (1982). Mutants of *Rhizobium japonicum* defective in nodulation. Arch. Microbiol. *132*, 219–224.

Stewart, F.M., and Levin, B.R. (1984). The population biology of bacterial viruses: why be temperate? Theor. Popl. Biol. *26*, 93–117.

Srinivasiah, S., Bhavsar, J., Thapar, K., Liles, M., Schoenfeld, T., and Wommack, K.E. (2008). Phages across the biosphere: contrasts of viruses in soil and aquatic environments. Res. Microbiol. *159*, 349–357.

Suttle, C.A. (2005). Viruses in the sea. Nature *437*, 356–361.

Taguchi, F., and Ichinose, Y. (2011). Role of type IV pili in virulence of *Pseudomonas syringae* pv. *tabaci* 6605: correlation of motility, multidrug resistance, and HR-inducing activity on a nonhost plant. Mol. Plant Microbe Interact. *24*, 1001–1011.

Tanaka, H., Negishi, H., and Maeda, H. (1990). Control of tobacco bacterial wilt by an avirulent strain of *Pseudomonas solanacearum* M4S and its bacteriophage. Ann. Phytopathol. Soc. Jpn. *56*, 243–246.

Thompson, J.N. (1999). Specific hypotheses on the geographic mosaic of coevolution. Am. Nat. *153*, S1–S14.

Torsvik, V., Øvreås, L., and Thingstad, T.F. (2002). Prokaryotic diversity-magnitude, dynamics, and controlling factors. Science *296*, 1064–1066.

Toth, I.K., Mulholland, V., Cooper, V., Bentley, S., Shih, Y.L., Perombelon, M.C.M., and Salmond, G.P.C. (1997). Generalized transduction in the potato blackleg pathogen *Erwinia carotovora* subsp. atroseptica by bacteriophage M1. Microbiology *143*, 2433–2438.

Varani, A.M., Monteiro-Vitorello, C.B., Nakaya, H.I., and Van Sluys, M.A. (2013). The role of prophage in plant-pathogenic bacteria. Ann. Rev. Phytopathol. *51*, 429–451.

Von Bodman, S.B., Bauer, W.D., and Coplin, D.L. (2003). Quorum sensing in plant-pathogenic bacteria. Ann. Rev. Phytopathol. *41*, 455–482.

Vos, M., Birkett, P.J., Birch, E., Griffiths, R.I., and Buckling, A. (2009). Local adaptation of bacteriophages to their bacterial hosts in soil. Science *325*, 833.

Wang, I.-N. (2006). Lysis timing and bacteriophage fitness. Genetics *172*, 17–26.

Waterbury, J.B., and Valois, F.W. (1993). Resistance to co-occurring phages enables marine *Synechococcus* communities to coexist with cyanophages abundant in seawater. Appl. Environ. Microbiol. *59*, 3393–3399.

Wei, Y., Kirby, A., and Levin, B.R. (2011). The population and evolutionary dynamics of *Vibrio cholerae* and its bacteriophage: conditions for maintaining phage-limited communities. Am. Nat. *178*, 715–725.

Weinbauer, M.G., and Rassoulzadegan, F. (2004). Are viruses driving microbial diversification and diversity? Environ. Microbiol. *6*, 1–11.

Whitchurch, C.B., and Mattick, J.S. (1994). Characterization of a gene, pilU, required for twitching motility but not phage sensitivity in *Pseudomonas aeruginosa*. Mol. Microbiol. *13*, 1079–1091.

Williams, H.T. (2013). Phage-induced diversification improves host evolvability. BMC Evol. Biol. *13*, 17.

Williamson, K.E., Wommack, K.E., and Radosevich, M. (2003). sampling natural viral communities from soil for culture-independent analyses. Appl. Environ. Microbiol. *69*, 6628–6633.

Williamson, K.E., Radosevich, M., Smith, D.W., and Wommacj, K.E. (2007). Incidence of lysogeny within temperate and extreme soil environments. Environ. Microbiol. *9*, 2563–2574.

Yong Yu, Z., ChunXiao, H., Jun, Y., and NianZhi, J. (2011). Interactions between marine microorganisms and their phages. Chinese Sci. Bull. *56*, 1770–1777.

Zhang, Q.G., and Buckling, A. (2012). Phages limit the evolution of bacterial antibiotic resistance in experimental microcosms. Evol. Appl. *5*, 575–582.

Index

Q

R

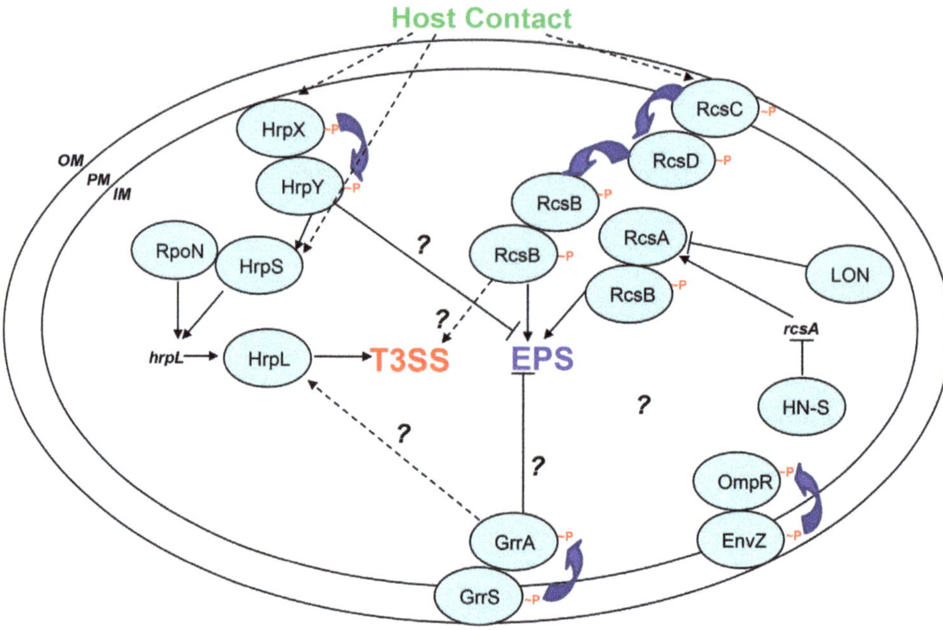

Figure 3.1 Simplified model of virulence factor regulation in *Erwinia amylovora*. Multiple signalling pathways coordinate the activity of pathogenicity factors in *E. amylovora* including the type III secretion system (T3SS) and exopolysaccharide (EPS) biosynthesis.

Figure 4.1 Symptoms of several diseases caused by *Acidovorax* species. Bacterial fruit blotch symptom in watermelon (A) and melon (B), caused by *A. citrulli*; bacterial brown stripe of maize, caused by *A. avenae* (C); leaf spot disease of *Cattleya*, caused by *A. cattleyae* (D); leaf symptoms caused by *Acidovorax* strains in the ornamental plants hydrangea (E), geranium (F) and lantana (G). [15](A) and (B), reproduced from Burdman and Walcott (2012) ; (C), kindly provided by Takao Tsukiboshi (National Institute of Livestock and Grassland Science, Japan); (D), kindly provided by Terry Bottom (St Augustine Orchid Society, FL, USA); (E–G), kindly provided by Michael Munster (North Carolina State University, NC, USA).

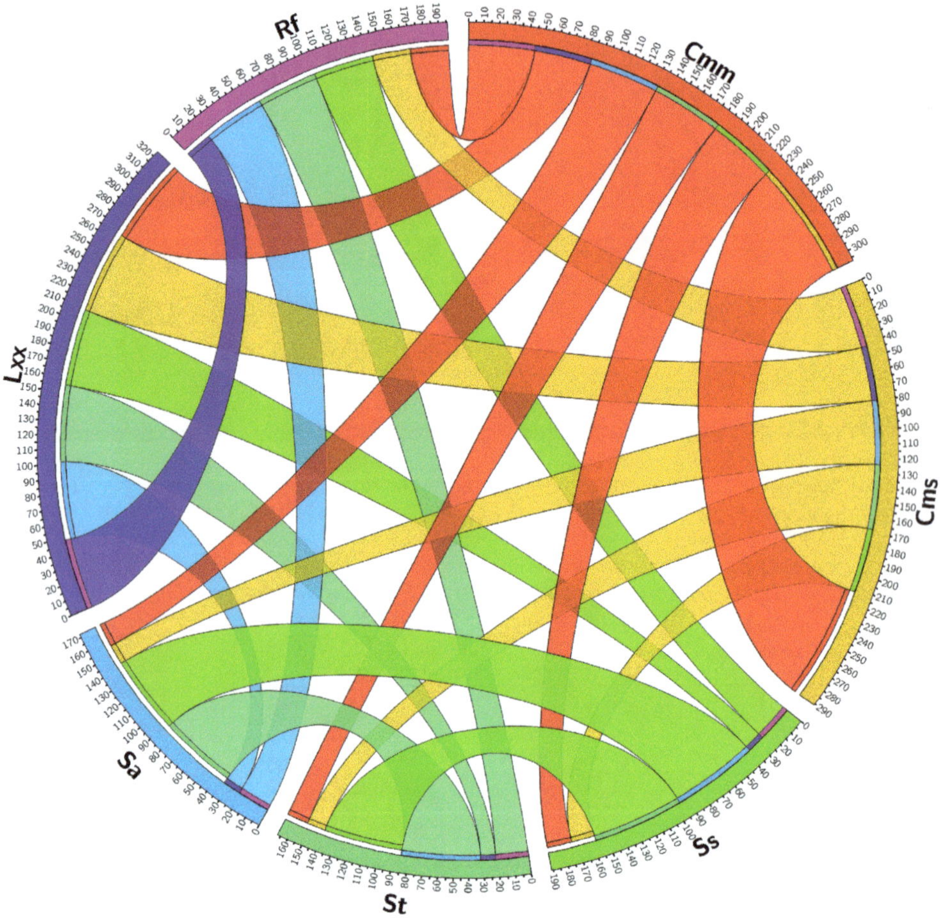

Figure 5.2 Visualization of genome-wide homology between Gram-positive plant pathogens. The outer track is expressed as a ratio of the total number of pairs of homologues divided by number of genes in a genome, e.g. there were three times as many pairs of homologues based on six pair wise comparisons, as genes in the genome of *Cmm*. The interior ribbons connect the genome comparisons. Ribbons vary in width to depict the extent of homology relative to genome size, e.g. pairs of homologues represented ~45% and ~25% of the *Cmm* and *Rf* genomes, respectively. Homologues were identified using reciprocal best BLAST hit analysis. Genomes were assigned arbitrary colours: red, *Clavibacter michiganesis* subsp. *michiganensis* (*Cmm*); yellow, *Clavibacter michiganensis* subsp. *sepedonicus* (*Cms*); lime green, *Streptomyces scabies* (*Ss*); mint green, *Streptomyces turgidiscabies* (*St*); light blue, *Streptomyces acidiscabies* (*Sa*); dark blue, *Leifsonia xyli* subsp. *xyli* (*Lxx*); purple, *Rhodococcus fascians* D188 (*Rf*). The figure was generated using Circos (Krzywinski *et al*., 2009).

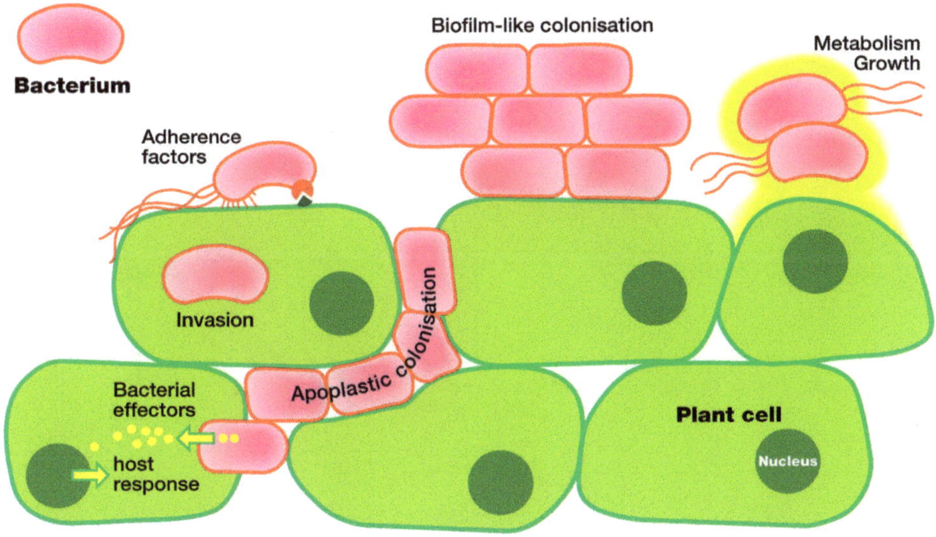

Figure 6.1 Cartoon showing the possible interactions between human-pathogenic bacteria (magenta) and plant hosts (plant cells in green).

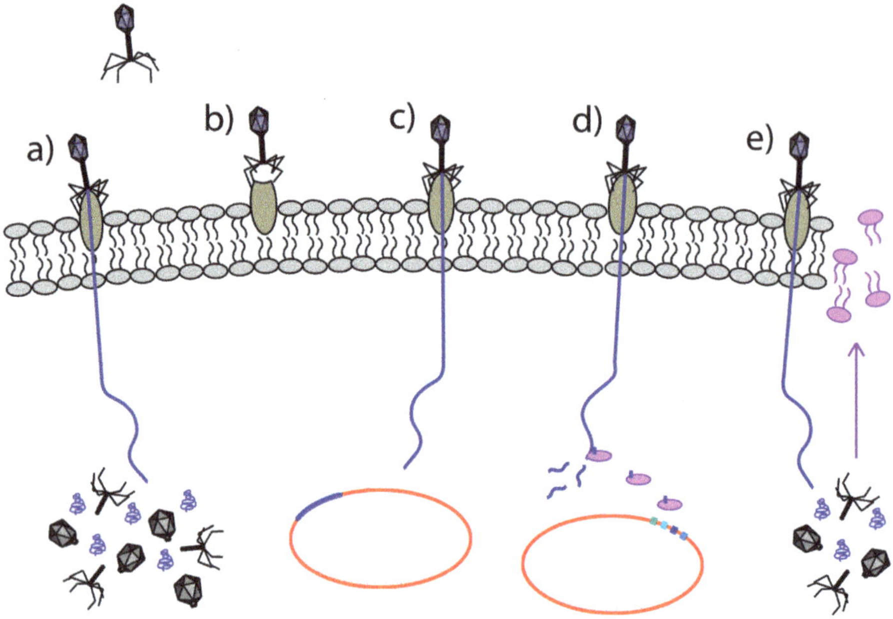

Figure 8.1 Bacteria–phage interactions are successful if (a) the phage is successfully able to bind to the bacterial cell via receptor-recognition, inject its RNA or DNA into the host cell and replicate using host cell machinery. (b) Bacteria can evolve resistance to a given phage by loss or alteration of the target receptor. (c) Upon successful infection, some phage genomes are integrated into the genome of their host, in which case they can replicate concurrently with their host and transmit vertically through a bacterial lineage (this is referred to as the lysogenic life cycle). However, bacteria have evolved a number of systems for degrading phage DNA upon successful binding. (d) For example, in some cases small segments of the phage genome are integrated into the host genome (inserted within spacers of the CRISPR region of the genome) and these segments are used to block future replication of that phage lineage. (e) Finally, many bacterial species have mechanisms that lead to cell death, for example by degrading the cell wall, upon successful phage infection, thereby protecting neighbouring cells from further infection. As discussed in the text, phages are capable of overcoming each of these resistance mechanisms via a diverse set of counter-adaptation.

www.ingramcontent.com/pod-product-compliance
Lightning Source LLC
LaVergne TN
LVHW051050210926
848486LV00001B/60
9781908230584